Language in Our Brain

The Origins of a Uniquely Human Capacity

Angela D. Friederici

Foreword by Noam Chomsky

The MIT Press
Cambridge, Massachusetts
London, England

© 2017 Massachusetts Institute of Technology

All rights reserved. No part of this book may be reproduced in any form by any electronic or mechanical means (including photocopying, recording, or information storage and retrieval) without permission in writing from the publisher.

This book was set in Syntax LT Std and Times New Roman by Toppan Best-set Premedia Limited. Printed and bound in the United States of America.

The cover displays the white matter fiber tracts of the human brain for the left and the right hemispheres provided by Alfred Anwander, Max Planck Institute for Human Cognitive and Brain Sciences, Leipzig, Germany.

Library of Congress Cataloging-in-Publication Data

Names: Friederici, Angela D., author.
Title: Language in our brain : the origins of a uniquely human capacity / Angela D. Friederici ; foreword by Noam Chomsky.
Description: Cambridge, MA : The MIT Press, [2017] | Includes bibliographical references and index.
Identifiers: LCCN 2017014254 | ISBN 9780262036924 (hardcover : alk. paper)
Subjects: LCSH: Cognitive grammar. | Cognitive learning. | Cognitive science. | Brain--Localization of functions. | Brain--Language. | Brain--Physiology. | Brain--Locations of functionality. | Psycholinguistics.
Classification: LCC P165 .F64 2017 | DDC 401/.9–dc23 LC record available at https://lccn.loc.gov/2017014254

10   9   8   7   6   5   4   3

Es ist recht unwahrscheinlich, daß die reine Psychologie zu einer wirklichen naturgemäßen Anschauung der Gliederung im Geistigen vordringen wird, solange sie der Anatomie des Seelenorgans grundsätzlich den Rücken kehrt.

It is rather unlikely that psychology, on its own, will arrive at the real, lawful characterization of the structure of the mind, as long as it neglects the anatomy of the organ of the mind.
—Paul Flechsig (1896), Leipzig

Contents

Foreword by Noam Chomsky   ix
Preface   xi
Acknowledgments   xiii

Introduction   1
    Language as a Uniquely Human Trait   1
    Language as a Specific Cognitive System   3
    Language as a Brain System   5

Part I   13

1 Language Functions in the Brain: From Auditory Input to Sentence Comprehension   15
    1.1   A Cognitive Model of Auditory Language Comprehension   15
    1.2   Acoustic-Phonological Processes   20
    1.3   From Word Form to Syntactic and Lexical-Semantic Information   27
    1.4   Initial Phrase Structure Building   32
    1.5   Syntactic Relations during Sentence Processing   43
    1.6   Processing Semantic Relations   56
    1.7   Thematic Role Assignment: Semantic and Syntactic Features   62
    1.8   Processing Prosodic Information   71
    1.9   Functional Neuroanatomy of Language Comprehension   82

2 Excursions   85
    2.1   Language Comprehension and Production: A Common Knowledge Base of Language   85
    2.2   Language Comprehension and Communication: Beyond the Core Language System   95

Part II   101

3 The Structural Language Network   103
    3.1   The Neuroanatomical Pathways of Language   103
    3.2   Pathways in the Right Hemisphere and Cross-Hemispheric Pathways   112
    3.3   The Neuroanatomical Pathway Model of Language: Syntactic and Semantic Networks   115

4 **The Functional Language Network** 121
   4.1 The Neuroreceptorarchitectonic Basis of the Language Network 121
   4.2 Functional Connectivity and Cortical Oscillations 125
   4.3 The Neural Language Circuit 134

Part III 143

5 The Brain's Critical Period for Language Acquisition 145
   5.1 Neurophysiology of Second Language Learning 146
   5.2 Critical and Sensitive Periods of Learning: Facts and Speculations 155
   5.3 Universality of the Neural Language Network 157

6 **Ontogeny of the Neural Language Network** 163
   6.1 Language in the First Three Years of Life 165
   6.2 Language beyond Age 3 183
   6.3 Structural and Functional Connectivity during Development 191
   6.4 The Ontogeny of the Language Network: A Model 196

Part IV 201

7 **Evolution of Language** 203
   7.1 Theories of Language Evolution 203
   7.2 Processing Structured Sequences in Songbirds 205
   7.3 Comparing Monkeys and Humans 207
   7.4 Paleoanthropological Considerations of Brain Development 219

8 **The Neural Basis of Language** 221
   8.1 An Integrative View of the Language Network 223
   8.2 Epilogue: Homo Loquens—More than Just Words 231

Glossary 233
Notes 241
References 245
Index 281

Foreword

Fifty years ago, Eric Lenneberg published his now-classic study that inaugurated the modern field of biology of language—by now a flourishing discipline, with remarkable richness of evidence and sophistication of experimentation, as revealed most impressively in Angela Friederici's careful and comprehensive review of the field, on its fiftieth anniversary.

Friederici's impressive study is indeed comprehensive, covering a rich range of experimental inquiries and theoretical analyses bearing on just about every significant aspect of the structure and use of language, while reaching as well to the ways language use is integrated with systems of belief about the world generally. Throughout, she adopts the general (and in my view highly reasonable if not unavoidable) conception that processing and perception access a common "knowledge base," core internal language, a computational system that yields an unbounded array of structured expressions. Her discussion of a variety of conceptions and approaches is judicious and thoughtful, bringing out clearly the significance and weaknesses of available evidence and experimental procedures. Her own proposals, developed with careful and cautious mustering and analysis of evidence, are challenging and far-reaching in their import.

Friederici's most striking conclusions concern specific regions of Broca's area (BA 44 and BA 45) and the white matter dorsal fiber tract that connects BA 44 to the posterior temporal cortex. Friederici suggests that "This fiber tract could be seen as the missing link which has to evolve in order to make the full language capacity possible." The conclusion is supported by evidence that this dorsal pathway is very weak in macaques and chimpanzees, and weak and poorly myelinated in newborns, but strong in adult humans with language mastery. Experiments reported here indicate further that the "Degree of myelination predicts behavior in processing syntactically complex non-canonical sentences" and that increasing strength of this pathway "correlates directly with the increasing ability to process complex syntactic structures." A variety of experimental results suggest that "This fiber tract may thus be one of the reasons for the difference in the language ability in human adults compared to the prelinguistic infant and the monkey." These structures, Friederici suggests, appear to "have evolved to subserve the human capacity to process syntax, which is at the core of the human language faculty."

BA 44, then, is responsible for generation of hierarchical syntactic structures, and accordingly has "particular characteristics at the neurophysiological level," differing from other brain regions at both functional and microstructural levels. More specifically, it is the ventral part of B44 in which basic syntactic computations—in the simplest case Merge—are localized, while adjacent areas are involved in combinatorial operations independent of syntactic structure. BA 45 is responsible for semantic processes. Within Broca's area neural networks are differentiated for language and action. These are bold proposals, with rich implications insofar as they can be sustained and developed further.

Friederici's extensive review covers a great deal of ground. To give only a sample of topics reviewed and proposed conclusions, the study deals with the dissociation of syntactic/semantic processing both structurally and developmentally. It reviews the evidence that right-hemisphere specialization for prosody may be more primitive in evolution, and that its contributions are integrated very rapidly (within a second) with core language areas for assignment and interpretation of prosodic structure of expressions. Experimentation reveals processes of production that precede externalization. The latter is typically articulatory, though since the pioneering work of Ursula Bellugi enriched by the illuminating work of Laura Ann Petitto and others, it is now known that sign is readily available for externalization and is so similar to speech in relevant dimensions that it is plausible, Friederici suggests, to conclude that there is "a universal neural language system largely independent of the input modality," modulated slightly by lifelong use of sign. Brain regions are specialized for semantic-syntactic information and syntactic processing of complex non-canonical sentences. Language mastery develops in regular and systematic ways that are revealed by mutually supportive behavioral and neurological investigations, with some definite semi-critical periods of sensitivity. By ages 2–3, children have attained substantial syntactic knowledge though full mastery of the language continues to early adolescence, as shown by behavioral studies and observation of neural domain-specificity for syntactic structures and processes.

The result of this wide-ranging exploration is a fascinating array of insights into what has been learned in this rapidly developing field and a picture of the exciting prospects that lie ahead.

Noam Chomsky
November 2016
Cambridge, Massachusetts

Preface

Language makes us human. It is an intrinsic part of us. We learn it, we use it, and we seldom think about it. But once we start thinking about it, language seems like a sheer wonder. Language is an extremely complex entity with several subcomponents responsible for the language sound, the word's meaning, and the grammatical rules governing the relation between words.

I first realized that there are indeed such subcomponents of language when I worked as a student in a clinic of language-impaired individuals. On one of my first days in the clinic I was confronted with a patient who was not able to speak in full sentences. He seemed quite intelligent, was able to communicate his needs, but did so in utterances in which basically all grammatical items were missing—similar to a telegram. It immediately occurred to me: if grammar can fail separately after a brain injury, it must be represented separately in the brain. This was in 1973. At this time structural brain imaging such as computer tomography was only about to develop and certainly not yet available in all clinics.

Years later in 1979, when I spent my postdoctoral year at the Massachusetts Institute of Technology (MIT) and at the Boston Veterans Hospital of the Boston University, the neurologist Norman Geschwind was one of the first to systematically relate sensory and cognitive impairments to particular brain sites in vivo by means of computer tomography. In his clinical seminars he examined a given patient behaviorally and from this predicted the site of the patient's brain lesion. Then the computer tomographic picture of the patient's brain lesion was presented. Norman Geschwind most of the time had made the correct prediction, thereby providing impressive evidence for a systematic relation between brain and behavior.

Today, more than 35 years later, our knowledge about the relationship between brain and cognitive behavior has dramatically increased due to the advent of new brain imaging techniques such as functional magnetic resonance tomography. This is in particular true for the domain of language thanks to studies that were guided and informed by linguistic theory.

Linguistics provides a systematic description of the three relevant language components: the sound of language, its semantics (dealing with the meaning of words and word combinations), and its syntax (dealing with the grammatical rules determining the combination of words). All these different components have to work together in milliseconds in order to keep track of online language use. If we want to understand language we first have to disentangle this complex entity into its relevant pieces—its subcomponents—and then see how they work together to make language use possible. It is like a mosaic, in that once all the pieces are in place a coherent picture will evolve.

In *Language in Our Brain* I will provide a description of the relevant brain systems in support of language with its subcomponents, how they develop during the first years of life, and, moreover, how they possibly emerged during evolution.

For this book I drew primarily from empirical data on the language-brain relationship available in the literature. Parts of it come from articles I have written together with my students and colleagues. It is the work and discussions with them on which the view laid down here is built. My thanks go to all who have been working with me over the years.

Acknowledgments

Had I not met Noam Chomsky in fall 1979 at a workshop at the MIT Endicott House, I would not have written this book. His idea of language as a biological organ, and Jerry Fodor's postulate of the modularity of mind—hotly debated at the time—left their trace in my subsequent work. I set out for the search of the neurobiological basis of language. In *Language in Our Brain* I present what we know today about the neural language network and its ontogeny and phylogeny.

On my scientific path I received support from a number of excellent teachers and colleagues. During my post-doctoral year at MIT this was Merrill Garrett, and back in Europe it was Pim Levelt, who offered me the chance to conduct my research at the Max Planck Institute for Psycholinguistics.

In 1994 the dream of my scientific life became true. The Max Planck Society offered an opportunity to me and to my dear colleague, the neurologist Yves von Cramon, to build up a new Max Planck Institute for Cognitive Neuroscience in Leipzig. Now we only had to turn the dream into reality, which with the help of the Max Planck Society we did. I am most grateful for this.

My sincere thanks go to my colleagues from inside and outside the institute who have read and critiqued portions of *Language in Our Brain*, each with their special expertise. I am deeply thankful to Pim Levelt and Marco Tettamanti for their feedback on large parts of the book. Special thanks go to Wolfgang Prinz, who as a non-expert in neuroscience gave many helpful suggestions that have greatly improved its clarity.

Many thanks go to the colleagues in my department who together with me wrote major review articles upon which parts of the book are based, and who also gave critical feedback on various parts of the book. These are, in alphabetical order: Alfred Anwander, Jens Brauer, Tomás Goucha, Thomas C. Gunter, Gesa Hartwigsen, Claudia Männel, Lars Meyer, Michael A. Skeide, and Emiliano Zaccarella.

Special thanks go to Emiliano Zaccarella, who constructed the glossary of this book. I am indebted to Andrea Gast-Sandmann, who produced and adapted the figures for this book. Many thanks go to Elizabeth Kelly for editing the text in English. I am indebted to Christina Schröder for her support throughout all stages of the emergence of this book. Moreover, I express my gratitude to Margund Greiner for her tireless commitment to bring the complete manuscript into a presentable form.

Introduction

Language as a Uniquely Human Trait

Humans are born to learn language. We learn our native language without any formal teaching or training, and we easily deal with language every day in all possible situations without even thinking about it. This means that—once learned—we use language automatically. This capacity is quite impressive, in particular when considering that no other creature or even machine is able to cope with language in a similar way.

What is it that puts us in the position to do this? What is the basis of this uniquely human trait?

An obvious answer is that the language faculty must be innate. This answer, however, does not mean much unless we define two things: First, what does the term *language* refer to, and second, what do we mean by *innate*? The first question appears to have a simple answer, as we all use language for communication, but this is not necessarily the correct answer. Because we are able to communicate without language by means of pointing, facial expressions, and iconic gestures it is clear that language and communication are not the same. Due to the fact that language is not identical with communication, but used for communication, it happens that researchers in the domain of language do not always agree on the topic of their research. Some researchers assume that language is a mental organ that refers to a set of finite computational mechanisms which allows us to generate an infinite number of sentences and take language to be distinct from communication (Chomsky, 1980, 1995a; Berwick and Chomsky, 2015). Other researchers see language as an epiphenomenon of the human capacity to share intentions in the use of communication (Tomasello, 1999, 2008; Tomasello, Carpenter, Call, Behne, and Moll, 2005). These different views are embedded in different beliefs of how language as a human faculty has evolved. Although both agree on a biological evolution of human cognition, one side believes that language develops in the use of social interaction and communication (Tomasello, 2008), whereas the other side views language as the result of a genetically determined neurobiological evolutionary step, the genetics of which still remain to be specified (Berwick and Chomsky, 2015).[1]

In *Language and Our Brain*, I start from the assumption that language is a biological system that evolved through phylogeny, and then I aim to specify the neurobiological basis for that system. Human language use is based on cognitive capacities that are partly present in other animals, except for the capacity necessary for all natural languages to build hierarchically structured sentences. Among those cognitive capacities non-specific to humans—and which are present in other animals such as non-human primates, dogs, or birds—are memory and attention, as well as the ability of associative learning and the capacity to recognize and memorize sequences across time. These cognitive abilities together with the human-specific, innate language faculty allow the acquisition and use of any natural language.

What does *innate* mean in this context? It cannot only refer to capacities present at birth, but it must also refer to those capacities that develop later in life, according to a fixed biological program, even if this requires language input during a critical period of development. "Abandoned children" such as Kaspar Hauser and the more recent, well-documented case of "Genie" (Curtis, 1977) who were both deprived of language input early in life, never developed full language capacity later in life (Blumenthal, 2003). Kaspar Hauser, a boy found when he was about 7 years old, did manage to learn vocabulary and some basic aspects of language, but like other "abandoned children" he never developed syntax. Genie, a girl discovered at age 13, who was isolated from human contact since age 2, was able to acquire vocabulary through intensive training. However, she was not able to learn the most basic principles of syntactic structure. This stands in stark contrast to the normal situation of language acquisition in which children with continuous language input throughout development normally gain the full language capacity, including syntax, seemingly following a fixed biological program. As we will see, this not only holds when the input is auditory language but also when the input is visual, as in sign language. Thus it appears that nature provides us with a biological program for language, but it needs input in order to develop.

In clear contrast to humans, the chimpanzees (our next relative) are unable to learn to combine words to construct larger utterances. It was shown that children by age 4, both hearing children and deaf children, construct utterances with a mean length of three to four words, whereas chimpanzees, even after four years of intensive training with meaningful symbols for communication, only demonstrate utterances with a mean length of little more than one word or symbol (Pearce, 1987). Thus, the ability of binding words to build a linguistic sequence is essentially non-existent in the chimpanzee. An interesting difference between the species was also shown for the processing of structured sequences in another study comparing humans and monkeys. This study revealed that cotton-top tamarins, although able to learn auditory rule-based sequences constructed according to a simple rule, were not able to learn sequences of the same length when constructed according to a more complex rule (Fitch and Hauser, 2004). Humans, in contrast, were able to learn both rule types within minutes. These results suggest dramatic differences between human and non-human primates when it comes to learning syntactically structured sequences. Why is that the case?

A possible answer may be found in neuroanatomical differences between the human and non-human primate brain, in particular, in those parts of the brain that constitute the language network in humans. And indeed, cross-species analyses of those brain regions that are responsible for language in humans, as well as those fiber bundles that connect these regions, reveal crucial differences between human and non-human primates (Rilling et al., 2008). I will discuss these differences in more detail at the end of this book. Before that, we will have to understand what language is, and moreover, we have to learn some basic neuroanatomy of language in the human brain.

In this introduction I will also briefly consider language as a specific cognitive system and language as a brain-based system. Only after these considerations will I discuss the enormous amount of detailed empirical data available on the relationship between language functions and their brain bases in humans, and then compare the brain structures in human and non-human primates.

Language as a Specific Cognitive System

Language is a cornerstone of human cognition. It serves human communication, but is not communication itself. Non-human creatures communicate—for example, whales and bees—but they do so by means other than language. In humans language serves communication, but in addition to language, humans use gestures and emotional tones accompanying speech as means of communication. Moreover, humans are able to infer the speaker's intentional meaning beyond the utterance given. Human language as such can be distinguished from other means of communication mainly by syntax, a specific cognitive system of rules and operations that permits the combination of words into meta-structures such as phrases and sentences. In addition to the core language system consisting of syntactic rules and an inventory of words, there are two interfaces: an external interface, that is, the sensory-motor interface subserving perception and production; and an internal interface, that is, the conceptual-intentional interface guaranteeing the relation of the core language component to concepts, intentions, and reasoning (Berwick, Friederici, Chomsky, and Bolhuis, 2013).

When producing or understanding a sentence, the different parts of the language system have to work together to achieve successful language use. Within the system, syntax is a central part that has been described well, although with some variations across linguistic theories (such as phrase structure grammar, Chomsky, 1957, 1965; head-driven phrase structure grammar, Pollard and Sag, 1994; and lexical functional grammar, Bresnan, 2001). One prominent theory (Chomsky, 1995b, 2007) proposed that syntax can be broken down into one single and basic mechanism called Merge, which binds elements into a hierarchical structure. This is a strong claim as it is assumed to hold for all natural languages.[2]

What does it mean? An example may be the best way to clarify. When we listen to a sentence and hear the word *the* (determiner) followed by the word *ship* (noun), the computation Merge binds these two elements together to form a minimal hierarchical structure. In this example, [*the ship*] is called a determiner phrase. Then when we hear the word *sinks* (verb) next, the computation Merge again binds two elements together: the element determiner phrase [*the ship*] and the element [*sinks*]. This leads to a hierarchically structured sentence: *The ship sinks*. Applying this computation recursively, over and over again, an infinite number of sentences of any length can be generated and mentally represented. As we will see, the basic computation Merge, like other more complex syntactic processes, has a well-defined localization in the human brain.

But language most generally is a system that subserves the mapping of structured sequences of words into meaning. This mapping requires us to identify not only the syntactic relations of words in a sentence but also the meanings of the words to establish "who is doing what to whom." In contrast to syntax with its strict rules, the representation of words and their meaning is not as easy to investigate neurophysiologically, perhaps because for each word there are two types of semantic representations: a linguistic-semantic representation on the one hand and a conceptual-semantic representation on the other hand (Bierwisch, 1982). These two types of representation are hard to separate empirically as they partly overlap. Compared to the quite rich conceptual-semantic representations with all their visual, tactile, olfactory, and episodic associations (Collins and Loftus, 1975; Miller, 1978; Jackendoff, 1983; Lakoff, 1987; Pustejovsky, 1995), linguistic-semantic representations are restricted to the linguistically relevant parts and thus underspecified (Bierwisch, 1982). According to this view a linguistic-semantic representation only contains a subset of the information present in the conceptual-semantic representation, just sufficient to build grammatically correct and parsable sentences.

What does that mean psychologically and neuroscientifically? Consider the word *chair* in the sentence *He sits on a chair*. In order to process this sentence the linguistic-semantic representation that a chair is "something to sit on" is sufficient to understand the sentence. The conceptual-semantic representation contains much more information about all possible forms of chairs in the world, in different historical contexts, as well as individual contexts. Under this view the conceptual-semantic representation in contrast to a linguistic-semantic representation may even include the individual's episodic memory, for example, *the chair I saw in a museum last week*. Given the large variability in the conceptual-semantic representation across individuals, it is conceivable that the brain basis of conceptual-semantics is less confined than the linguistic-semantic representation.[3] To grasp the full meaning of a word thus often requires the inclusion of further aspects outside the core language system, such as the speaker's intention, and/or the speaker's situational or cultural background. Considering this, it is not surprising that comprehension sometimes fails—at least on the first attempt—but it is surprising that interpersonal communication most of the time is successful.

Language as a Brain System

The brain itself is a very complex system, and the description of the relation between language functions and the brain remains a big challenge. The brain consists of gray matter and white matter. The gray matter is composed of about 100 billion neuronal cells that are interconnected via trillions of synapses. Each neuron has a number of connections via which it receives signals from other neurons (these are the dendrites), and it also has connections via which it forward signals to other neurons (these are the axons). The axons come into contact with other neurons via synapses at which the transmission of the signals is realized by neurotransmitters. The white matter, in contrast, contains only few neuronal cells, being composed of fiber bundles connecting adjacent brain regions by short-range fiber bundles, or connecting more distant parts of the brain by long-range fiber bundles that guarantee communication between these. In their mature state, these fiber bundles are surrounded by myelin, which serves as insulation and enables rapid propagation of the signal. Both the gray and the white matter are the basis for all cognitive abilities, including language.

The brain's functioning, however, is not yet completely understood. This holds for all different neural levels, from the single neurons and the communication between them up to the level of local circuits and the level of macrocircuits at which neuronal ensembles or even entire brain regions communicate. But our knowledge about the brain has increased considerably over the past centuries and decades, allowing us to provide a first description of the neural language network. Today we are able to bring together data from different neuroscientific levels of the language-related brain regions ranging from the cellular and molecular level, to neural circuits, and up to the systems level represented by larger neural networks consisting of distant brain regions. All these levels of analyses should lead to a physiologically integrative view on language in the human brain.

A Brief History of the Language-Brain Relationship

Historically,[4] Franz Gall (1758–1828) was the first to suggest a relation between cognitive functions and particular brain regions, and he postulated that language is located in the left frontal lobe. In 1836 Marc Dax presented a talk and, based on his observations that lesions of the left hemisphere lead to language impairments, suggested an involvement of the left hemisphere for language (Dax, 1836). The first documented empirical proof that language is indeed represented in particular parts of the brain was a clinical case in the late nineteenth century. This first famous case was a patient of the French scientist Paul Broca. The patient was described to suffer from a severe deficit in language production as he was only able to utter the syllable *tan*. Broca described the language behavior of this patient well, but in order to report a language-brain relationship, the neurologists at that time had to wait until the patient died and a brain autopsy was conducted. During autopsy, neurologists

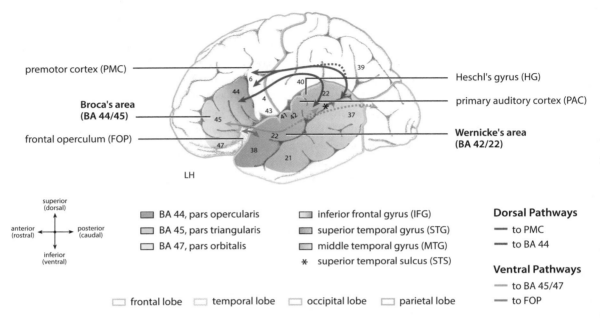

Figure 0.1
Neuroanatomy of language. Anatomical and cytoarchitectonic details of the left hemisphere (LH). Top: The different lobes (frontal, temporal, parietal, occipital) are marked by colored borders. Major language-relevant gyri (inferior frontal gyrus (IFG), superior temporal gyrus (STG), middle temporal gyrus (MTG)) are color-coded, superior temporal sulcus (STS) located between STG and MTG is marked by an asterisk. Numbers indicate language-relevant Brodmann areas (BA) that Brodmann (1909) defined on the basis of cytoarchitectonic characteristics. The coordinate labels (see bottom left) superior (dorsal)/inferior (ventral) indicate the position of the gyrus within a lobe (e.g., superior temporal gyrus) or within a BA (e.g., superior BA 44). The horizontal coordinate labels anterior (rostral)/posterior (caudal) indicate the position within a gyrus (e.g., anterior superior temporal gyrus). Broca's area consists of the pars opercularis (BA 44) and the pars triangularis (BA 45). Located anterior to Broca's area is the pars orbitalis (BA 47). The frontal operculum (FOP) is located ventrally and medially to BA 44, BA 45. The premotor cortex (PMC) is located in BA 6. Wernicke's area is defined as BA 42 and BA 22. The primary auditory cortex (PAC) and Heschl's gyrus (HG) are located in a lateral to medial orientation in the temporal lobe. White matter fiber tracts, i.e., the dorsal and ventral pathways connecting the language-relevant brain regions are indicated by color-coded arrows (see bottom right). Reprinted from Friederici (2011). The brain basis of language processing: From structure to function. *Physiological Reviews*, 91 (4): 1357–1392.

located the brain lesion of "Monsieur *Tan*" in the left inferior frontal gyrus, today called Broca's area (Broca, 1861) (see figure 0.1). Some years later, Carl Wernicke from Breslau described a number of patients who were characterized by a deficit in language comprehension (Wernicke, 1874). These patients were found to have lesions in the left temporal cortex, today called Wernicke's area (see figure 0.1). Based on these descriptions, Broca's area was seen to support language production, and Wernicke's area was taken to support language comprehension. These brain lesions were caused by a stroke—in these cases an ischemic stroke in which a blood vessel carrying blood to a particular brain area is blocked, leading to death of neuronal cells in the respective area. Later these language deficits were called Broca's and Wernicke's aphasia.

Initially, the respective language deficits were categorized on the basis of the patient's language output and comprehension during clinical observations. Patients with a primarily deficit language output often spoke in a telegraphic style leaving out all the function words, such that sentences sounded like a telegram as if grammar was gone. This symptom was called agrammatism (Pick, 1913). In contrast, patients who mainly appeared to have a comprehension deficit as described by Wernicke were shown to produce sentences that were not easy to understand for others. These patients often produce utterances in which constituents are incorrectly crossed or unfinished. For example, consider the following utterance in German and its literal English translation: *und ich nehme auch an dass das mit dem Sprechen oder was sonst überall, also das da, das ist es, sonst anders als die Beine so was hat ich keine /* and I suppose also that this with the speaking or what else overall, this one, this is it, different to the legs that one I did not have. This symptom was called paragrammatism (Kleist, 1914; Jakobsen, 1956).

Until the late 1970s and early 1980s, neurologists were not able to look into a patient's brain in vivo. Only with the advent of new imaging techniques did the possibility emerge to describe the relation between language behavior and brain lesion in a living patient. Together with the advancement in neuroimaging methods, crucial developments in linguistic and psycholinguistic theorizing occurred. This led to a revision of the classical view of Broca's area being responsible for language production, and Wernicke's area being responsible for comprehension.

This classical view of the function of Broca's area and Wernicke's area was revised when more fine-grained tests based on linguistic and psycholinguistic theories were applied, and the outcome of those tests required a functional redefinition of these brain regions (Zurif, Caramazza, and Myerson, 1972; Caramazza and Zurif, 1976). It was found that Broca's aphasics not only had problems in language production—they spoke in an agrammatic telegraphic style leaving out all function words—but they also demonstrated problems in language comprehension when confronted with grammatically complex constructions whose comprehension depended on function words. These were, for example, passive sentences—*The boy was pushed by the girl*—which were misinterpreted when only processing the content words *boy pushed girl*. The observation led to the view that Broca's area subserves grammatical processes during both production and comprehension. A similar shift in functional interpretation took place with respect to Wernicke's area. The finding that Wernicke's patients not only showed deficits in comprehending the meaning of utterances, but also displayed problems in word selection during production, led to the view that the temporal cortex where Wernicke's area is located supports lexical-semantic processes. Thus, it was linguistic and psycholinguistic theorizing that brought on new insights into the language-brain relationship. These studies provided first indications that different subcomponents of language, such as the grammar and the lexicon, might have different localizations in the brain.

With the development of new neuroscientific methods in the late twentieth century, the relation between language and the brain could be observed in living persons while processing language. Today the language network can be described to consist of a number of cortical regions in the left and the right hemisphere. These interact in time under the involvement of some subcortical structures that are not specific for language but may serve as a relational system between the language network and its sensory input systems (e.g., auditory cortex for auditory language and visual cortex for sign language) and its output systems (e.g., motor cortex for articulation and for signing). In this book I primarily focus on the cortical regions of the neural language network, which I will discuss with respect to their particular language function in the adult brain and the developing brain.

A Brief Neuroanatomy of Language
In order to map the language network in the human brain, we need a coordinate system that subdivides the brain into relevant parts and applies labels to these parts. Unfortunately, different neuroanatomical coordinate systems and labels have been and still are used nowadays. Therefore, I will present them all here. One way to describe the neuroanatomical location of a given function in the brain is to name the different gyri in the cortex: inferior frontal gyrus in the frontal lobe or superior temporal gyrus and middle temporal gyrus in the temporal lobe (see figure 0.1). This model represents the anatomical details of the left hemisphere, the dominant hemisphere for language.[5]

The brain regions that have been known for more than a century to be centers of language processing are Broca's area and Wernicke's area in the left hemisphere. These two brain regions have been described with respect to their neuroanatomical microstructure at the cytoarchitectonic level and more recently also at the neuroreceptorarchitectonic level. These microstructural analyses allow the parcellation of larger cortical areas, which is much more finely grained than any lesion-based functional parcellation, since brain lesions usually cover larger parts of cortical tissue.

The initial and still valid parcellation is the *cytoarchitectonic* description of the cortical structure, and was provided in 1909 by Korbian Brodmann (Brodmann, 1909). He only gave it for the left hemisphere because it took him many years to present this analysis, as he was looking at the cortex millimeter by millimeter under the microscope. He analyzed the six layers of the cortex with respect to the type of neurons and their density. As a result, he identified cytoarchitectonically different regions, which he numbered, and are known today as Brodmann areas (BA) (see figure 0.1). This numbering is still used to indicate the location of a functional activity in the brain. Based on the cytoarchitectonic analysis by Brodmann, Broca's area, for example, was subdivided into a more posterior part (BA 44) and a more anterior part (BA 45). This cytoarchitectonic subdivision of Broca's area into two subparts was later confirmed by objective cytoarchitectonic analyses (Amunts et al.,

1999). The subdivision into BA 44 and BA 45 coincides with the neuroanatomical separation into pars opercularis and pars triangularis, respectively. Wernicke's area is located in the temporal cortex and refers to BA 42/22. Cytoarchitectonically, BA 22 has been differentiated from the primary and secondary auditory cortex (BA 41, 42) and the inferiorly located middle temporal gyrus (Brodmann, 1909).

In the twenty-first century, novel methods of parcellating different brain regions have been developed. One method is *receptorarchitectonic* parcellation and the other method is connectivity-based parcellation. The receptorarchitectonic analysis takes into consideration the type and density of neurotransmitters in a given region (Zilles and Amunts, 2009; Amunts et al., 2010;). Neurotransmitters are biochemicals that transfer signals between neurons and thereby transport information from one neuron to the next. The receptorarchitectonic analysis also indicates a global distinction between BA 44 and BA 45, an even more fine-grained differentiation within each of these areas, and, moreover, a distinction between lateral area 45 and 44 and the more ventral-medial area, the frontal operculum (Amunts et al., 2010; for details see legend of figure 0.2).

The *connectivity-based* parcellation approach refers to an analysis that differentiates cortical areas on the basis of their white-matter fiber connections to other areas in the brain (Johansen-Berg, 2004; Anwander, Tittgemeyer, von Cramon, Friederici, and Knösche, 2007). With the use of connectivity-based parcellation analysis, Broca's area has also been divided into BA 44 and BA 45, both of which are separated from the more ventrally

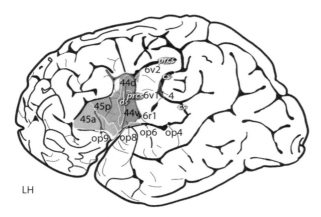

Figure 0.2
Receptorarchitectonic parcellation of left IFG. Receptorarchitectonic differences of areas within the inferior frontal gyrus (IFG) analyzed in post mortem brain. Extent of delineated areas projected to the lateral surface of an individual left hemisphere (LH) of a post mortem brain. Different receptor binding sites were studied. The color coding indicates receptorarchitectonically defined borders. Area 44 can be subdivided into 44d (dorsal) and 44v (ventral). Area 45 can be subdivided receptorarchitectonically into an anterior (45a) and a posterior (45p) part. Area 6 can be subdivided into three subparts. Other abbreviations: op, Operculum (numbering indicates different subparts); prcs, precentral sulcus; cs, central sulcus. Reprinted from Amunts et al. (2010). Broca's region: novel organizational principles and multiple receptor mapping. *PLoS Biology*, 8: e1000489.

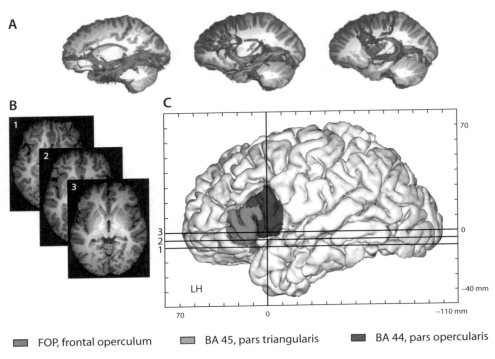

Figure 0.3
Structural connectivity-based parcellation of left IFG. Parcellation of the inferior frontal gyrus (IFG) based on the areas' connectivity of the white matter fiber tracts. (A) Parcellation results for the inferior frontal gyrus into three clusters for the group average, pars opercularis, BA 44 (purple); pars triangularis, BA 45 (orange); and frontal operculum, FOP (red). (B) Delineates these areas for one participant in three different axial levels on the Talairach-scaled cortical surfaces of the left hemisphere (LH). (C) Shows the location of the axial slices at three different levels that equidistantly cut through the parcellated areas; locations of axial slices are indicated by solid horizontal lines (numbered 1–3). Note the intersubject variability with respect to differences in shape and size of both areas, but the invariable principle order of the three clusters. Adapted from Anwander et al. (2007). Connectivity-based parcellation of Broca's area. *Cerebral Cortex*, 17 (4): 816–825, by permission of Oxford University Press.

located frontal operculum (Anwander et al., 2007; figure 0.3). Using a connectivity-based approach, BA 22 in the superior temporal gyrus can be parcellated into three subareas (Raettig, Kotz, Anwander, von Cramon, and Friederici, submitted). Based on this approach the superior temporal gyrus and the superior temporal sulcus can be subdivided into three parts: a posterior, middle, and anterior subpart.

The combined data from different analyses provide a strong empirical basis for the view that particular brain regions can be separated from each other at different neuroscientific levels: at the lower levels on the basis of cyto- or receptorarchitectonic parameters, and at higher levels on the basis of structural connectivity. As we will see, these separations at the different neuroscientific levels are most relevant for the differentiation of functional subcomponents in the cognitive language system.

The brain regions that constitute the major parts of the functional language network in the left hemisphere—in particular Broca's area and Wernicke's area—are displayed and color-coded in figure 0.1. These regions subserve the processing of words and sentences across all languages in the world. Homolog areas in the right hemisphere (not displayed in figure 0.1) are also part of the language network and subserve the processing of prosodic information, such as a sentence's melody. The different brain regions in the network are connected by fiber bundles, which provide pathways that guarantee information transmission from one region to the other (schematically indicated by the colored arrows in figure 0.1). These pathways exist within each hemisphere and connect frontal and temporal language regions, and, moreover, there are fiber bundles that connect the left and right hemispheres allowing information transfer between the two hemispheres.

From this short description it already becomes clear that the relation between language and the brain is a complex one, in particular when considering that the different parts of the network have to work in real-time to make language processing possible.

Language and the Brain: The Story to Be Told

My goal in this book is to provide a coherent view on the relation between language and the brain based on the multifold empirical data coming from different research fields: electrophysiological and functional neuroimaging data of the adult language system, neuroscientific data of the developmental language system, and data on the brain structure and its maturation in humans as well as in non-human primates.

I have organized these data in four larger parts at the end of which I define the anatomical model outlined in figure 0.1 with respect to language functions. In part I (chapters 1 and 2) I describe the basic language functions and their brain basis from auditory input to sentence comprehension, both with respect to "where" and "when" different subprocesses take place. I dissect the complex language system into its subparts and then put the bits and pieces together again to provide then a functional neuroanatomical model of language comprehension. I follow chapter 1 with an excursion (chapter 2) that goes beyond the issue of language comprehension. I provide empirical data in support of the view that there is a common knowledge base for language comprehension and production. Moreover, I briefly discuss aspects of communication that are not part of the core language system. In part II (chapters 3 and 4) I provide the description of the language networks connecting the different language-relevant brain region, both structurally and functionally. Here I delineate the spatial and temporal characteristics of and within the neural language network. I will propose a dynamic neural language circuit describing the interaction of the language-relevant areas in the adult brain. In part III (chapters 5 and 6) I discuss the brain basis of language acquisition, looking into language development during early childhood and into aspects of second language learning. Here we will learn about the neurobiological conditions of language development. A neurocognitive model of the ontogeny of language capturing these data is proposed. In part IV (chapters 7 and 8), first I tackle questions about the evolution of language and the possible underlying neural constraints, and then bring together the material covered in the different chapters. At the end I propose an integrative view on the neural basis of the core language system.

As the book unfolds I hope to guide the reader through a rich database for language and the brain by starting with a purely functional model which is then developed into a neurocognitive model becoming ever more complex as additional data are discussed. We will see that the different language-relevant brain regions alone cannot explain language, but that the information exchange between these supported by white matter fiber tracts is crucial, both in language development and evolution.

In order to ease the recognition of language-relevant brain regions and the pathways across the figures representing variations of the model presented in figure 0.1, I have kept the colors constant. Moreover, I have also adapted the colors of the figures coming from different published papers for coherence.

I

The miracle of language often occurs to us only when language does not function anymore, such as after brain damage, or when language acquisition becomes really tedious, such as when learning a second language. In everyday language, however, we never think about the efficiency of the language system that allows us to produce and understand complex sentences in milliseconds. We just use it. Within a very short time frame we process the sound of language, the meaning of words, and the syntactic structure of an utterance.

Psycholinguistic models assume that these processes, which are concerned with different information types, take place in different subsystems that work together in time in a partly cascadic and partly parallel manner. During language comprehension this means that once the acoustic information is perceived, the different subprocesses dealing with phonological, syntactic, and semantic information have to be performed. These processes constitute what is called the "core language system" (Berwick, Friederici, Chomsky, and Bolhuis, 2013; see also Fedorenko and Thompson-Schill, 2014). Comprehension during interpersonal communication, however, may also involve situational and emotional aspects that are not considered part of the core language system since these, as we will see, can act independently of language.

Chapter 1 focuses on the cognitive structure of these different processes during language comprehension and their respective neural basis. An excursion (chapter 2) has two purposes. It will discuss a common knowledge base of syntactic rules and words for language comprehension and language production. Moreover, it will briefly touch upon aspects of communication that go beyond the core language system.

1

Language Functions in the Brain: From Auditory Input to Sentence Comprehension

To comprehend an utterance means to interpret what has been said. Comprehension is achieved once linguistic meaning is successfully mapped onto and integrated into existing world knowledge. In order to reach this goal, the language processing system has to compute a number of different types of information including acoustic, phonological, syntactic, and semantic information—that is, information about the speech sounds, the grammatical structure of an utterance, and the relation of meaning between different words. Linguistic theories generally assign these different information types to distinct subcomponents of language. Phonetics and phonology deal with sounds of speech and language, respectively. Phonetics is about the physical aspects of sounds, whereas phonology is about the abstract aspects of sounds independent of their physical realization in speech. Speech, moreover, carries physical information concerning units larger than a single sound, such as words and phrases realized as stress/accentuation or intonation at a suprasegmental level. This is also considered part of phonology. Syntax as an important subcomponent deals with the grammatical structure of phrases and sentences. Semantics as the next subcomponent deals with the meaning of words and word combinations. The computation of these different information types has to take place within milliseconds in order to guarantee comprehension online. This speed is necessary as an average speaker talks with a speech rate of about 300 syllables per minute, that is 1 syllable in 200 milliseconds (ms), which is incredibly fast.

Chapter 1 focuses on the core language system in the mature brain and describes the procedural way from auditory input up to language comprehension based on a neurocognitive model that defines the different subprocesses and its neural basis, both with respect to where in the brain these subprocesses take place and what their temporal relation is.

1.1 A Cognitive Model of Auditory Language Comprehension

The language processing model sketched in figure 1.1 assumes different processing stages from auditory input up to interpretation (for versions of this model see Friederici, 2002, 2011). At the first stage there are the acoustic-phonological processes, which are dealt

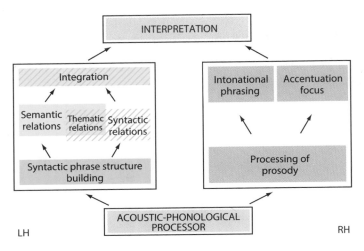

Figure 1.1
Cognitive model of auditory language comprehension. The model represents the different subprocesses during comprehension from auditory input to interpretation. Different color-coded boxes represent subprocesses assumed to take place in the left hemisphere (LH) and right hemisphere (RH). For the details of these subprocesses see text. Adapted from Friederici (2011). The brain basis of language processing: From structure to function. *Physiological Reviews*, 91 (4): 1357–1392.

with by the auditory cortex in both hemispheres. The output of this initial process is then further processed by the left hemisphere and the right hemisphere according to their specifications—segmental sounds in the left hemisphere and suprasegmental parameters in the right hemisphere. In the left hemisphere, three processing stages deal with syntactic and semantic information before interpretation and integration into the existing world knowledge is possible. In the right hemisphere, at least two separate aspects of prosodic information in speech have to be dealt with: the first is the processing of sentence melody and intonation, which can signal the beginning or end of a phrase in a sentence, and the second is the processing of accentuation relevant for thematic focus. During auditory speech comprehension, the different subsystems within one hemisphere (as well as across the two hemispheres) work together to achieve smooth comprehension.

These different processes are described in the functional cognitive model shown in figure 1.1 (Friederici, 2011). The processes are assumed to run in a partly parallel, but cascadic manner. That means that each subsystem forwards its output to the next subsystem as soon as possible, thereby causing several subsystems to work in parallel.[1] Each of the subsystems corresponds to local cortical networks in specialized brain areas (Friederici, 2011), which together form large-scale dynamic neural networks supporting language comprehension (Friederici and Singer, 2015).

The challenge of an adequate description of the neural basis of language is to identify the different processing subsystems in the brain not only with respect to where they are located in the brain but, moreover, how these subsystems interact in time. Language-related activation, like any other brain activation, is based on the activation of neurons and neuronal ensembles resulting in electrical activity unfolding in milliseconds that can be measured at the scalp by means of electroencephalography. Since each electrical current has its magnetic field, brain activity can also be measured with magnetoencephalography. Moreover, because neuronal activity is dependent on cerebral blood flow, measurements of changes in the blood flow can be taken as an indicator for brain activity that can be measured by functional magnetic resonance imaging and by near-infrared spectroscopy. Here I will take the following approach: For each of the subprocesses postulated in the functional auditory language comprehension model, I will describe the neurotemporal parameters as reflected by the time-sensitive electroencephalography or magnetoencephalography measures and specify the neurospatial parameters of one or more brain areas as reflected by changes in the blood flow using functional magnetic resonance imaging, by near-infrared spectroscopy, or by calculating the location of the source of a given effect observed in electroencephalography or magnetoencephalography data. I will briefly describe these neuroscientific methods, as they are a necessary part of understanding the data underlying the language-brain relationship and the resulting neurocognitive model of language comprehension.

Traditionally, the language-brain relationship was investigated through work with patients whose brain lesions were causing particular language impairments. In the late 19th century lesions were diagnosed postmortem, but thanks to the availability of computer tomography in the late 20th century lesions could be identified in vivo. The advent of magnetic resonance imaging in the past decades has led to fundamental knowledge about the representation of different aspects of language in the lesioned and intact brain. Due to the fact that brain lesions, in particular vascular lesions caused by a stroke, are not always well circumscribed but rather concern the entire territory of a given blood vessel and are thus often unspecific, in this book I will not cover the entire literature on patient-related language studies. However, I will report on patient studies that together with other neuroscientific studies provide essential information for an adequate description of language in the brain.

Neuroscientific Methods
There are two time-sensitive neurophysiological methods, *electroencephalography* and *magnetoencephalography*. Electroencephalography records electrical activity in a non-invasive manner as the brain generated electrical activity is measured at the outside of the brain, namely at the scalp. Electroencephalography registers neural oscillations that are reflected in different frequency bands. In neurocognitive research, electroencephalography is used to measure brain activity time-locked to a particular stimulus, provided to the individual either auditorily or visual, called *event-related brain potentials* (ERP). The ERP is a quantification of electrical activity in the cortex in response to a particular type of stimulus

event with high temporal resolution in the order of milliseconds. As the signal in response to one event is very small, the electrical activity has to be averaged over events of similar type in order to reach a larger signal against the ongoing brain activity not related to the stimulus. Average electrocortical activity appears as waveforms in which so-called ERP components have either positive or negative polarity relative to baseline, have a certain temporal latency in milliseconds after stimulus onset, and have a characteristic but poorly resolved spatial distribution over the scalp. Both the polarity and the time point at which the maximum ERP component occurs, as well as partly its distribution, are the basis for the names of the different ERP components. For example: negativity (N) around 400 ms is called N400, and positivity (P) around 600 ms is called P600.

For the speech and language domain there are at least five ERP components that must be considered. The temporal aspects that will enter the neurocognitive model later are taken from ERP studies. These studies often include violations of respective information types in sentences, because the brain reacts most strongly to unexpected events and violations in the input. The language-relevant ERP components are ordered in time and reflect the different processes as described in the functional model in figure 1.1. The first is the N100 (negativity around 100 ms), which has been associated with acoustic processes; the second is the ELAN (early left anterior negativity, between 120 and 200 ms), which reflects phrase structure building processes; the third is the LAN (left anterior negativity, between 300 and 500 ms), which reflects morphosyntactic processes; the fourth is the N400 (negativity around 400 ms), which reflects lexical-semantic processes; and the fifth is the P600 (positivity after 600 ms), which is associated with late syntactic integration processes. These ERP components are discussed in detail below (examples of ERPs are given in figures 1.4, 1.14, and 1.17).

Magnetoencephalography is a related neurophysiological method that records magnetic fields induced by electrocortical activity. Magnetoencephalography provides information about the amplitude, latency, and topography of language-related magnetic components with a temporal resolution comparable to ERPs but with an improved spatial resolution due to a high number of registration channels.

Functional magnetic resonance imaging, a technique to localize activity related to particular functions in the brain, is widely used for neurolinguistic experiments. It has replaced the partly invasive positron emission tomography by a non-invasive, state-of-the-art method for functional-anatomical reconstruction of the language network in the order of submillimeters. However, the temporal resolution of magnetic resonance imaging is limited as it measures the hemodynamics of the brain activity taking place in the order of seconds. Functional magnetic resonance imaging reveals precise information about the location and the magnitude of neural activity changes in response to external stimulation but also about intrinsic fluctuations at rest, that is, in the absence of external stimulation. These neural activity changes are reflected in blood-oxygen-level-dependent (BOLD) signal changes based on the effect of neurovascular coupling.

Near-infrared spectroscopy allows more flexible recording of the BOLD response since the monitoring system is mounted directly on the participant's head, which means that the participant does not have to lie still, as is the case during functional magnetic resonance imaging. This advantage made it an important method for language acquisition research in infants and young children. However, the spatial resolution of near-infrared spectroscopy is much lower than that of magnetic resonance imaging whereas its temporal resolution is just as poor. For this reason this technique is mainly used with very young participants.

In addition to these methods that measure brain function, there is a method that allows us to register aspects of the brain's structure. *Structural magnetic resonance imaging* provides detailed morphometric and geometric features of the brain's gray and white matter such as its volume, density, thickness, and surface area. Diffusion-weighted magnetic resonance imaging, especially diffusion tensor imaging, is used to reconstruct the trajectory and quantify tissue probabilities of white matter fiber bundles interconnecting brain areas.

Each of the mentioned non-invasive methods provides either fine-grained temporal or spatial information, but not both. Currently, the best approach for gaining reliable knowledge about the brain-function relation is to combine high temporal and spatial resolution. Invasive techniques, particularly intracranial electrophysiology, overcome this need to combine techniques but are exclusively feasible in clinical settings. Based on the available non-invasive techniques such as electroencephalography, magnetoencephalography, near-infrared spectroscopy, and functional magnetic resonance imaging methods, we are able to specify the neural correlates of those regions that support the different subprocesses described in the neurocognitive model, from the acoustic input to the level at which comprehension is achieved.

Summary and Plan for the First Chapter

In section 1.1, I presented the cognitive model of language comprehension with its functional processing components and processing steps, from auditory input to comprehension.[2] I also presented the neuroscientific methods that are used to measure the brain's functional activation and its structure. In the following sections the different functional components represented by boxes in the cognitive model will be described in more detail with respect to their brain location and their activation as it unfolds in time. The acoustic-phonological processes will be discussed in section 1.2, followed by access to syntactic and semantic information encoded in the word in section 1.3, initial phrase structure building in section 1.4, the processing of syntactic, semantic, and thematic relations and their integration in sections 1.5, 1.6, and 1.7. The processing of prosodic information will be reviewed in section 1.8. In section 1.9, I will present a brain-based model in which the different functional components displayed in the cognitive model of auditory language comprehension are localized anatomically.

1.2 Acoustic-Phonological Processes

The comprehension of spoken language begins with the acoustic-phonological analysis of the speech input. This step precedes any further language process and takes place during the first couple of milliseconds after speech onset. The process is reflected in particular ERP components that make it possible to describe the temporal course of acoustic-phonological processes and, moreover, the localization of these in the brain.

The Temporal Course: EEG/MEG Studies
Each language contains different speech sounds relevant for the recognition of words. Those speech sounds that are crucial for the differentiation of words in a given language are called phonemes in linguistic theory. For example, in English the two speech sounds /l/ and /r/ differentiate the words *low* and *row*, and, therefore, are phonemes in English. In Japanese, this difference between /l/ and /r/ does not distinguish between words; these sounds are not relevant in Japanese and thus are not phonemes in Japanese. For each language the identification of phonemes is a crucial first step during speech perception.

Neuroscientists have recognized an ERP effect that occurs around 100 ms after speech onset and reflects the identification of phonemes. This effect, called N100, is a negative deflection in the waveform of the ERP (Obleser, Lahiri, and Eulitz, 2003). A second ERP component discussed in the context of acoustic-phonological processing occurs shortly after 100 ms is the so-called mismatch negativity, which has been shown to reflect the discrimination of acoustic and phoneme categories (Näätänen et al., 1997). Note that this ERP component is not specific to language, but reflects the discrimination of auditory categories. It can, however, be used to investigate linguistic aspects such as phoneme discrimination using phonemes or syllables as stimulus material (for a review, see Näätänen and Alho, 1997; Phillips, 2001; Näätänen, Paavilainen, Rinne, and Alho, 2007; Winkler, Horvath, Weisz, and Trejo, 2009).

Studies exist to indicate that language-specific representations are built at the phoneme and syllable level shortly after language is perceived (Dehaene-Lambertz, Dupoux, and Gout, 2000; Phillips et al., 2000). A study on phoneme processing in word-like units comparing French and Japanese listeners found that during speech processing, the auditory input signal is directly parsed into the language-specific phonological format of the listener's native language. This allows fast computation of the phonological representation from the speech input and may thus facilitate access to the phonological word form stored in the lexicon. The lexicon is generally thought to be an inventory of all words, and that attached to each word there is information about its phonological word form, about its syntactic category (noun, verb, etc.), and about its meaning. The phonological form of a word, however, is the means by which the lexicon can be accessed (Dehaene-Lambertz et al., 2000).

Although early speech-related ERP effects have long been reported to occur around 100 ms after speech onset for phonetic and phonemic aspects of spoken language (see below), more recent work even suggests very early effects before 100 ms. During speech processing this has been reported for the recognition of a word's lexical status, for example, "Is this element a word of my language or not"? (MacGregor, Pulvermüller, van Casteren, and Shtyrov, 2012) as well as for the recognition of a word's syntactic category, for example, "Is the word's syntactic category correct in this phrasal context"? (Herrmann, Maess, Hahne, Schröger, and Friederici, 2011). These early ERP effects challenge the view of a strictly incremental speech processing system in which a subprocess can only start when the previous one is finished. They rather suggest a partly parallel working comprehension system in which even partial information from the acoustic-phonological processor is forwarded to the next processing level to allow early word-related processes.

Localization: fMRI Studies
The obvious neural candidate to support the acoustic-phonological analysis of speech is the auditory cortex and adjacent areas. And indeed the N100 and the mismatch negativity in response to phonemes have been located in or in the vicinity of the auditory cortex (Diesch, Eulitz, Hampson, and Ross, 1996; Poeppel et al., 1997), thereby indicating that these processes take place early during speech perception in this region. This is compatible with results from neuroimaging studies on phoneme processing that localize the N100 component for vowels and consonants in the so-called Heschl's gyrus and the planum temporale, regions which as parts of the auditory cortex are both located in the superior part of the temporal lobe (for anatomical details see figure 0.1) (Obleser et al., 2003; Obleser, Scott, and Eulitz, 2006; Shestakova, Brattico, Soloviev, Klucharev, and Huotilainen, 2004). But the auditory cortex as such must be a general processing system as it has to deal with any type of auditory input.

In order to describe to what extent phoneme perception is located in the domain-general auditory region or a specific region for processing speech, subregions in the superior temporal lobe had to be empirically evaluated both neuroanatomically and functionally. Such data could inform functional models as to whether these should assume a language-specific component for phoneme processing. In a first attempt to specify subregions in the auditory cortex and adjacent areas in the human brain, researchers have primarily investigated non-language animals. For non-human primates neuroanatomical data identified a core region in Heschl's gyrus, as well as a surrounding belt and parabelt region (Scott and Johnsrude, 2003; Rauschecker and Scott, 2009). In humans, the primary auditory cortex is located on the superior surface of the temporal lobe bilaterally in the Heschl's gyrus, as part of the primary auditory cortex (BA 41) (see figure 0.1). In addition three regions can be identified adjacent to Heschl's gyrus: a region located posteriorly, called the planum temporale; a region located anterolaterally, called the planum polare; and a region located lateral to Heschl's gyrus, in the superior temporal gyrus, extending inferiorly to the superior temporal sulcus. The superior temporal sulcus is the sulcus between the superior temporal gyrus and the middle temporal gyrus. All these regions are involved in the acoustic analysis of speech.[3]

Functionally, all three regions—Heschl's gyrus, the planum polare, and the planum temporale—are involved in speech analysis, but Heschl's gyrus serves a more general auditory function. When processing speech, a primary step is to differentiate speech from non-speech acoustic signals, and, for a description of the neuroanatomical basis of speech comprehension, it would be of major interest to identify where different subprocesses in the auditory processing stream take place. We know that any primary auditory analysis is computed in Heschl's gyrus because functional neuroimaging studies show that Heschl's gyrus is activated by any type of sound (Mummery, Ashburner, Scott, and Wise, 1999; Johnsrude, Giraud, and Frackowiak, 2002). The region lateral to Heschl's gyrus in the superior temporal gyrus extending into the superior temporal sulcus has been found to respond to both variations of frequency and spectral information in non-speech sounds (Hall et al., 2002) as well as to acoustic speech parameters (Binder et al., 2000) and is therefore not specialized for speech. Functional imaging studies, moreover, have shown that the planum temporale, like Heschl's gyrus, does not react specifically to speech sounds, at least when compared with equally complex non-speech sounds (Démonet et al., 1992; Zatorre, Evans, Meyer, and Gjedde, 1992; Wise et al., 2001). However, in a time-sensitive functional magnetic resonance imaging paradigm investigating the information flow from Heschl's gyrus to the planum temporale, it has been demonstrated that Heschl's gyrus is active prior to the planum temporale (Zaehle, Wustenberg, Meyer, and Jancke, 2004). From these data it can be concluded that Heschl's gyrus supports the processing of sound signals per se, whereas the planum temporale may be involved in a later process of categorizing sound signals. Therefore, the planum temporale has been proposed as the region that segregates and matches spectrotemporal patterns, and serves as a "computational hub," categorizing and gating the information to higher-order cortical areas for further processes (Griffiths and Warren, 2002).

During speech processing the perception of phonemes—consonants, in the example here—two temporal regions were also observed to be active (Obleser, Zimmermann, Van Meter, and Rauschecker, 2007). One region was found to process the basic acoustic characteristics of the signal, independent of speech, and a second region was found to differentiate between speech and non-speech sounds. Given their respective responsibilities, these findings suggest hierarchical steps in processing acoustic information (Kumar, Stephan, Warren, Friston, and Griffiths, 2007). The functional magnetic resonance imaging finding is also consistent with magnetoencephalographic evidence locating the relatively early N100 response to consonants in Heschl's gyrus and planum temporale (Obleser et al., 2006). Moreover, it is compatible with patient evidence showing that lesions in the posterior superior temporal gyrus lead to word deafness, the inability to process and understand words, as well as to deficits in the perception of non-speech sounds (Pinard, Chertkow, Black, and Peretz, 2002). These data suggest that during auditory language processing two brain regions are involved, a first one that processes auditory information independent of whether it is speech or not, and a second region that categorizes the acoustic information into non-speech and speech sounds.[4]

Left and Right Hemisphere
Auditory processing is subserved by the auditory cortices in both hemispheres. Functionally, primary auditory cortices in the left and the right hemispheres respond to speech and tonal pitch, but they appear to have different computational preferences. While the left primary auditory cortex reacts specifically to speech sound characteristics, the right primary auditory cortex reacts to characteristics of tonal pitch (Zatorre, Belin, and Penhune, 2002). The relative specialization of the two auditory cortices in the left and the right hemisphere for these stimulus types has been explained as being based on the different sensitivity of these regions to temporal and spectral characteristics present in the input. The left hemisphere has been described to be specialized for rapidly changing information with a limited frequency resolution whereas the right hemisphere has been described to be specialized for stimuli with reverse characteristics. The former system would be ideal for the perception and recognition of speech sounds, as the determination of these (i.e., phonemes in a sequence) requires a system with a time resolution of 20–50 ms. The latter system would be able to deal with suprasegmental information (i.e., prosody) requiring a system with a time resolution of 150–300 ms. Hickok and Poeppel (2007) proposed that the left and right hemisphere generally work at different time scales reflected in different frequency band–related brain waves in the electroencephalographic signal, leading to a relative lateralization of functions with the left hemisphere primarily working in the gamma band range, and the right hemisphere in the theta band range (Giraud et al., 2007).

It is still an open question how information associated with these different time scales is processed and integrated during online speech perception. There is behavioral evidence from psychophysics that phoneme-sized (10–40 Hz) and syllable-sized (2–10 Hz) information are bound to the different time scales (Chait, Greenberg, Arai, Simon, and Poeppel, 2015). An interesting question is, how can these different time scales relevant for speech processing that are present in parallel be reflected in the brain's activity? Giraud and Poeppel (2012), who argue for a principled relation between time scales in speech and frequencies of cortical oscillatory activations, provide an answer to this question. Oscillations of cortical activations come in different frequency bands: delta (1–4 Hz), theta (4–8 Hz), beta (13–30 Hz), and gamma (30–70 Hz). In their study, Giraud and Poeppel (2012) showed that phonemic, syllabic, and phrasal processing, which take place in parallel, are reflected in nested theta-gamma band oscillation patterns in the auditory cortex. Thus, the brain's solution to parallel processes may be to nest different frequency bands that reflect different aspects of processing—for example phoneme processing and syllable processing—into each other. (For more details on oscillatory activations, see section 4.2.)

When considering functional levels of speech perception, psychologists have proposed "intelligibility" (i.e., to identify what has been said) in its most general sense as a next relevant level. The methodological approach used to investigate processes at this level is the manipulation of the acoustic signal by spectrally rotating normal speech to render the speech signals less intelligible (Blesser, 1972). Studies using such manipulations have consistently shown that the anterior superior temporal sulcus is systematically activated as a function of intelligibility (Scott, Blank, Rosen, and Wise, 2000; Crinion, Lambon-Ralph, Warburton, and Wise, 2003; Narain et al., 2003; Obleser, Zimmermann, et al., 2007; Obleser and Kotz, 2010). In contrast the posterior superior temporal sulcus was found to be equally activated by normal speech and less intelligible speech (Shannon, Zeng, Kamath, Wygonski, and Ekelid, 1995). This led to the idea that the posterior superior temporal sulcus is involved in the short-term representation of sequences of sounds that contain some phonetic information (without necessarily being intelligible) (Scott et al., 2000), whereas the anterior superior temporal sulcus is involved in processes necessary for the identification of speech.

A clear functional differentiation into different subregions within the superior temporal gyrus/superior temporal sulcus of the left hemisphere into three parts—a posterior, a middle, and an anterior part—has been demonstrated by Giraud and Price (2012). The middle portion of the superior temporal sulcus/superior temporal gyrus was found to respond to general sounds as well as speech sounds, whereas the anterior and posterior portions of the superior temporal gyrus respond to speech only (see figure 1.2). These two latter regions may have different functions, as proposed above.

This functional differentiation is interesting in light of data from a connectivity-based parcellation study which parcelled the superior temporal gyrus into three subparts based on their structural connectivity—that is, their structural connections to other regions in the brain. With this parcellation-by-tractography approach, the superior temporal gyrus and the superior temporal sulcus can also be divided into three parts: a part posterior to Heschl's gyrus, a middle part, and a part anterior to Heschl's gyrus (Raettig, Kotz, Anwander, von Cramon, and Friederici, submitted; figure 1.3). These three subparts are connected via white matter fiber bundles to different regions in the frontal cortex, in particular BA 44, BA 45, and BA 47, which subserve different language functions, as we will see later. This tractography-based parcellation mirrors the subdivision of the superior temporal cortex into three parts based on functional relevance, in that only the posterior and the anterior part supported speech processing, while the middle part subserved auditory processing in general (Giraud and Price, 2001).

Based on a meta-analysis of different studies investigating word, phrase, and sentence processing, the superior temporal gyrus has been interpreted to reflect a functional posterior-to-anterior gradient for speech processing: when speech stimuli become more complex, more anterior portions of the superior temporal gyrus are recruited (DeWitt and Rauschecker, 2012).

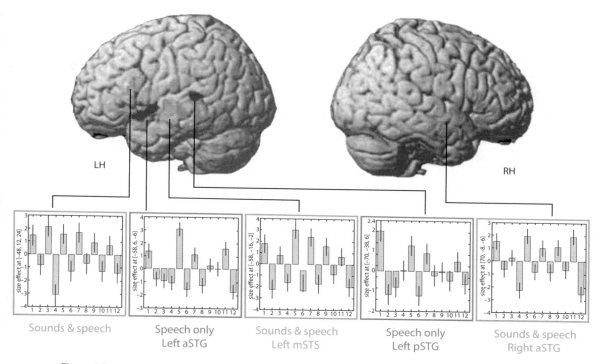

Figure 1.2
Brain activations for sounds and speech. Activations common to sounds and speech relative to noise (orange) and specific to speech (red) rendered onto left and right hemispheres of a template brain. Histograms illustrate relative blood flow response in each activated area for 12 conditions including words, syllables, and non-language sounds. Adapted from Giraud and Price (2001). The constraints functional neuroimaging places on classical models of auditory word processing. *Journal of Cognitive Neuroscience*, 13 (6): 754–765. © 2001 by the Massachusetts Institute of Technology.

Summary

The processing of language-relevant sounds during acoustic speech input involves several processing steps. As a first processing step during auditory language comprehension, the brain performs an acoustic analysis in an auditory cortical network starting at the primary auditory cortex, which then distributes the information in two directions: posteriorly toward the planum temporale and posterior superior temporal gyrus, and anteriorly toward the planum polare and anterior superior temporal gyrus. The planum temporale in the posterior superior temporal gyrus has been suggested as the general auditory "computational hub" from which information is gated to higher-order cortical regions (Griffiths and Warren, 2002). Thus in this section we have learned that during the initial processing steps of speech recognition, physical acoustic and phonological categorization processes are presented differently in the brain. These observations can be mapped onto the functional

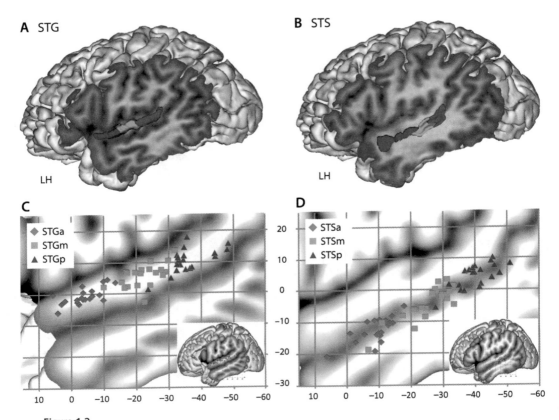

Figure 1.3
Connectivity-based parcellation of left STG/STS. Parcellation of the (A) superior temporal gyrus (STG) and (B) superior temporal sulcus (STS) of the left hemisphere (LH) for one representative participant into 3 different areas—anterior (a, red); middle (m, orange); and posterior (p, purple)—superimposed on an individual brain. The STS areas are plotted on top of the anatomical slice for better visualization. Intersubject variability of the parcellation result of (C) left STG and (D) left STS. Each participant is represented by the respective symbol (diamond, cube, triangle) within each of the 3 areas. Centers-of-mass of the individual anterior (red), middle (orange), and posterior (purple) subdivisions of the STG/STS (in MNI coordinates originated by the Montreal Neurological Institute) displayed on the MNI 152 standard brain, but note the general subdivision of the regions into 3 clusters. Adapted from Raettig et al. (submitted). The language connection: a structural connectivity-based parcellation of the left superior temporal cortex.

differentiation of acoustic (non-speech and speech) versus speech-related processes relevant for any psycholinguistic modeling.

1.3 From Word Form to Syntactic and Lexical-Semantic Information

Once acoustic-phonological processes have identified initial speech sounds (i.e., phonemes), access to a particular word in the lexicon—that is, the inventory of words—can begin. The entry of a given word in the lexicon stores different types of information relevant for further processing: in addition to the word form, it stores information about its syntactic word category (e.g., noun, verb) and meaning. During lexical access, various processing steps have to be performed. As a first step the word form must be identified. The processing system has to check whether the combination of identified phonemes in a given sequence is legal in the target language. If no match between the phoneme sequence and a lexical entry is found, the system reacts with an error signal, indicating "no entry found." If a match is found, access to the word in the lexicon can start and syntactic and semantic information can be retrieved.

This two-step process is suggested by a meta-analysis on brain imaging studies contrasting the processing of real words against words that are phonologically possible words in a language, but do not exist in the lexicon of that language (Davis and Gaskell, 2009). The meta-analysis revealed two effects that the authors related to two subprocesses reflected in two different brain activation patterns. Pseudowords—words that are phonologically possible words of a given language, but are not real words—elicit solid activation in the superior temporal gyrus close to the auditory cortex, indicating an initial first search and checking process of possible word forms. Real words, in contrast, elicit more activation in the anterior, posterior, and inferior regions of the temporal cortex, suggesting a second step during which lexical-semantic and word category information, encoded in the lexical entry, are processed.

Lexical Access

The time course of lexical access can be modeled to start as soon as the acoustic-phonological processes lead to the identification of a first phonological legal syllable that can trigger lexical access. A number of behavioral studies suggest that an initial syllable of a possible word builds the cohort for lexical access (Marslen-Wilson, 1987) and that this cohort is reduced in size once contextual information from the prior sentence part is available (Zwitserlood, 1989). Neurophysiologically this process is reflected in the brain by a number of ERP effects which can best be described when considering the stimuli material. Consider, for example, a sentence such as: *To light up the dark she needed her can_*, which should be completed by the word *candle*. In an ERP experiment, such sentence fragments were presented auditorily, and then a visual word completion was shown that either completed the word fragment *can_* correctly (*candle*) or not (*candy*) (Friedrich and Kotz, 2007). The incorrect completion that did not match the contextually indicated meaning (*candy*) led to a left and a right lateralized positivity, which may be interpreted as a general surprise reaction. The fully matching word compared to the mismatching word led to a negativity around 400 ms (N400), an ERP component known to indicate lexical-semantic processes. These two ERP effects suggest that there are two processes at work: early bottom-up processes as reflected in the positivities, and lexical-semantic integration processes indicated by the negativity. A similar negativity with a preceding positivity was even observed out of sentential context at the single word level when the initial syllable (*nün_*) presented prior to a target word did not match the initial syllable of the perceived target word (as *mün_* in *Münze*; Engl. coin) (Friedrich, Schild, and Röder, 2009). Therefore, it appears that the speech processing system is very sensitive using relevant contextual information already early during processing.

Another study has shown that at the word level even prosodic factors can assist or hamper lexical access. Consider, for example, a compound word like *wheelchair*. Within this compound the first part-word *wheel* of the compound word *wheelchair* is pronounced differently from the way it is pronounced as a separate word *wheel*. When presenting *wheel* spoken as a word separate from the second part of the compound, *chair,* the lexical access is hampered as compared to *wheelchair*, when spoken as a compound (Isel, Gunter, and Friederici, 2003). This prosodic difference is due to the fact that the acoustic envelope of *wheel* spoken as part of the compound *wheelchair* is acoustically different from *wheel* spoken as a single word. The processing system thus already takes into account the prosodic information from the beginning of the compound word when accessing the lexicon. These findings impressively demonstrate the effective use of phonetic and prosodic information for lexical access and word recognition. It is this fine-scaled interplay of different processes that makes word recognition fast.

Besides information about of word's phonological form, the lexical entry contains other important information, namely information about the word's *syntactic word category* (noun, verb, adjective, determiner, etc.) and information about its meaning, that is, *lexical-semantic information*. Moreover, in the case of verbs, the lexicon entry contains syntactic information about the *verb's argument structure*—that is, information about how many arguments the particular verb can take: one argument (*he cries*), two arguments (*he sees Peter*), or three arguments (*he sends the letter to Peter*). Once these information types are accessed, they will be made available for further processing at a subsequent level where syntactic, semantic, and thematic relations are considered (see figure 1.1).

When thinking about word processing during language comprehension, a word's syntactic category may not be the first that comes to one's mind. However, this information is highly relevant during language processing, as it is crucial to know whether a given word is a noun or a verb. This is the case because word category information guides the buildup of syntactic structures (noun phrase or verb phrase) during comprehension. Moreover, verb-argument information encoded in the verb determines the sentence structure, in particular with respect to how many argument noun phrases it takes. I will take up these issues in the next sections but will now first turn to the processing of lexical-semantic information at the single word level.

Lexical-Semantic Information at the Word Level

The comprehension of a single word and its meaning is associated with the neural substrate of the left temporal pole and adjacent anterior temporal cortex (Mesulam, Thompson, Weintraub, and Rogalski, 2015). The functional role of the anterior temporal lobe within the language network, however, is still under discussion (Rogalsky, 2015), as neuroimaging studies show systematic activation in the anterior temporal cortex not only for single word processing but also as a function of sentence processing. It is likely that the left anterior temporal lobe is necessary for word comprehension and also conditional for sentence comprehension when meaningful units are built on the basis of semantic combinatorics (Vandenberghe, Nobre, and Price, 2002; Pallier, Devauchelle, and Dehaene, 2011).

Neuroscientific studies on semantic processes on the word level are manifold. They include neurophysiological (electrophysiological, magnetophysiological, and intracranial recording) measures as well as functional magnetic resonance imaging approaches. For reviews on studies pertaining to single word processing, I refer to Démonet, Thierry, and Cardebat (2005) and Price (2010). Many of the word-level studies are problematic, however, because they often did not vary their stimulus material in a very systematic, theory-driven way. The reasons for this may be multifold. Unfortunately, until today solid empirical evidence has not uniformly supported or accepted any theory of semantics. This situation is even further complicated by the fact that currently no empirical findings provide clear information about the relation between a word's form and its lexical-semantic information and the relation between the lexical-semantic information and the conceptual-semantic information, that is, the associated semantic memory. This relation has been reflected upon by different researchers, resulting in quite different outcomes. Why should that be?

As I mentioned in the introduction to this book, some researchers argued for a representation of a word's semantics as part of the linguistic representation and distinguish this from conceptual semantics, with the latter including memory representation that comprises all possible semantic associations (Bierwisch, 1982). Others, in contrast, did not distinguish between lexical semantics and conceptual semantics (Jackendoff, 1983). The language-independent conceptual-semantic knowledge is traditionally thought to be represented in semantic networks (Collins and Loftus, 1975) in which nodes represent particular semantic features of a concept (e.g., *bird* has the features *is animate, has wings, has feathers, can fly*) and are arranged in a way which permits the encoding of hierarchical relations (with category name *animal* being higher in the hierarchy than the specific name *canary*). In order to generate a coherent sentence, however, not all the features need to be activated. For the correct use of the word *bird* in a sentence (e.g., *Birds fly*), the activation of the semantic features *animal* and *can fly* might suffice. According to some models, conceptual-semantic memory representations and the corresponding lexical-semantic representations can be described as ensembles of semantic features, with the word in the lexicon representing a smaller set of features (Bierwisch, 1982) and the conceptual-semantic structure in memory representing a richer set of features (Collins and Loftus, 1975; Miller, 1978; Jackendoff, 1983). Such features could be thought of as the relating elements between a lexical representation and a conceptual representation. There are also different views on the mental representation of semantics (see Lakoff, 1987; Pustejovsky, 1995), but none of these are well investigated at the neuroscientific level.

Neuroscientifically, the anterior temporal lobe has been discussed as the semantic "hub" in a memory system which is modality invariant and thus domain general, taking modality-specific features such as visual features, sound, olfactory, and motoric features as secondary associations into account (Patterson and Lambon Ralph, 2016). This view is evidenced in findings from patients with semantic dementia indicating a deficit across different semantic categories and modalities (Bozeat, Lambon Ralph, Patterson, Garrard, and Hodges, 2000). These data, however, leave open how the information is transmitted from the domain-specific systems to the domain-general, modality-invariant memory system in the anterior temporal lobe. Another open question concerns the relation between the domain-general memory system and the lexicon, as the inventory of words.

One model on semantic processing assumes that more posterior portions of the middle temporal cortex support semantic processing during language comprehension (Lau, Phillips, and Poeppel, 2008). Moreover, the medial temporal lobe together and the adjacent hippocampus have been considered to play a major role in semantic processing, in particular for the learning of novel words and their subsequent recognition (Breitenstein et al., 2005; Davis and Gaskell, 2009; Takashima, Bakker, van Hell, Janzen, and McQueen, 2014).

With respect to the medial temporal lobe and the hippocampus there is even evidence for semantic processes at the level of single cell recordings. This method makes it possible to measure the activation of a single cell in response to a given stimulus (i.e., a picture or a word). As this method can only be applied in clinical settings, the respective studies are very rare. The results from these studies show that semantic aspects are processed in the medial temporal lobe. Neurons in this area were found to be activated by a visually displayed picture of an object or a familiar person, or even by the person's name (Quian Quiroga, Reddy, Kreiman, Koch, and Fried, 2005). For example, a neuron was identified that responds with activation to different photographs of the famous actress Halle Berry and also to her name written in block letters. The very same neuron does not react to photographs of other famous women or to another female name. This surprising finding suggests a high degree of conceptual abstraction already at the neuronal level. These results have been interpreted as evidence for one crucial aspect of the theory of sparse coding, that is, the representation of specific contents by a small number of highly specialized neurons (Olshausen and Field, 2004). However, one has to admit that the statistical considerations of the probability of encountering such neurons by chance is so low that even representations of highly abstract, modality-invariant concepts must still involve a large number of neurons whose activation is bound temporarily into functionally coherent ensembles. If this assumption is valid the sparseness of coding is progressively achieved by iterative recombination of feature-specific neurons that can be part of different concepts at different points in time. This is suggested by findings from intracranial recordings in the right anterior hippocampus suggesting that certain neurons respond to semantic categories rather than specific members of that category. Quian Quiroga and colleagues (2005) reported that a single neuron reacted not only to the Tower of Pisa but also to the Eiffel Tower, suggesting responsiveness to the semantic feature *tower*.[5]

These findings can be directly related to cognitive feature-based semantic theories modeling the relation between semantic memory and word representation. The assumption is that both semantic engrams and word representations share a number of common features. Empirical support for the view that a word is represented as an ensemble of features comes from behavioral and electrophysiological studies. At the behavioral level it has been found that the time taken to recognize a word depends on the number of semantic features (semantic richness) a word carries (Sajin and Connine, 2014). At the neurocognitive level, Li, Shu, Liu, and Li (2006) showed that electrophysiological brain responses increase as a function of the number of mismatching semantic features by which a given word differs from the best-fit into the prior sentential context (see section 1.6).

At the neuronal level, this would theoretically imply that a neuron representing one basic semantic feature might participate in various ensembles of different word representations, with semantically related words that have overlapping semantic features leading to partly overlapping ensembles. Such a view is in principle compatible with a behavioral effect long known in the psychological literature as a *priming effect*, referring to the observation that semantically related words (*tiger* and *lion*) are easier to process when they are presented with a short delay. This allows preactivation of a number of neurons in the ensemble representing the respective overlapping semantic features. When presenting two semantically related words with no delay, however, the perception of the second word is inhibited. In this case neurons necessary to process the first word are still engaged and are therefore not yet available for the second word, thereby leading to an interference effect (Levelt et al., 1991). These effects are probably caused by the hysteresis of the neurons encoding the overlapping semantic features. This time-sensitive interference effect of lexical-semantic processing has been located by means of magnetoencephalography in the left temporal cortex (Maess, Friederici, Damian, Meyer, and Levelt, 2002), a cortical region known to be involved in processing semantic aspects at various levels of the semantic hierarchy (Tyler et al., 2004).

The next obvious question is how to model the relation between the different brain systems dealing with semantic aspects at the word level, as for example between the hippocampus/medial temporal lobe and the anterior and posterior temporal cortex. Such models still have to be developed.

For the time being we can only speculate and say that the hippocampus and the neighboring structures of the medial temporal lobe appear to play a crucial role. The hippocampus is known to support memory (Milner, Squire, and Kandel, 1998; Xu and Sudhof, 2013). Neurons that react to specific objects or persons have been identified even in the human medial temporal lobe (Quian Quiroga et al., 2005). This latter finding has been interpreted as evidence for one crucial aspect of sparse coding, that is the representation of specific contents by a small number of highly specialized neurons (Olshausen and Field, 2004). Another, by no means contradictory, interpretation may be that a particular neuron represents only one of many features in an ensemble of neurons, which then make up a more complex concept.

Summary
In the context of the cognitive model of auditory language comprehension, a first step toward comprehension is the access to the lexicon and the information encoded in the lexical entry. Neuroscientifically the temporal cortex, in particular the temporal gyri, together with the medial temporal lobe and the hippocampus play a major role in this process.

1.4 Initial Phrase Structure Building

The cognitive model displayed in figure 1.1 assumes that the processing of the different information types follows a strict time course—with word category information processes taking place prior to the processing of syntactic, semantic, and thematic relations. The investigation of these processes in principle requires studies at the phrase or sentential level as syntactic, and thematic aspects only emerge at these levels. The majority of these studies suggest that word category information is necessary for building up an initial phrase structure.

In psycholinguistic literature, it has long been debated whether it is valid to assume a first syntactic processing stage during which an initial phrase structure is built on the basis of word category information, for example noun, verb, preposition, determiner (Frazier and Fodor, 1978), or not (Marslen-Wilson and Tyler, 1980, 1987; MacDonald, Pearlmutter, and Seidenberg, 1994;). The idea of syntax-first models (Frazier and Fodor, 1978; Friederici, 2002; Bornkessel and Schlesewsky, 2006) is that during this initial stage a local phrase structure can be built quite fast and independent of semantic information, allowing the incoming information to be efficiently chunked and structured.

However, the non-expert reader may not think about syntax as the first relevant information in the context of sentence processing, but may rather think of lexical-semantic and meaning information when considering language comprehension. Moreover, it appears that non-experts think about words as the building blocks of language rather than of syntax as providing the rules of how words in sentential context come together. But we have to consider that the processing system at least during auditory language comprehension has to cope with the incoming speech in an exceptionally fast manner. In order to do so, the system appears to initially structure the incoming information into larger chunks and only then further processes are pursued. Chunking of the incoming information is facilitated by word category information. For example, once the listener identifies a determiner (e.g., *the*), the system knows that the determiner is the beginning of a determiner phrase that has a certain grammatical structure ending with a noun (e.g., *the ship, the old ship*). This knowledge can be used to identify the beginning and the end of a syntactic phrase.

What are syntactic phrases, in general? In the linguistic analysis a phrase is a group of words (or even a single word) that functions as a constituent in a sentence. Different linguistic theories differ in their syntactic details. Here we start from the assumption that a phrase is a unit within the syntax hierarchy, and that there are at least the following phrase types: noun phrase (NP), also called determine phrase (DP); verb phrase (VP); prepositional phrase (PP); adjective phrase (AP); and possibly adverbial phrase (AdvP).[6] These phrasal types can be seen as the building blocks of language.

A crucial issue for processing is when and under which conditions the information about a word's category becomes available. Given the findings from a number of studies, the assumption appears plausible that during lexical access word category information is listed high in the lexical entry, since it becomes available fast. The early availability of this information could be due to the fact that word category information is limited in size. There are only a few different syntactic word categories in natural language (e.g., noun, verb, preposition, adjective, adverb, conjunction), whereas lexical-semantic information is much richer (all the words an individual ever learned). The early availability of syntactic word category information makes good sense because it is the relevant information for the buildup of syntactic phrase structure, which is assumed to constitute an initial step during sentence comprehension.

This may differ from language to language, however, because word category information of verb and noun can be marked either by inflectional morphology, such as in English or German (*refine* vs. *refinement*, *veredeln* versus *Veredelung*), or only by the lexical entry itself (*eat* vs. *meal*). The time point at which online processing of word category information becomes available could play a crucial role when modeling auditory language comprehension.

Temporal Course: EEG and MEG Studies
The cognitive model of auditory language comprehension proposed in section 1.1 assumes such a first syntactic stage during which word category information is processed. At this stage the language processing system has to check whether the word category of the incoming word allows the buildup of a phrase or not. Once a phrase structure can be built, further processes at the next processing stages can take place. Here I will describe the temporal

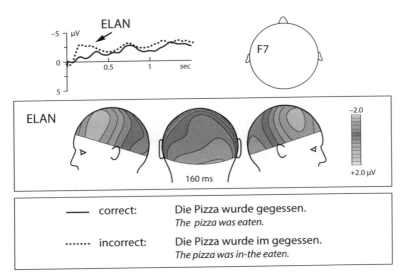

Figure 1.4
Syntactic violation effect: ELAN. ELAN stands for *early left anterior negativity*. It is an event-related brain potential observed in response to a syntactic violation (here word category violation in an incorrect sentence compared to incorrect sentences recorded at the left anterior electrode F7). For example sentences see box at the bottom of the figure (more negative-going wave form for the incorrect compared to the correct sentence). Top: wave form for correct and incorrect condition for the sentence final word. Middle: voltage maps displaying the activation difference between incorrect and correct condition around 160 milliseconds for the sentence final word. Bottom: example sentences. Adapted from Hahne and Friederici (2002). Differential task effects on semantic and syntactic processes as revealed by ERPs. *Cognitive Brain Research*, 13 (3): 339–356, with permission from Elsevier.

parameters of phrase structure building and its neural basis. The temporal aspects are taken from ERP studies. Evidence from ERP research indicates that a word category violation led to an early syntax effect in brain response. This ERP component is the early left anterior negativity (ELAN), which occurs in response to a word category violation 120–200 ms after word onset or after the part of the word that provides the word category information (see figure 1.4) (Neville, Nichol, Barss, Forster, and Garrett, 1991; Friederici, Pfeifer, and Hahne, 1993; Hahne and Friederici, 2002; Kubota, Ferrari, and Roberts, 2003; Isel, Hahne, Maess, and Friederici, 2007; for a review, see Friederici and Weissenborn, 2007).[7]

The initial phrase structure is usually built based on the word category information (e.g., determiner phrase, verb phrase). These phrases are the building blocks for larger sentence structures. Within the neurocognitive model of language comprehension (Friederici, 2002), this process takes place in the initial processing phase. The buildup of local phrase structure has been shown to be highly automatic in the adult listener as it is independent of attentional processes (Hahne and Friederici, 1999). Moreover, it is independent of the instructions given to the listener, because the ELAN component is even present when the listener is instructed to ignore the syntactic error in the sentence (Hahne and Friederici, 2002). This indicates that the process reflected by the ELAN is quite automatic.

The earliness of the ERP effect was attributed to the ease with which word category information can be extracted from the stimulus, whether it pertains to the word's shortness (e.g., function word, as in Neville et al., 1991) or its morphological markedness (e.g., inflection, as in Dikker, Rabagliati, and Pylkkänen, 2009). The ELAN component has been reported mostly in the auditory domain for connected speech (but see Dikker et al., 2009), and in rapid visual sentence presentation (see Neville et al., 1991). Varying the visual contrast of the words during sentence reading indicated that the visual input must fall into an optimal perceptual range to guarantee the ELAN effect of fast initial syntactic processes (Gunter, Friederici, and Hahne, 1999). The finding, however, that the ELAN effect was reported mainly for the auditory domain has raised an important question. To what extent might this component be due to prosodic deviances also present in connected speech when inducing a word category violation? A study focusing on this very question systematically induced syntactic and prosodic violations independently, and demonstrated that the prosodic contour cannot account for the early syntactic effect (Herrmann, Maess, and Friederici, 2011). In an additional ERP study again varying syntactic and prosodic violations in a highly complex design we found that prosodic violations alone elicit a different ERP component, namely a right hemispheric anterior negativity (Eckstein and Friederici, 2006). Because of these latter findings and because of the reported ELAN effects for syntactic violations in visual studies with fast and optimal visual input, it appears that this early ERP effect reflects the recognition of a syntactic violation leading to the impossibility of building up a syntactic phrase (Friederici and Weissenborn, 2007).

The speed of this initial phrase structure–building process may be surprising. However, the process of building up a local structure, such as a determiner phrase (determiner plus noun, *the car*) or a prepositional phrase (preposition plus determiner phrase, *in the car*), on the basis of word category information could be performed quickly once the possible minimal local structures in a given language are learned and stored as a fixed template. If such templates are established, the initial step of the phrase structure–building process could be viewed as a fast template-matching process taking place very early in comprehension (Bornkessel and Schlesewsky, 2006). During this process, templates of local phrase structures are activated (e.g., a preposition would activate a template of a prepositional phrase), against which the incoming information is checked. If the incoming information does not match the template, a phrase structure violation is detected and any further processes are not syntactically licensed.

This view would predict that sentences containing a phrase structure violation should hamper any further processing at the next processing level, including semantic processes. ERP studies can address this issue as, in addition to initial phrase structure–building processes (ELAN), lexical-semantic processes and semantic integration processes have been related to a particular ERP component. This is a negativity of around 400 ms called

N400. At the sentence level, N400 effects show up whenever a semantic violation occurs (discussed in section 1.6). If the syntax-first prediction holds, and if a syntactic violation hampers the subsequent lexical-semantic processes, no N400 effect should be observed but instead only a syntactic effect should be visible in a sentence containing both a syntactic and a semantic violation at the same word. The syntactic violation should hinder any further lexical-semantic processes. We demonstrated that this is indeed the case when combining a word category violation with a semantic violation in one sentence, since the syntactic violation (ELAN) should disallow semantic processes (no N400) (Hahne and Friederici, 2002; Friederici, Gunter, Hahne, and Mauth, 2004). This even holds when a word category violation is combined with a violation of the verb-argument information, which again is processed at a later processing stage (Frisch, Hahne, and Friederici, 2004). Together these data indicate that syntactic phrase structure violations are processed prior to semantic and thematic information and can block higher-level processes, thereby providing strong evidence for models assuming an initial syntactic processing phase (Bornkessel and Schlesewsky, 2006; Friederici, 2002). Although this conclusion was called into question on the basis of an experiment using Dutch language material (Van den Brink and Hagoort, 2004), we have argued that the data from this study do not speak against the model's assumption (Friederici and Weissenborn, 2007). This is because in the Dutch study the syntactic word category information (noun or verb) of the critical word was provided in the word's inflectional ending and thus only became available after the semantic information that was provided in the word stem preceding the ending. A review of the literature on the timing of syntactic information and semantic information across the different languages reveals that the absolute timing of the syntax-initial and other processes may vary from language to language, but that the order of these processes in time is fixed and the same across the different languages with syntactic word category information being processed first (Friederici and Weissenborn, 2007).

Since the ELAN was not found in every study across different languages, it is important to discuss what process this ERP component might reflect. The combined findings suggest that it reflects the detection of an outright phrase structure violation, that is, the impossibility of building up a phrase structure from the input. This is a fast process most likely to be visible when word category information, on which phrase structures are built, is unambiguously available from the input. What is the evidence for this claim? One study (Friederici, Hahne, and Mecklinger, 1996) clearly demonstrated that the processing system only reacts to outright phrase structure violations, but not to structures that only occur rarely or are ambiguous. The respective study used German as the test language in which the word *wurde* is ambiguous, it can be an auxiliary (*was*) or it can be a main verb (*became*). The occurrence of the word *wurde* as the auxiliary *was* is much more frequent than its occurrence as the main verb *became*. Now, a sentence like *The metal wurde* (meaning *was*) *refined* should be easy to process, but how about a sentence in which *wurde* is more ambiguous, such as: *The metal wurde refinement*? If *wurde* is read as *was* it should elicit an ELAN, since the auxiliary *was* requires a past participle verb form (*refined*), but not a noun (*refinement*). If, however, the processing system also takes into consideration the *became*-meaning of *wurde,* then a noun (*refinement*) as the next word is possible and would not lead to a phrase structure violation because a noun is a syntactically valid continuation (*The metal became refinement standard*). Syntactically this latter sentence is correct, although its semantics is strange. The data from this German experiment show that the parsing system does not signal an ELAN in case the crucial word of the phrase (e.g., *wurde*) is ambiguous, even under the condition that the correct reading probability is very low (Friederici et al., 1996). These findings indicate that the ELAN component is a reflection of a rule violation rather than an expectancy violation. For the time being, however, it must remain open whether the ELAN effect reflects initial phrase structure building in general or a reflection of a phrase structure template error-detection process.

Localization of Phrase Structure Building: A Multimethod Approach
Where in the brain does the initial phrase structure building process take place? There are several ways to approach this question, but not all are optimal as the effect is only visible in a certain time window. One way to localize language processes online is to apply electroencephalography measures in patients with circumscribed brain lesions and to use an ERP design known to elicit certain language-related components. Using this approach, it was found that the ELAN component is absent in patients with left frontal cortical lesions (including lesions of the left basal ganglia), but is present in patients who only suffer from left basal ganglia lesions. This indicates that the left frontal cortex plays a crucial role in the generation of the ELAN (Friederici, von Cramon, and Kotz, 1999). Interestingly, the ELAN also appears to be affected in patients with lesions in the left anterior temporal lobe, but not in patients with lesions in the right temporal lobe, suggesting that the left anterior temporal cortex in addition to the left frontal cortex may be involved in the early structure building processes reflected in the ELAN (Friederici and Kotz, 2003).

A second way to localize the ELAN effect is to use magnetoencephalography, as it provides good topographic resolution (depending on the number of channels), although the method inherently has to deal with the so-called inverse problem, as it has to calculate the neural generators of the processes based on data recorded over the scalp. With the use of magnetoencephalography, the ELAN effect has indeed been localized in the anterior temporal cortex and the inferior frontal cortex (Knösche, Maess, and Friederici, 1999; Friederici, Wang, Herrmann, Maess, and Oertel, 2000) or solely in the temporal cortex (Herrmann, Maess, Hasting, and Friederici, 2009; Gross, et al., 1998) for auditory language experiments. In a visual experiment, however, a morphosyntactic violation effect was localized in the visual cortex, at least for sentences in which the word category information was morphologically marked (Dikker, Rabagliati, and Pylkkänen, 2009). These data raised the question of whether or not clearly marked syntactic word category violations are detected in the sensory cortices, whether visual or auditory (Dikker et al., 2009; Herrmann et al., 2009). Evidence from an auditory magnetoencephalography experiment, however, revealed syntactic effects located in the anterior superior temporal gyrus anterior to the auditory cortex, but not in the primary auditory cortex itself (Herrmann, Maess, Hahne, et al., 2011) (see figure 1.5). These data demonstrate that at least in the auditory domain the ELAN effect is not located in the sensory auditory cortex.

The localization of the ELAN effect by means of magnetoencephalography suggests an involvement of the anterior superior temporal gyrus and possibly the ventral inferior frontal cortex. Specifying the localization of the ELAN effect remains difficult, as localization by means of functional magnetic resonance imaging has major drawbacks due to the method's low resolution in time. Since syntax-first models (Friederici, 2002, 2011; Bornkessel and Schlesewsky, 2006) assume that the initial syntactic processing phase is immediately followed by a phase during which syntactic and semantic relations are processed jointly to achieve thematic assignment, it is not easy to separate these two stages of syntactic processing that take place within milliseconds. In principle there are several ways to approach this problem, but each of these has its limitations.

A third way is to use functional magnetic resonance imaging and compare the brain activation pattern for natural sentences that contain only a syntactic violation with sentences that contain a semantic violation. In this study we found that the middle and posterior superior temporal gyrus were activated for both the semantic and the syntactic violation condition, but that the anterior superior temporal gyrus and the frontal operculum were selectively activated only for sentences with syntactic violations (Friederici, Rüschemeyer, Hahne, and Fiebach, 2003). Thus, the latter regions may be crucial for processing syntactic violations, but not necessarily specific for structure building processes.

A fourth way to investigate the localization of phrase structure building is to compare brain activation pattern for full sentences to non-structured word lists under the assumption that during sentence processing the initial stage of phrase structure building is mandatory and should be observable, but not when a word list is processed. Activation of the frontal operculum was observed in a study comparing sentences to word lists without function words (Friederici, Meyer, and von Cramon, 2000). However, this pattern was not reported for functional magnetic resonance imaging studies that compared normal sentences to mixed word lists containing content words and function words. These other studies used mixed word lists, which led to the situation that local structure building was partly possible due to syntactically legal combinations of two or three words in the list, for example, the combination of determiners and nouns or adjectives and nouns (Stowe et al., 1998; Vandenberghe, Nobre, and Price, 2002; Humphries, Binder, Medler, and Liebenthal, 2006; Humphries, Love, Swinney, and Hickok, 2005; Snijders et al., 2009). Interestingly, Vandenberghe and colleagues (2002) report activation in the frontal operculum for sentence conditions providing word category information when comparing these with control conditions of unpronounceable letter sequences (providing no word category information). Thus, these studies suggest that the frontal operculum is involved when

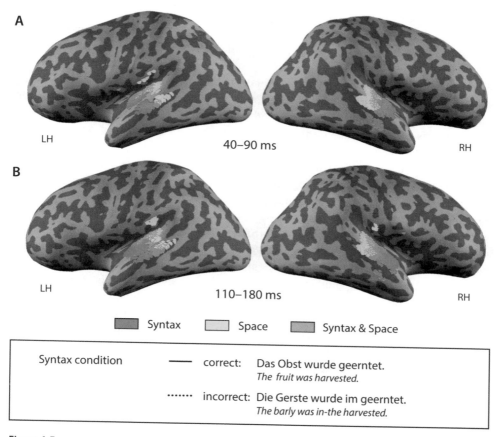

Figure 1.5
Effects of syntax and auditory space violations. Display of the location of the grand average activations for different violation types: for syntax violation only condition (red) as in sentences displayed in the box at the bottom; for a non-linguistic auditory space (left side / right side) violation only condition (yellow), and for the joint syntax + auditory space violation condition (orange). Brain activity is displayed on the inflated surface of the brain, with dark gray representing sulci and light gray representing gyri. Results of the local maxima analysis in the temporal cortex for the two time windows (40–90 ms) and 110–180 ms after stimulus onset. Activations for the syntax violation condition (red) are located anterior to the auditory space condition (yellow). Bottom: example sentences. Adapted from Herrmann, Maess, Hahne, et al. (2011). Syntactic and auditory spatial processing in the human temporal cortex: An MEG study. *NeuroImage*, 57 (2): 624–633, with permission from Elsevier.

processing sequences allowing the combination of words versus unrelated word lists or letter sequences. However, it remains to be determined what particular role the frontal operculum plays in combinatorial processes more generally. I will come back to this at the end of the section.

Yet another way of investigating structure building is to use artificial grammars that lack semantic information (Petersson, Forkstam, and Ingvar, 2004; Friederici, Bahlmann, Heim, Schubotz, and Anwander, 2006). In the context of artificial grammar learning, it is frequently discussed to what extent artificial grammar studies can tell us something about natural language processing (Petersson, Folia, and Hagoort, 2012). The discussion centers on the issue of whether those artificial grammar sequences that used nonsense syllables are processed according to phonological principles or indeed according to syntactic principles. For the latter position the argument is that only once the experimental conditions in functional magnetic resonance imaging studies force the buildup of a syntactic hierarchy, syntactic principles are at work. The only artificial grammars that go beyond using syllable sequences are those artificial grammars that entail syntactic categories, such as nouns and verbs, as well as function words (e.g., determiners and prepositions).

Both types of artificial grammars have been used to determine the neural basis of grammar processing. One of the artificial grammar studies used syllable sequences in which an element of category A (a certain syllable type) was always followed by an element of category B (another syllable type): for example, ABABAB (see figure 1.6). A violation was created by having an A syllable followed by another A syllable in the sequence. The processing of this syntactic error in the artificial grammar sequence led to activation in the frontal operculum (Friederici, Bahlmann, et al., 2006). In a second condition of this study the syllable sequence follows an A^nB^n rule creating long-distance nested dependencies between the A and B elements. Violations of these sequences led to an activation in BA 44 in addition to the frontal operculum (Friederici, Bahlmann, et al., 2006). On the basis of these findings, we suggested that the frontal operculum together with the anterior superior temporal gyrus supports some aspects of local structure building, whereas BA 44 is involved in the processing of more complex dependencies. However, the interpretation with respect to the frontal operculum has to be treated with caution, as it has been argued that processing artificial grammar sequences of the ABABAB type does not necessarily require the buildup of a syntactic hierarchy between element A and element B, but could simply be processed on the knowledge that every A element is followed by a B element.

A most recent study using natural language, however, indicates that the frontal operculum supports the combination of two subsequent elements rather than hierarchy building as such (Zaccarella and Friederici, 2015a, see below). In a functional magnetic resonance imaging study we investigated the structure building process at its most fundamental level, that is, when two words hierarchically bind together to form a phrase. This process, called Merge in linguistics, is the most basic syntactic computation at the root of every language (Chomsky, 2013). For example, this basic computation binds two words, for instance, *this*

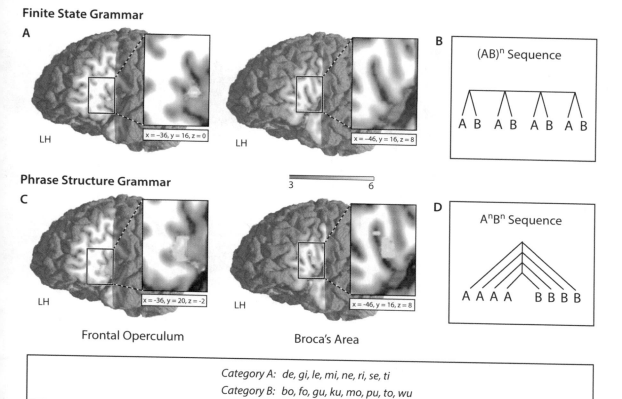

Figure 1.6
Brain activation as a function of grammar type. Statistical parametric maps of the group-averaged activation during processing of violations of two different types of artificial grammar (P < 0.001, corrected at cluster level). (A) The contrast of incorrect vs. correct sequences in the Finite State Grammar (FSG) is shown for the frontal operculum (left) and Broca's area (right). (B) The FSG sequence structure. (C) The contrast of incorrect vs. correct sequences in the Phrase Structure Grammar (PSG) is shown for the frontal operculum (left) and Broca's area (right). (D) The PSG sequence structure. Bottom: Members of the two categories (A and B) were coded phonologically with category "A" syllables containing the vowels "i" or "e" (de, gi, le, ri, se, ne, ti, and mi) with category "B" syllables containing the vowels "o" or "u" (bo, fo, ku, mo, pu, wo, tu, and gu). Adapted from Friederici, Bahlmann, et al. (2006). The brain differentiates human and non-human grammars: Functional localization and structural connectivity. *Proceedings of the National Academy of Sciences of the USA*, 103 (7): 2458–2463. © (2006) National Academy of Sciences, U.S.A.

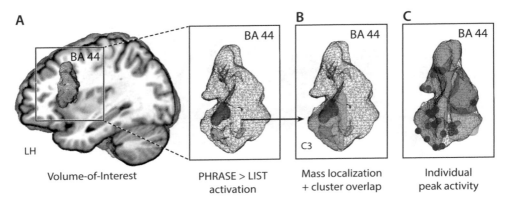

Figure 1.7
Brain activation for phrase structure building. Brain activation of phrase structure building during Merge computation. Volume of interest analysis is BA 44. (A) Activation contrast between a determiner phrase building condition (*the pish*) compared the processing of a two-word list condition (*pish, the*) within BA 44 extracted from the whole brain (green part), (B) localization within BA 44 (C3 cluster according to Clos et al., 2003). (C) Individual peak activity distribution within BA 44. Different clusters are color-coded. Significant accumulation of individual peaks (blue dots) are located in cluster C3 (P < 01). Adapted from Zaccarella and Friederici (2015a). Merge in the human brain: a sub-region based functional investigation in the left pars opercularis. *Frontiers in Psychology*, 6:1818. doi:10.3389/fpsyg.2015.01818

and *ship*, to form a minimal hierarchical phrase, namely a determiner phrase, for instance, *this ship*. Multiple applications of Merge can lead to quite complex sentence structures. Here, however, the focus is on a single Merge process, necessary to build a determiner phrase like *this ship*. When comparing this condition that allows determiner phrase building with a condition that only presents two words as a list, for example, *cloud ship*, a clear activation in the most ventral anterior portion of the BA 44 was found (Zaccarella and Friederici, 2015a, see figure 1.7). This activation was clearly confined in one out of five predefined subregions in BA 44 (Clos, Amunts, Laird, Fox, and Eickhoff, 2013) with a high consistency across individuals. This result adds a crucial piece of information to the question of where in the adult human brain initial phrase structure building processes are located. From these data we can conclude that the frontal operculum is responsible for the combining of two elements independent of syntactic phrasal aspects, whereas syntactic phrase structure building is subserved by BA 44 in Broca's area.

The interpretation that BA 44 comes into play when processing syntactic structures has been supported by a number of studies, for complex artificial grammars (Bahlmann, Schubotz, and Friederici, 2008; Uddén et al., 2008; Bahlmann et al., 2009), for artificial grammars with syntactic categories (Opitz and Friederici, 2003, 2007), and for hierarchy building in natural languages (see section 1.5).

The functional difference in the frontal operculum and BA 44 is interesting because the frontal operculum and BA 44 can also be distinguished at the level of brain evolution and at the level of the brain's receptorarchitectonics. The frontal operculum is a phylogenetically older brain region than Broca's area including BA 44 and 45 (Sanides, 1962). Moreover, the functional differentiation between the frontal operculum and BA 44 is mirrored in a receptorarchitectonic differentiation as revealed by receptorarchitectonic analyses. As I have already pointed out in the introduction, the *receptorarchitectonics* of the cortex refers to the density of neurotransmitters present in a given region (Zilles and Amunts, 2009; Amunts et al., 2010). Using this analysis a receptorarchitectonic distinction was found between the lateral area 44 and the more ventral-medial frontal operculum (Amunts et al., 2010). This receptorarchitectonic difference between BA 44 in Broca's area and the adjacent frontal operculum has been shown to be functionally relevant because these two regions support different subfunctions during language processing (Friederici, Bahlmann, et al., 2006; Zaccarella and Friederici, 2015a).

Summary

The anterior superior temporal gyrus and the frontal operculum seem to support the combinatorics of two elements independent of the combination's underlying syntactic structure whereas the ventral portion of the inferior frontal gyrus, in particular the most ventral part of BA 44, supports initial phrase structure building. These regions seem to constitute a functional neural network subserving initial phrase structure building. In sum, in this section I discussed those neuroscientific studies relevant for the initial phrase structure building as assumed by syntax-first models. The available data indicate that such an initial syntax-related process can be demonstrated at neurobiological level.

1.5 Syntactic Relations during Sentence Processing

Sentence comprehension crucially depends on the identification of syntactic, semantic, and thematic relations in order to establish "who is doing what to whom." In the cognitive model outlined in section 1.1, these processes are assumed to take place in parallel, but in separate subsystems (figure 1.1).

The fact that we are able to understand syntactic relations even in a meaningless sentence, as in Noam Chomsky's famous 1957 example *"Colorless green ideas sleep furiously,"* already indicates that syntactic relations can be identified and processed independent of meaningful semantic relations. Testing the independence of these processes (in principle), and thus identifying their respective neural networks, requires particular manipulations in a given experiment because in normal sentences syntactic information comes together with semantic information. The approaches that have been taken to experimentally disentangle syntactic from semantic processes are twofold: studies have either manipulated the complexity of the syntactic structure in the sentences, keeping everything else constant, or they have systematically stripped away the semantic information from the sentence until the syntactic skeleton is uncovered.

In this part of the book I draw from studies that have been carefully designed, on the basis on linguistic theories, to investigate syntactic processes. I will describe several of these studies in some detail and then come to the conclusion that Broca's area and the posterior superior temporal gyrus/superior temporal sulcus constitute the neural network supporting syntactic processes.[8] The stimulus material usually varies only one aspect of syntax in order to receive a precise response from the brain that varies its activation according to the stimulus material. The more precise the variation in the sentence material, the more precise will be the answer of the brain. This in turn will allow us to dissect the syntactic processes and define their brain basis. In the remainder of this section I will discuss the relation between the sentence material used in the different studies and the respective brain activation pattern by taking the following approach. The sentence material used in the studies discussed in this section is always presented together with the brain activations in the respective figures. The notations of the sentences presented in respective boxes matches those used in the text.

Varying Syntactic Complexity[9]
The term *syntactic complexity* is used for different types of linguistic phenomena, for sentences deviating from canonical subject-first word order, and for sentences with varying degrees of hierarchical embedding and their relative interplay with verbal working memory during processing.

A common way to vary syntactic complexity is to work with sentences that do not follow the canonical subject-verb-object word order. Among those sentence constructions that deviate from the canonical word order, "scrambling" is a classical example. In such constructions one or more arguments (noun phrases) are shifted to a different position within the sentence. For example, the object noun phrases can be shifted in front of the subject noun phrase (see figure 1.8; sentences a2–a3), leaving a gap in their original position (shift is indicated by arrow and gap by "_"). Because of this shift, the new sentences (a2–a3) now have a non-canonical structure. The processing of such non-canonical structures is more complex and more difficult to process than canonical sentences. The non-canonical structures need be rearranged according to the ordering rules imposed by the language, before the interpretation of verb-argument relationships becomes possible (Friederici, Fiebach, Schlesewsky, Bornkessel, and von Cramon, 2006). This study based on functional magnetic resonance imaging data for such sentences found that argument reordering activates the inferior frontal gyrus, in particular BA 44 in Broca's area, as a function of the number of permutations, with highest activity for multiple reordering and lowest activity for null reordering (see figure 1.8).

This result is in line with other functional magnetic resonance imaging studies investigating argument reordering in non-canonical sentences, which found activation in the inferior frontal gyrus, or BA 44 in particular (inferior frontal gyrus: Röder, Stock, Neville, Bien, and Rösler, 2002; Grewe, Bornkessel, Zysset, Wiese, and von Cramon, 2005; BA 44:

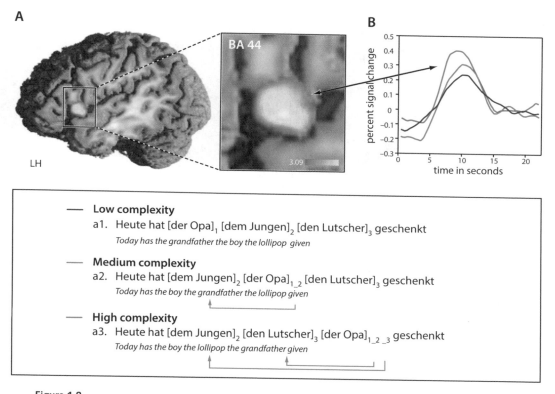

Figure 1.8
Brain activation of BA 44 as a function of syntactic complexity. Complexity effect operationalized as different degrees of syntactic complexity in BA 44, identified in a parametric analysis. For the parametric variation from low complexity, medium complexity, to high complexity, see example sentences at the bottom of the figure. (A) Activated area (BA 44) and blow up of this area displayed in parasagittal view. (B) Trial-averaged brain activation in BA 44 for the three conditions from low to high complexity. P uncorrected < 0.001. Bottom: Example sentences. Adapted from Friederici, Fiebach, et al. (2006). Processing linguistic complexity and grammaticality in the left frontal cortex. *Cerebral Cortex*, 16 (12): 1709–1717, by permission of Oxford University Press.

Friederici, Fiebach, et al., 2006; Meyer, Obleser, Anwander, and Friederici, 2012). Accordingly, there is strong evidence that syntactic complexity (requiring elements in a sentence to be reordered) and activity in Broca's area are directly related.

Syntactic movement is a similar phenomenon also resulting in sentences with noncanonical word order present in languages with an otherwise fixed word order, such as English. With respect to movement operations, special attention has been paid to relative clause constructions and their internal architecture. Relative clauses can either appear to the right of the main clause (see figure 1.9; sentences b1–b2) or be nested within it (b3–b4). Similar to scrambling constructions, the relationship between the moved noun phrase (*the*

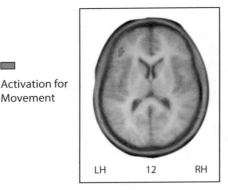
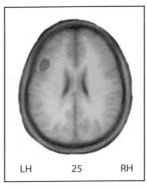

Activation for Movement

Activation for Movement and Embedding

Movement, short distance
b1. Derek is [the boy]₁ [who ₋₁ is chasing the tall girl]

Movement, long distance
b2. Derek is [the boy]₁ [who the tall girl is chasing ₋₁]

Embedding, short distance
b3. [The boy]₁ [who ₋₁ is chasing the tall girl] is Derek

Embedding, long distance
b4. [The boy]₁ [who the tall girl is chasing ₋₁] is Derek

Figure 1.9
Brain activation clusters for different types of syntactic structures. Activation clusters for movement and embedding in the left inferior frontal gyrus (IFG; Broca's area) overlaid on a group average. (Top left) The red color corresponds to those areas that activated for movement type only. (Top right) The pink color corresponds to those areas that activated for both syntactic structures, movement and embedding. The clusters are thresholded at $P < 0.05$ voxel-wise and $P < 0.05$ map corrected for multiple comparisons. Bottom: Example sentences. Adapted from Santi and Grodzinsky (2010). fMRI adaptation dissociates syntactic complexity dimensions. *NeuroImage*, 51 (4): 1285–1293, with permission from Elsevier.

boy) and its original position (gap indicated by "_") also needs to be established in order to achieve comprehension. This is more difficult for object-relative clauses (b2–b4) than for subject-relative clauses (b1–b3), and more difficult for nested center-embedded relative clauses (b3–b4) than for right-branching relative clauses (b1–b2). By separating the processing difficulty into filler-gap linking (movement: subject vs. object-relative clause) and relative-clause complexity (embedding: right branching vs. center embedding), it has been shown that movement activates BA 44 and BA 45 within Broca's area, while clause complexity activates BA 44 only (Santi and Grodzinsky, 2010). Studies on movement in Hebrew also revealed activation in left BA 45 and in the left posterior superior temporal sulcus, suggesting a certain universality of the language-brain relationship (Ben-Shachar, Hendler, Kahn, Ben-Bashat, and Grodzinsky, 2003; Ben-Shachar, Palti, and Grodzinsky, 2004).

The issue has been raised that processing a non-canonical sentence requires working memory in order to identify the relation between the moved noun phrase and its original position (gap), and that this working-memory process, rather than reordering as such, might cause the activation in Broca's area (Rogalsky and Hickok, 2011). The issue was investigated in a study using German as the test language. In this study we systematically varied the factor of *ordering* (object-first vs. subject-first German sentences) and the factor of *storage* in working memory (one vs. four phrases intervene between the critical argument and the verb) in a crossed design using functional magnetic resonance imaging (Meyer et al., 2012). The factor of ordering object-first sentences compared to subject-first sentences activated BA 44 in Broca's area. The factor of storage reflecting verbal working memory activated left temporo-parietal regions. Interestingly, the left temporo-parietal activation correlated with listeners' verbal working memory capacity as measured independently by the digit span test, whereas BA 44 activation did not. These results suggest that verbal working memory during sentence processing relies on temporo-parietal regions, while BA 44 in Broca's area appears as a distinct neural correlate of reordering constituents in a sentence. This result is in line with a transcranial magnetic stimulation study in which inhibitory stimulation in healthy participants was applied over left BA 44 in the inferior frontal gyrus and over left BA 40 in the parietal cortex when processing sentences varying in complexity and length (Romero Lauro, Reis, Cohen, Cecchetto, and Papagno, 2010). Stimulation over BA 44 reduced accuracy of processing syntactically complex sentences only, whereas stimulation over BA 40 reduced accuracy not only of syntactically complex sentences, but also for long though syntactically simpler sentences.

Another way to investigate syntactic complexity is to work with hierarchically embedded sentences. Varying the amount of hierarchical embedding, a functional magnetic resonance imaging study found the syntactic hierarchy effect in BA 44 to be independent of sentence-related working-memory processes (see figure 1.10; syntactic hierarchy: high c_1–c_2 vs. low c_3–c_4; working memory distance: long c_1–c_3 vs. short c_2–c_4) (Makuuchi, Bahlmann, Anwander, and Friederici, 2009). Note that syntactic hierarchy activates BA 44 only during native language processing whereas less automatic second language processing recruits more anterior regions (Jeon and Friederici, 2013).

It is important to note that the effect of working memory in the special case of syntactic dependency was located in the inferior frontal sulcus, whereas general verbal working memory is located in the temporo-parietal cortex (e.g., Jonides et al., 1998; Gruber and von Cramon, 2003). Thus there appear to be two working-memory systems with possibly different functions that can be active during sentence processing. Studies report that the

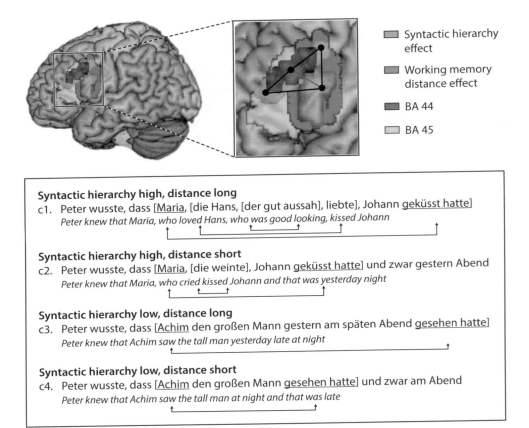

Figure 1.10
Brain activation for syntax and working memory. Activation in the left inferior frontal cortex for sentences with different degrees of hierarchical embeddings. For example sentences see box at the bottom of the figure. A region-of-interest analysis revealed the following activations: Syntactic hierarchical embedding (orange) in BA 44 (color-coded in green) and working-memory distance (blue) in the left inferior frontal sulcus projected onto the surface of the left hemisphere of the brain. Effect of Psychophysiological Interaction (PPI) cluster (purple) indicates that the left inferior frontal sulcus is highly coupled with BA 44. $P < 0.05$ corrected for whole brain as the search volume. Bottom: Example sentences. Detail perspective with Jülich cytoarchitectonic maximum probability maps for BA 45 (yellow) and BA 44 (green). Adapted from Makuuchi et al. (2009). Segregating the core computational faculty of human language from working memory. *Proceedings of the National Academy of Sciences of the United States of America*, 106 (20): 8362–8367.

working-memory system in the parietal cortex has been activated during processing non-canonical sentences with a large distance between the moved noun phrase and its original position (gap) (Grossman et al., 2002; Meyer et al., 2012). In this case the moved noun phrase had to be kept in verbal working memory until its original position was encountered. Fiebach, Schlesewsky, and Friederici (2001, 2002) have documented this process by using a combined functional magnetic resonance imaging and ERP approach. Therefore, it appears that there are two working memory systems that are relevant for language processing. It is assumed that the working memory system in the prefrontal cortex is syntax-specific, whereas the working memory system in the parietal cortex is not specific for syntax, but covers verbal working memory more generally (Caplan and Waters, 1999).

Stripping away Semantics: The Syntactic Skeleton

In the approaches of uncovering sentence-level syntactic processes a number of studies varied the presence or absence of semantic information available in the stimulus material. This leads to differential findings depending on the amount of semantic information present in the stimulus materials and partly on tasks used in the different studies.

Many imaging studies on syntactic aspects during sentence processing report major activation only in BA 44 (as in Friederici, Fiebach, et al., 2006; Newman, Ikuta, and Burns, 2010), but some also report BA 44 and additional activation in BA 45 (Kinno, Kawamura, Shioda, and Sakai, 2008; Tyler and Marslen-Wilson, 2008; Tyler et al., 2010; Fedorenko, Behr, and Kanwisher, 2011; Pallier et al., 2011). Such reporting discrepancies occur because these studies differed in the stimulus material, in the tasks, or even the different experimental languages. In the following we will try to disentangle these factors. Concerning the factor of language we can state that English and German—although both Indo-European languages—differ quite dramatically with respect to the required syntactic processes. In German, word order is relatively free, with constituents, for example noun arguments, being able to occupy different positions in the sentence. The flexibility of word order is compensated by rich morphosyntactic information—word elements that convey the grammatical relations between constituents (inflectional morphology)—which allow us to determine "who is doing what to whom." In English, morphosyntactic information is comparatively scarce, and therefore English speakers can mainly rely upon word order to identify the syntactic roles of constituents (Thompson, 1978) and possibly on semantic information. Accordingly, it may not be surprising that German studies show a clear activation of BA 44 for various syntactic manipulations (Friederici, Fiebach, et al., 2006; Makuuchi et al., 2009; Obleser, Meyer, and Friederici, 2011), whereas English studies frequently show the activation of BA 44 and additionally BA 45 (Caplan, Chen, and Waters, 2008; Santi and Grodzinsky, 2010; Tyler et al., 2010). English studies only see a clear BA 44 activation for syntactic processes in a strictly controlled experiment in which the syntactic parameters are crucial for sentence understanding (Newman et al., 2010). The differences between the studies depend not only on the language types but also on how syntactic processes are operationalized.

The factor of stimulus material becomes obvious in a recent study by Goucha and Friederici (2015) who designed a German study similar to a prior English study by Tyler and colleagues (2010). In the English study participants were presented with three conditions: (1) a normal sentence condition; (2) a nonsense sentence condition, called anomalous prose, in which all content words of a normal sentence were replaced by semantically implausible words of the same lexical category; and (3) a condition consisting of the same word material in a permutated order that disrupted syntactic structure consisting of content and function words, called random word order. The original English functional magnetic resonance imaging study reported widespread activation in the left hemisphere for their syntax-based contrast. But this activation pattern could have resulted from the additional semantic processes possibly triggered by semantic information available in all their conditions due to the presence of content words or parts of it. In this German study we therefore progressively removed the semantic information from the sentences in order to identify a representation of pure syntactic processing. We constructed conditions similar to the English study (Tyler et al. 2010) and, in addition, included pseudowords devoid of any word meaning. Thus the German study contained no semantic information in pseudoword counterparts or sentences, and the only syntactic information appeared in sentences.

In the study by Goucha and Friederici (2015) the deletion of the semantic information was realized in two steps: first, real words (word stems) were replaced by pseudowords leaving morphological elements of words conveying aspects of meaning intact: for example, the prefix *un-* in *unhappy* would stay in place (condition d1>d2 in figure 1.11); in a second step, even these morphological elements, called derivational morphology, were replaced by nonsense syllables (condition d3>d4 in figure 1.11). For real word conditions the German study found broad activation in the left hemisphere, including the large part of inferior frontal gyrus (BA 44/45/47), the anterior temporal lobe, and the posterior superior temporal gyrus/superior temporal sulcus. In the German study for pseudoword sentences, in which derivational morphology conveying semantic meaning like *un-* in *unhappy* or *-hood* in *brotherhood* was still present (d1>d2), a similar activation pattern still involving BA 45 in addition to BA 44 was found. Thus, when derivational morphology was present, BA 45 known to activate as a function of semantic aspects showed increased activation. But in pseudoword sentences in which derivational morphology was absent and only inflectional morphology (e.g., the verb ending *-s* in *paints*) was present (d3>d4), only BA 44 was found to be active. These findings confirm BA 44 as a core area for the processing of pure syntactic information and show that any additional semantic information can lead to the recruitment of BA 45. This nicely demonstrates that the brain represents the linguistic distinction between syntactic aspects and meaning aspects as separate activation patterns.

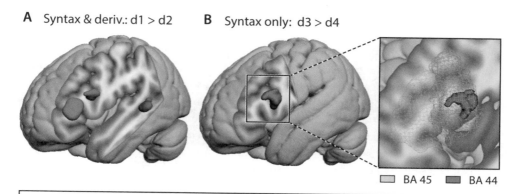

Figure 1.11
Brain activation for syntax and derivational morphology. Brain activation for pseudoword sentences with and without derivational morphology. For examples see box at the bottom of the figure, different conditions are labeled from d1 to d4. Only significant brain activations are displayed. (A) Contrast condition d1 > d2 i.e., with derivational morphology—activation clusters in the language network: BA 45, BA 44 and pSTG/STS; (B) Contrast condition d3 > d4, i.e., without derivational morphology—activation clusters in Broca's area: BA 44 (red), detail perspective with Jülich cytoarchitectonic maximum probability maps for BA 45 (yellow) and BA 44 (green). Bottom: Example sentences. Adapted from Goucha and Friederici (2015). The language skeleton after dissecting meaning: a functional segregation within Broca's area. *NeuroImage*, 114 (6): 294–302, with permission from Elsevier.

And how about other languages? A functional magnetic resonance imaging study in Japanese on syntactic processes also investigated pseudoword sentences in which all content words were replaced by pseudowords leaving all grammatical particles and morphosyntactic elements in place (Ohta, Fukui, and Sakai, 2013). These morphosyntactic elements, realized as inflections at nouns and verbs in natural Japanese sentences, allow identification of the grammatical and, moreover, the thematic relations in a sentence and may do so even in pseudoword sentences. Varying the syntactic structure, the authors report increased activation in BA 44/45 for nested sentences compared to simple sentences. The Japanese material, unlike the German material (Goucha and Friederici, 2015), allowed the identification of thematic roles. This may have led to partial involvement of BA 45. The coordinates of the activation reported, however, strongly suggest a major involvement of BA 44, thereby indicating again BA 44 as a region subserving syntactic processes.

The findings across the different languages show a systematic involvement of BA 44 for the processing of syntactic aspects, but sometimes activation is also reported on the border of BA 45. It is interesting to note, functionally, that BA 45 may subdivide into a more posterior portion and a more anterior portion. If BA 45 is involved in syntactic processes it is mainly the more posterior portion bordering BA 44. The more anterior portion of BA 45 bordering BA 47 is rather seen to be activated together with BA 47 for semantic processes. Therefore, it appears that BA 45 is either a region in which both semantics and syntax overlap, or a region that is functionally subdivided. Both views are possible in principle. However, there is evidence that area 45 can also be subdivided receptorarchitectonically into two portions: a more anterior area 45a bordering BA 47 and a more posterior area 45p bordering BA 44 (Amunts et al., 2010). This subdivision may be of particular functional importance when considering the language processes.

Building Up Syntactic Structures: From Constituents to Sentences
The approach to strip away the semantics is one way to uncover syntax. Another way is to break down syntax into its most basic computation and build up from there. The most fundamental syntactic process is a binary process that syntactically binds two elements together hierarchically to form larger structures. This process, called Merge, briefly discussed in the previous section, is assumed to be at the basis of syntactic processes in all languages (Chomsky, 2013; Berwick et al., 2013). As I explained in the introduction, the process is the most basic syntactic computation that combines two elements (e.g., a determiner and a noun) into a new structure (e.g., a determiner phrase). The recursive application of Merge allows the buildup of complex syntactic structure. For example, the determiner *the* merges with the noun *ship* to build a determiner phrase (*the ship*). This determiner phrase is now a new element that merges with the verb *sinks* to make up the sentence, for instance: *The ship sinks*. Multiple Merge processes now can lead to sentences like *The ship sank late at night during the heavy storm*.

The computation of a single Merge process has been localized in the most ventral anterior part of BA 44 when operationalized as the buildup of a determiner phrase (Zaccarella and Friederici, 2015a). Taking this idea further, an additional study (Zaccarella, Meyer, Makuuchi, and Friederici, 2015) investigated conditions involving two Merge processes, either when building a prepositional phrase, like *on the ship*, or when building a sentence, like *The ship sinks*. Both conditions, which contain the same number of words, led to activation in BA 44 and in the posterior superior temporal sulcus, when compared to word lists. The sentence condition additionally showed activation in BA 45. The BA 45 activation was related to the verb, present in the sentence condition only. The BA 44 activation present in both conditions was taken to reflect syntactic structure building processes, that is, Merge.

A recent study was able to track the processing of linguistic structures for different levels of syntactic hierarchy by oscillatory frequency measures (Ding, Melloni, Zhang, Tian, and Poeppel, 2015). It was shown that different levels of linguistic input from syllable sequences, to a phrase and to a sentence, elicited different frequency patterns in the magnetoencephalography signal even when keeping the number of syllables constant. The data suggest that neural processing reflects the mental construction of hierarchical linguistic structures depending upon whether a given syllable sequence perceived made up a phrase or sentence in the perceiver's native language. Additional electrocorticography measures revealed that these processes involve the middle and posterior superior temporal gyrus bilaterally and the left inferior frontal gyrus. These data clearly demonstrate the online construction of linguistic hierarchies.

The systematic buildup of hierarchical structures was investigated in a French study with sequences of words that made up either a few or progressively more constituents (Pallier et al., 2011). In this study Broca's area was found to increase its activation with the increase of the number of constituents even when using pseudoword sentences. This activation is reported for the posterior portion of BA 45 bordering BA 44 (with 80% in BA 45 and 20% in BA 44). The finding that this activation cluster increases with constituent size is observed for both conditions—for natural language and for pseudoword sentences—may be explained by the fact that the stimulus material in the pseudoword sentence condition contained a number of real words such as personal pronouns (*you, they*), possessive pronouns (*your, his*) and adverbs (*very, ahead*), which have a referential character and thereby represent a semantic aspect of language. In this study three regions increased their activity as a function of constituent size: BA 45, BA 47, and the posterior superior temporal sulcus. The study by Pallier and colleagues (2011) demonstrates that the posterior superior temporal sulcus and the inferior frontal gyrus are involved in processing constituent structure, but also suggest a close relation between constituent structure and syntactic phrase structure.

The Functional Network of Processing Complex Sentences
Thus when it comes to processing sentences, a second region often reported to be activated in addition to Broca's area is the posterior superior temporal gyrus/superior temporal sulcus. This region is suggested to subserve language comprehension as an interface mechanism for different types of information, which can be both syntactic and semantic in nature (Friederici, 2011). This suggestion is based on the observation that the posterior superior temporal gyrus was found active as a reflection of structural complexity for natural language sentences in which semantic information was available, making sentence interpretation possible (Friederici, Makuuchi, and Bahlmann, 2009) but not in hierarchical artificial grammar experiments where no semantic information was available (Bahlmann et al., 2008).

More specifically, the posterior superior temporal gyrus may play a central role in the assignment of meaning to a sentence establishment of verb-argument hierarchies at the sentential level. Its major function is to map syntactic structures to sentential meaning. A number of neuroimaging studies support this view. They reported activation in the posterior temporal gyrus or the posterior superior temporal gyrus/middle temporal gyrus for different conditions during sentence processing: for the processing of syntactic complexity (Ben-Shachar et al., 2004; Kinno et al., 2008; Friederici et al., 2009; Newman et al., 2010; Santi and Grodzinsky, 2010; see figure 1.11); for the processing of syntactic anomaly (Suzuki and Sakai, 2003); for the processing of the verb-argument violation (Friederici et al., 2003; Friederici, Kotz, Scott, and Obleser, 2010; predictability: Obleser and Kotz, 2010); and when the presence/absence of semantic information was manipulated with syntactic information being present (Goucha and Friederici, 2015).

The functional network for the processing of syntactically complex sentences involves two regions, namely Broca's area—in particular BA 44—and the posterior superior temporal gyrus/superior temporal sulcus. This present description of the language networks is compatible with other models in the field, although some of these deviate in some relevant details. Other models describe the functions of particular regions somewhat differently: either they consider the larger Broca's region to support syntactic processes and syntax-semantic integration (unification) (Hagoort, 2005), or they consider Broca's area to support verbal rehearsal during syntactic processes rather than syntactic processing (Rogalsky and Hickok, 2011). However, a meta-analysis over more than 50 studies revealed a functional separation of syntactic and semantic processes in the left inferior frontal gyrus, studies with higher syntactic demands show stronger activation in BA 44, whereas studies with higher semantic demands show stronger activation in BA 45/47 (Hagoort and Indefrey, 2014). This meta-analysis therefore confirms a functional distinction between BA 44 and BA 45/47 as proposed earlier (Bookheimer, 2002; Friederici, 2002) and as evidenced by the findings discussed here.

Above I discussed a number of different studies on syntactic processing using different approaches. It appeared that BA 44 was found to be active whenever syntactic hierarchy building was involved—even at the lowest level for two-word combinations. A recent meta-analysis across 18 studies on syntactic processing found two main activation clusters in the inferior frontal cortex (BA 44 and BA 45) and in the temporal cortex (BA 22, BA 39, and temporal pole) (Zaccarella and Friederici, 2016). An interesting difference was found when subdividing the studies into (a) those that had used a control condition, including word-lists of both function words and content words, thereby allowing buildup of minimal syntactic structures, and (b) those that used word-lists consisting of only either function words or content words disallowing any structure building. The activation for the studies with mixed word list control conditions involved BA 45 and temporal regions. The cluster for the studies with control conditions restricted to only one word class show a more constrained pattern that was limited to BA 44 and the posterior superior temporal sulcus (see figure 1.12). These data clearly indicate that BA 44 and the posterior superior temporal cortex are the main regions constituting the syntactic network.

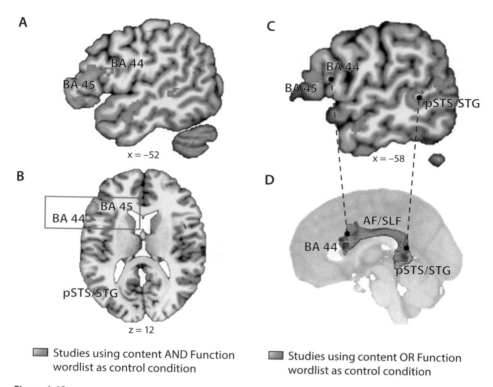

Figure 1.12

Meta-analysis of brain activation for syntactic studies. Functional clusters as a function of control condition used. Subsets of studies based on the type of word-list control condition used in the corresponding experiment: studies using content words AND function words (green-to-yellow), studies using content words OR function words (blue-to-turquoise). (A) Sagittal plane at x = − 52. (B) Transverse plane at z = 12. Red rectangle indicates Broca's region to show activation split between the two subsets of studies in BA 44 (content OR function words) and BA 45 (content AND function words). (C) Sagittal plane at x = − 58 showing activation overlap between the two subsets in pSTS/STG. Brodmann area (BA) 44/45; posterior superior temporal sulcus/gyrus (pSTS/STG; BA 22) (D) White matter fiber tract connecting BA 44 and posterior temporal cortex. Adapted from Zaccarella and Friederici (2016). The neuroanatomical network of syntactic Merge for language. MPI CBS Research Report 2014–2016 (see http://www.cbs.mpg.de/institute/research-reports).

Summary

This section discussed the many neuroscientific studies on syntax processing. The data suggest that the basic syntactic computation of binding two elements into a phrase (called Merge) assumed by linguistic theory can be evidenced at the neurobiological level in a confined brain region, BA 44. Moreover, the studies indicate that the processing of syntactically complex sentences involves the superior part of BA 44 and the posterior superior temporal gyrus/superior temporal sulcus. The left temporo-parietal cortex comes into play whenever working memory is required during sentence processing. These areas work together to achieve the comprehension of complex sentences.

1.6 Processing Semantic Relations

Sentence understanding crucially depends on the extraction of the sentence's meaning, provided by the meaning of the different words and the relation between them. An almost infinite number of word combinations—although guided by syntactic rules—allow us to construct complex concepts that go beyond the concept represented by a single word. This processing domain has partly been discussed under the heading of combinatorial semantics (Jackendoff, 2002). Here I will not describe Jackendoff's approach in particular, I will rather describe electroencephalography, magnetoencephalography, and functional magnetic resonance imaging studies that investigated semantic processes at the phrasal or sentential level, thereby going beyond the level of word semantics.

The semantic network as such is hard to localize in the human brain, as semantically triggered activations are reported for a large number of regions in the left hemisphere. A meta-analysis of over 120 neuroimaging studies on semantics including sentence level and word level studies by Binder, Desai, Graves, and Conant (2009) is titled "Where Is the Semantic System?" The regions reported to be activated cover large portions of the left hemisphere. The reason for this may be grounded in the circumstance that semantic processes even at the word level immediately trigger associations in different domains of semantic memory also involving sensory and motor aspects (Davis, 2015). These associations vary from person to person as each individual person may have learned a given word in different situations thereby triggering different aspects of their individual memory. I briefly discussed this in section 1.3.

Research studies describe the general semantic network to typically include inferior frontal, temporal, and parietal brain regions such as the left inferior frontal gyrus (Hagoort, Baggio, and Willems, 2009), the left temporal lobe (Humphries, Binder, Medler, and Liebenthal, 2007; Bemis and Pylkkänen, 2013), and the angular gyrus (Humphries et al., 2007; Obleser, Wise, Dresner, and Scott, 2007). On the basis of the available neuroscientific data, Lau et al. (2008) proposed a neuroanatomical model for semantic processing of words in context (figure 1.13). This model includes different regions in the temporal cortex, the inferior frontal gyrus, and the angular gyrus.

These regions were found to be activated in different studies but were not simultaneously reported in each of the studies. Early functional magnetic resonance imaging studies reported BA 45 and BA 47 in the inferior frontal gyrus as regions that support semantic language-related processes. BA 45/47 activation was observed in judgment tasks when participants were asked to judge whether two sentences had the same meaning (Dapretto and Bookheimer, 1999), or in tasks that required strategic and executive aspects of semantic processing, such as word retrieval and categorization (Fiez, 1997; Thompson-Schill, D'Esposito, Aguirre, and Farah, 1997). Based on these and later studies, which mainly reported temporal activation for semantic processes, it was concluded that in the semantic domain the inferior frontal gyrus mainly supports strategic and controlled processes.

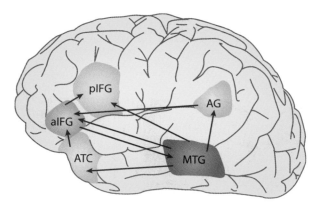

Figure 1.13
A neurofunctional model for semantic processing of words in context. According to the model, lexical representations are stored and activated in the middle temporal gyrus (MTG) and in the nearby superior temporal sulcus and inferior temporal cortex, and are accessed by other parts of the semantic network. The anterior temporal cortex (ATC) and angular gyrus (AG) are involved in integrating incoming information into current contextual and syntactic representations. The anterior inferior frontal gyrus (aIFG) mediates controlled retrieval of lexical representations based on top-down information, and the posterior IFG (pIFG) mediates selection between highly activated candidate representations. Adapted by permission from Nature Publishing Group: Lau, Colin, and Poeppel. 2008. A cortical network for semantics: (De)constructing the N400. *Nature Reviews Neuroscience*, 9 (12): 920–933.

Semantic-related activations in the temporal cortex were mainly reported for sentence processing studies in the anterior temporal lobe (Mazoyer et al., 1993; Vandenberghe et al., 2002; Humphries et al., 2006) and also in the posterior superior temporal gyrus (Obleser and Kotz, 2010) as well as the angular gyrus (Humphries et al., 2007; Obleser, Wise, et al., 2007). However, a recent meta-analysis across sentence processing studies suggests an involvement of BA 45/47 when it comes to processing semantic aspects (Hagoort and Indefrey, 2014). Here I will discuss those studies that found activations as a function of processes at the phrasal or sentential level.

A large number of the studies investigating semantic processes at the sentential level were initially conducted using electroencephalography and later magnetoencephalography, for historical reasons. Electroencephalography as a technology was available more than a decade earlier than functional magnetic resonance imaging technology. One of the first ERP papers on language processing (Kutas and Hillyard, 1980) identified a specific ERP component that was correlated with the processing of semantic information. This ERP component is a centro-parietal negativity around 400 ms, called N400. The example in figure 1.14 demonstrates the N400 in response to a semantic violation in sentence context.

The N400 has been viewed as a correlate of lexical-semantic processing since its first discovery (Kutas and Hillyard, 1980). Later it was interpreted to reflect difficulty in

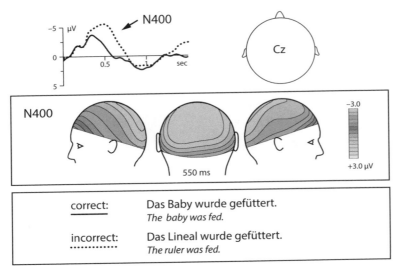

Figure 1.14
Semantic violation effect: N400. Event-related brain potentials in response to a semantic violation in an incorrect sentence compared to a correct sentence. For example sentence see box at the bottom of the figure. N400 stands for a central negativity (more negative-going wave form for the incorrect compared to the correct sentence) around 400 ms. Top: Wave form recorded at the centrally located electrode Cz for correct and incorrect condition for the sentence final word. Middle: Voltage maps displaying the activation difference between incorrect and correct condition at 550 ms for the sentence final word. Bottom: Example sentences. Adapted from Hahne and Friederici (2002). Differential task effects on semantic and syntactic processes as revealed by ERPs. *Cognitive Brain Research*, 13 (3): 339–356, with permission from Elsevier.

lexical-semantic integration (Hagoort and van Berkum, 2007' Hagoort, 2008), or to reflect lexical preactivation (Kutas and Federmeier, 2000; Lau et al., 2008; Stroud and Phillips, 2012). The N400 effect was observed under five different constellations at the word level and at the sentence level: the N400 amplitude was found to increase (1) when a word does not have a lexical status (i.e., a non-word or a pseudoword); (2) when the second word of a word pair does not fit the first word semantically; (3) when in a sentence the selectional restriction of verb-argument relations is violated; (4) when a word does not fit the preceding sentence context with respect to world knowledge or is simply unexpected; however, (5) N400 amplitude decreases for words as the sentence unrolls due to increased predictability of the next word. Therefore, the N400 is seen an indicator of lexical processes, lexical-semantic processes, semantic contextual predictability, and predictability due to world knowledge (Hagoort, Hald, Bastiaansen, and Petersson, 2004). Moreover, the N400 effect varies during sentence processing when semantic inferential processes that go beyond the meaning of the words are required. For example the processing of the final word in the sentence, *The journalist began the article* compared to the sentence, *The journalist wrote the article* leads to a long-lasting negative shift, longer in duration than the classical N400 effect (Baggio, Choma, van Lambalgen, and Hagoort, 2009). It thus reflects processes relevant to language comprehension at different levels, but also those that concern world knowledge.

The classical N400 effect is usually broadly distributed across the scalp (see figure 1.14). Its main generators were localized in respective magnetoencephalography studies in the vicinity of the auditory cortex (Halgren et al., 2002). Additional information concerning the location of the N400-related processes comes from brain imaging studies. There are a number of functional magnetic resonance imaging studies that used sentence materials identical to those used in a prior ERP study reporting N400 effects. Functional magnetic resonance imaging studies that worked with semantic violations similar to those presented reported activation in the middle and posterior portion of the superior temporal gyrus (Friederici et al., 2003; Brauer and Friederici, 2007).

Functional magnetic resonance imaging studies that used correct sentences in which the final word of the sentence was highly predictable from the prior context reported activation increase for low predictable compared to high predictable words in the left posterior superior temporal gyrus/superior temporal sulcus (Obleser and Kotz, 2010; Suzuki and Sakai, 2003) or in the left angular gyrus (Obleser, Wise et al., 2007). The assumed relation between predictability and the left angular gyrus was investigated in a study using transcranial magnetic stimulation by which normal functioning of a given brain region is interrupted (Hartwigsen, Golombek, and Obleser, 2015). Applying transcranial magnetic stimulation over the left angular gyrus during the comprehension of degraded speech revealed that this region is crucial for predictions at the sentential level. As the predictability effect mainly relies on the relation between the verb and the noun in a given sentence and reflects the aspect of thematic role assignment, the respective studies will be discussed in section 1.7.

Most of the semantic studies beyond the word level used sentence material that was constructed empirically, but not necessarily according to any linguistic semantic theory. Often, sentences were used in which a given word did not fit the prior context based on word knowledge or frequency of co-occurrence. Such sentence material was constructed on the basis of a so-called cloze procedure applied to an independent group of participants. The procedure requires individuals to complete a sentence fragment with the first word that came to their mind. Words that were frequently given as a response were defined as *high cloze* words, and words that were given rarely as *low cloze* words. The experimental sentences were constructed from these empirical data. The lack of theoretical rigor could be one reason for the observed variation in the localization of the brain activations. Another reason may simply be due to the fact that sentences used in different studies are of different length, contain different contextual constraints, or differ with respect to other parameters difficult to control, such as the various associations an individual may make when hearing longer sentences.

In order to address this problem, Bemis and Pylkkänen (2011, 2013) implemented a novel approach in which semantic combinatorics were broken down to a two-word phrase level. In this paradigm the same noun is presented either in a two-word phrase (*red boat*) or in the context of an unpronounceable consonant string (*xkq boat*) or a word list (*cup, boat*). When a combination of the first and the second element (word) was possible, the left anterior temporal lobe was seen to increase its activation as measured by magnetoencephalography, in both reading and listening (Bemis and Pylkkänen, 2013). Given that combinations such as *red boat* may not only trigger semantic combinatorics, but also the buildup of a syntactic phrase, additional experimentation had to disentangle these two aspects. An additional magnetoencephalography study showed that activation in the left anterior temporal lobe varies as a function of the conceptual specificity of adjective-noun combinations in which the syntax was kept constant (Westerlund and Pylkkänen, 2014). Moreover, activation in the left anterior temporal lobe was found across different types of composition (Del Prato and Pylkkänen, 2014) as well as across different languages, such as English and Arabic (Westerlund, Kastner, Al Kaabi, and Pylkkänen, 2015). From these findings the authors conclude that the anterior temporal lobe is primarily responsible for basic semantic composition.

Another region that appears to be involved in semantic processes is the angular gyrus in the posterior temporal cortex (Bemis and Pylkkänen, 2013). This is also suggested by studies that used functional magnetic resonance imaging to investigate the neural basis of combinatorial semantics. One study (Price, Bonner, Peelle, and Grossman, 2015) used two-word combinations that were either meaningful (*plaid jacket*) or not (*moss pony*). A region-of-interest analysis was performed for the left anterior temporal lobe and the left and right angular gyrus, since these regions had been suggested as relevant regions for combinatorial semantics (Bemis and Pylkkänen, 2013). This analysis revealed a major involvement of the angular gyrus, rather than the anterior lobe.

The involvement of the angular gyrus in combinatorial semantic processes is also suggested in an additional functional magnetic resonance imaging study (Molinaro, Paz-Alonso, Duñabeitia, and Carreiras, 2015) using adjective-noun pairs of prototypically varying degrees (*dark cave, bright cave, wise cave*). Brain activation data revealed the involvement of a number of regions, the anterior inferior frontal gyrus (BA 45/47), middle frontal gyrus (BA 9), anterior temporal cortex, middle temporal gyrus, angular gyrus, and fusiform gyrus. A region-of-interest analysis focused on inferior frontal gyrus, middle frontal gyrus, anterior temporal cortex, posterior middle temporal gyrus, and fusiform gyrus, because these are the regions that are considered to constitute the neural semantic network according to Lau and colleagues (2008). A subsequent functional connectivity analysis for these regions revealed a strong coupling between them all except for the angular gyrus. In the Molinaro and colleagues (2015) study the angular gyrus only comes into play for the contrastive condition (*bright cave*). For this condition a strong functional

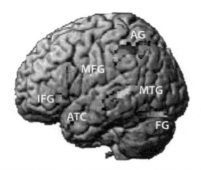

Figure 1.15
The semantic network for sentence processing. Brain activation for processing sentences with three types of adjective-noun pairs compared to a null-event control condition. IFG, inferior frontal gyrus; MFG, middle frontal gyrus; AG, angular gyrus; ATC, anterior temporal cortex; MTG, posterior middle temporal gyrus; FG, fusiform gyrus. Adapted from Molinaro et al. (2015). Combinatorial semantics strengthens angular-anterior temporal coupling. *Cortex*, 65: 113–127, with permission from Elsevier.

coupling between the angular gyrus and the anterior temporal cortex was observed (see figure 1.15). Therefore, these two regions are taken to be involved in semantic combinatorics. Molinaro and colleagues (2015) view the anterior temporal lobe to be devoted to high-level abstract semantic representations with tight connections to the language processing network, whereas the angular gyrus is taken to be mainly devoted to process combined semantic representations that are complex and anomalous. Regarding the respective role the anterior temporal pole and the angular gyrus play, it is interesting to see that the anterior temporal lobe is activated prior to the angular gyrus (Bemis and Pylkkänen, 2013), thereby supporting the view that the anterior temporal lobe represents more basic combinatorial processes and the angular gyrus more complex ones.

The magnetoencephalography and the functional magnetic resonance imaging studies together suggest different regions in the service of semantic combinatorial processes; these are the anterior temporal lobe and the angular gyrus, respectively. Both regions seem to play their part, but it is still open what functional part they play during combinatorial semantics. It appears that the anterior temporal lobe is involved in basic combinatorial processes, whereas the angular gyrus is involved in combinatorial semantic processes creating more complex semantic concepts.

Summary

In this section we have looked at the possible neurobiological basis of semantic processes and relations. A number of brain regions have been reported to support semantic processes: the anterior temporal lobe, the posterior temporal and middle temporal gyrus, the angular gyrus, and BA 47/45 in the inferior frontal gyrus. From the available data it appears that the anterior temporal region subserves basic semantic composition, the angular gyrus is involved in combinatorial semantics, whereas BA 47/45 comes into play for controlled executive semantic processes. Psycholinguistic models may want to take the neurobiological differentiation between semantic composition, combinational processes, and controlled executive processes into account.

1.7 Thematic Role Assignment: Semantic and Syntactic Features

A crucial part of sentence comprehension involves the assignment of thematic relations between the verb and its argument noun phrases in order to understand "who is doing what to whom." A number of features have to be processed to achieve this, one being a semantic determination of whether the actor of an action is animate (most actors are, except for robots doing an action). Syntactic features such as verb agreement and case marking (nominative, accusative) also code the likelihood of being an actor (doer) or a patient (undergoer) of the action, and are used as well for thematic role assignment. Neurolinguistic models assume that processes of thematic role assignment take place after the phase of initial structure building. In Friederici's model of language comprehension (Friederici, 2002, 2011) this constitutes the second phase. Bornkessel and Schlesewsky (2006) subdivide this phase into two subphases: one during which relevant features are extracted, and one during which computation takes place. For details of their model, see their review (Bornkessel and Schlesewsky, 2006).

The cognitive model outlined in section 1.1 (figure 1.1) postulates parallel processes in different subsystems that follow the initial phase of syntactic phrase structure building. These concern the establishment of syntactic relations and semantic relations discussed earlier and of thematic relations, which will be discussed in this section. The assignment of thematic relations mainly depends on different types of information available from nouns and verbs and their inflectional endings but can be influenced by semantic aspects of agency. These processes appear to work in parallel, but they may also interact when it comes to assigning the thematic roles online in order to determine "who is doing what to whom."

Sentence comprehension crucially depends on the identification of the grammatical subject and thereby the possible actor in a sentence. Languages differ in their cues upon which thematic roles can be assigned (MacWhinney, Bates, and Kliegl, 1984; Bates and MacWhinney, 1987). In a language with a fixed word order, like English, one can rely on positional information to assign the actor role (e.g., the first noun phrase is likely to be the actor). In a language with relatively free word order, like German, morphosyntactic information such as case marking can signal actorhood. Nominative case marked in the article preceding the noun indicates the grammatical subject and thereby most likely the actor. Thematic role assignment (e.g., actor, patient) can also be based on semantic features (e.g., animate, inanimate) as animate nouns are more likely to be the actor of an action.

Here I will first review the major ERP findings observed across different languages. The focus is on ERP components often reported in the literature, namely: (a) the N400 around 400 ms, which not only signals purely semantic processes but, moreover, semantic-thematic processes; (b) the LAN, a left anterior negativity between 300 and 500 ms, which was found for morphosyntactic processes; and (c) the P600 around 600 ms as a general marker reflecting syntactic and semantic integration processes.

Semantic and Verb Argument Features during Thematic Role Assignment

When mapping language input onto meaning, the verb plays a central role. Linguistic theory has defined the requirements of a verb with respect to its argument nouns. The verb determines its argument nouns both with respect to semantic aspects, called selectional restrictions, (Chomsky, 1965), and also with respect to structural aspects, called argument structure (Chomsky, 1980). Selectional restrictions refer to the semantic constraints a verb requires for its possible arguments. For example, the verb *drink* requires an object noun that has the semantic feature "fluid, drinkable," for example, *he drinks the beer*. A verb's argument structure determines how many arguments a verb can take: one argument (*he cries*); two arguments (*he visited his father*); three arguments (*he gave a book to his father, he gave his father a book*); and, moreover, the type of phrase a verb can take (noun phrase, prepositional phrase). According to the linguistic theory (Chomsky, 1980) a sentence is ungrammatical and unparsable if the verb's argument structure does not match the number and type of arguments in the sentence. Based on these theoretical considerations, we would expect violations of semantic and syntactic aspects to lead to different ERP patterns, respectively. We will see that this is indeed the case.

In the context of thematic role assignment, the semantic feature *animacy* in the noun argument plays a major role because animate nouns are more likely to assume the role of an actor than inanimate nouns. Since the prototypical actor is animate, this information may help to assign the role of the actor, but not always. For example in the sentence that follows, *The tree hit the man when falling*, the animacy-strategy could lead to an initial misassignment of the tree's role, as tree is an inanimate noun. Nevertheless, the parsing system has to assign thematic roles online as the sentence is perceived in order to keep the working memory demands low, even if initial assignments must be reanalyzed later in the sentence.

Selectional restriction violations tapping semantic aspects have been tested in many languages including English, German, Dutch, French, Spanish, and Chinese (for a review, see Lau et al., 2008). Most interestingly, it has been shown that the amplitude of the N400 increases systematically as a function of the number of semantic features violating the

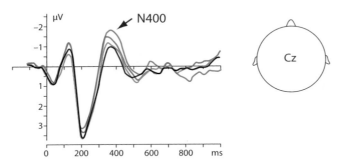

Figure 1.16
Effects of degree of semantic feature violations. Event-related brain potentials (ERPs) for sentences with different degrees of feature violations between the verb and its object noun. For examples see box at the bottom of the figure. Displayed are the ERP grand averages for the four different conditions recorded at the centrally located electrode Cz. The amplitude of the N400 increases as the "inappropriateness" of the target object noun in verb-noun combinations increases. The increase of "inappropriateness" is defined as the number of mismatching semantic features of the noun. Semantic features are, for example, ± human, ± animate, etc. The most inappropriate noun for the verb (*hire*) in this case is an inanimate object noun (*wire*) leading to the largest amplitude of the N400. Adapted from Li et al. (2006). Mental representation of verb meaning: Behavioral and electrophysiological evidence. *Journal of Cognitive Neuroscience*, 18 (10): 1774–1787. © 2006 by the Massachusetts Institute of Technology.

relation (in sentences) between the verb and its noun argument. The following type of sentence fragment was used: *For the sake of his safety, the millionaire decided to hire a ...* which was then completed by different nouns that varied from being an appropriate object to a totally inappropriate object (Li et al., 2006) (see figure 1.16). The appropriate object noun condition, *bodyguard* (semantic features are + animate, + human, + guard) carries no violation; the inappropriate human object noun, *baby* (semantic features are + animate, + human, − guard) carries a one-feature violation; the inappropriate non-human object, *hen* (semantic features are + animate, − human, − guard) carries a two-feature violation; and the inappropriate inanimate object noun, *wire* (semantic features are − animate, − human, − guard) carries a three-feature violation. For examples see figure 1.16. This impressively demonstrates that the N400 is modulated by a theoretically defined number of semantic word features and the resulting selectional restriction violations. As the N400 is taken to reflect semantic processing aspects, these data support the view that the processing of selectional restriction falls within the semantic domain.

Violations of the number of arguments in a sentence (for example when too many arguments for a given verb are present, e.g., *Anna knows that the inspector the banker departed*) lead to a combination of a N400 and a P600 effect, referred to as N400–P600 pattern (Friederici and Frisch, 2000). The N400–P600 pattern suggests that in this case the processing system registers a misfit between the verb and the noun both at the semantic and at the syntactic level, suggesting that the verb-argument structure processing involves both aspects. If there are too many or too few noun arguments, the sentence's meaning cannot be constructed, nor can the overall sentence's syntactic structure. In contrast, when a number of arguments are correct, but the case marking of the object noun phrase (direct object vs. indirect object) is incorrect, a combination of a LAN effect and a P600 effect is observed, this is referred to as a LAN–P600 pattern. The processing of case as a morphosyntactic process is reflected in the LAN followed by a P600 assumed to reflect processes of syntactic reanalysis (see also section 1.8). The combined results reveal that during sentence processing different aspects of the verb-argument relation have their different, quite specific neural traces. Given that the N400 and the LAN fall within the same time window (300–500 ms), these data further show that semantic and thematic relations involving the verb's selectional and verb-argument restrictions are processed in parallel as assumed in the functional language processing model (see figure 1.1).

Thus, when it comes to processing a verb with its argument nouns in a sentence, violations of number and type of arguments (structural domain) and violations of selectional restrictions (semantic domain) lead to different ERP patterns. The former is reflected in a biphasic pattern, N400–P600 or LAN–P600, whereas the latter is reflected in an N400. For variations of the particular realization of argument-related negativity as a function of different language typologies, see Bornkessel and Schlesewsky (2006).

Syntactic Features during Thematic Role Assignment
In parallel to the processing of information regarding a verb's selectional restrictions (semantic domain) and its verb-argument structure (structural domain), morphosyntactic information provided by a verb's inflection (number and person) is essential for the assignment of grammatical roles in a sentence. While this information is less important for sentence interpretation in languages with fixed word order, like English, it is crucial for languages with free word order, like German. In a free word order language, morphosyntactic features can signal thematical roles.

In English, subject-verb number agreement—plural (*we see*) versus singular (*he sees*)— determines who is the subject of the action. But in a free word order language, such as German, the assignment of subject or object role in a noun-verb-noun sequence is sometimes only possible under two conditions: if (1) a subject and object noun either differ in number marking (plural vs. singular) as in the object-first sentence 1 below. The principle of subject-verb agreement says that since the noun phrase *the boy* and the *verb* agree in number *the boy* must be the subject of the sentence. Assignment of subject and object in

German is also possible if (2) the noun phrases or one of them are case marked (nominative vs. accusative) as in the object-first sentence 2. Case marking allows assignment of a syntactic role: nominative assigns the subject role, and accusative assigns the object role.

Object-first sentence

(1) *Die Männer* [plural] *grüßt* [singular] *der Junge* [singular].

The men greet the boy [actor] [literal].

The boy greets the man [non-literal].

Object-first sentence

(2) *Den Mann* [singular, accusative] *grüßt der Junge* [singular, nominative].

The man greets the boy [actor] [literal].

The boy greets the men [non-literal].

But if the two noun phrases in a sentence carry the same number marking and ambiguous case marking as in sentence 3, the sentence is completely ambiguous and roles are assigned by a default "subject-first" strategy.

Ambiguous sentence

(3) *Die Männer* [plural] *grüßen* [plural] *die Jungen* [plural].

The men greet the boys.

When processing an ambiguous sentence often a subject-first strategy is applied, taking the first noun as the actor. There are a number of languages in which thematic roles (actor, patient, etc.) can be determined by case (nominative case assigns the actor, accusative case assigns the patient, etc.), thereby allowing the assignment of "who is doing what to whom."

Violations of subject-verb agreement (singular versus plural) in an inflecting language usually induce a LAN between 300 and 500 ms ([German], Penke et al., 1997; [Italian], Angrilli et al., 2002; [Spanish], Silva-Pereyra and Carreiras, 2007). In a fixed word order language such as English, a LAN is found less systematically (LAN reported in Osterhout and Mobley, 1995), but not in other studies (Kutas and Hillyard, 1983; Osterhout and Nicol, 1999). It has been argued that the presence or absence of the LAN should be viewed as a continuum across different languages, and that the likelihood of observing this effect increases with the increasing amount of overt and clear morphosyntactic marking in a given language (Friederici and Weissenborn, 2007).

However, it is not purely the amount of morphosyntactic marking that determines the possible presence of an LAN effect, but also whether the morphosyntactic information is crucial for the assignment of syntactic roles. In some languages, determiner-noun agreement with respect to gender (masculine, feminine, neuter) is crucial for role assignment, and in others it is not. If this information is not crucial for the assignment of grammatical relations between a verb and its arguments in sentences as in German or French, a violation of gender agreement between determiner and noun does not lead to a strong LAN effect. In these languages the agreement-relation between the verb and its subject noun does not depend on syntactic gender, that is, on whether the noun is masculine (German: *der Mann*, the man) or feminine (German: *die Frau, the woman*). But in languages in which gender agreement is relevant for the assignment of grammatical roles, as in Hebrew, in which gender agreement between subject noun and verb inflection is obligatory, the LAN is clearly present (Deutsch and Bentin, 2001). This leads to the conclusion that, whenever morphosyntactic marking is crucial for the assignment of grammatical relations in a sentence, a LAN is observed.

Thus, different languages provide different types of information that are relevant for the assignment of thematic roles. Bornkessel-Schlesewsky and colleagues (see Bornkessel-Schlesewsky, 2011; Tune et al., 2014; Muralikrishnan, Schlesewsky, and Bornkessel-Schlesewsky, 2015) provide compelling evidence that there are cross-linguistic differences in how semantic and syntactic information are used for thematic role assignment with the goal to achieve comprehension.

Integration

Models that consider the time course of language processes have assumed a late processing phase, around and beyond 600 ms after the critical event, during which different information types are mapped onto each other to accomplish comprehension (Friederici, 2002; Friederici and Kotz, 2003; Bornkessel and Schlesewsky, 2006). It is proposed that this takes place during a final processing phase (called Integration in figure 1.1) during which—if necessary—processes of syntactic reanalysis and repair take place. These processes are reflected in a late centro-parietal positivity in the ERP, called P600 (see figure 1.17).

The P600 displayed in figure 1.17 was found in response to a syntactic violation. The P600, however, has been observed in reaction to a number of different experimental sentential variations and has, therefore, received various functional interpretations.

This component, first observed for the processing of syntactic anomalies (Osterhout and Holcomb, 1992), was found for the processing of temporarily ambiguous sentences right at the point of disambiguation resolution when reanalysis was necessary (Osterhout, Holcomb, and Swinney, 1994), and also after a syntactic violation when repair was required (Hagoort, Brown, and Groothusen, 1993; Hahne and Friederici, 2002). In the past decade P600 effects were observed in response to a number of different language stimuli with partly different distributions over the scalp. The P600 has been reported for ambiguous sentences (Osterhout, Holcomb, and Swinney, 1994), for complex sentences (Friederici, Hahne, and Saddy, 2002), but also as part of a biphasic ELAN–P600 pattern in response to

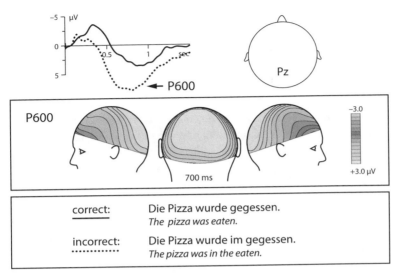

Figure 1.17
Late syntactic effect: P600. Event-related brain potentials in response to a syntactic violation (here word category violation), incorrect compared to correct sentences recorded at the centro-parietal electrode Pz. P600 stands for a positivity (more positive-going wave form for incorrect compared to correct sentences) around 600 ms. Top: Wave form for correct and incorrect condition for the sentence final word. Middle: Voltage maps displaying the activation difference between incorrect and correct condition at 700 ms for the sentence final word. Bottom: Example sentences. Adapted from Hahne and Friederici (2002). Differential task effects on semantic and syntactic processes as revealed by ERPs. *Cognitive Brain Research*, 13 (3): 339–356, with permission from Elsevier.

syntactic phrase structure violations (Friederici et al., 1993; Hahne and Friederici, 1999). This biphasic pattern is believed to reflect two phases in syntactic processing, with the ELAN indicating syntactic violation detection and the P600 indicating processes of reanalysis and repair. A direct comparison of the P600 topography for correct sentences requiring a reanalysis and for incorrect sentences requiring a repair revealed a differential pattern of distribution with a more fronto-central distribution for the syntactic reanalysis P600 and a centro-parietal distribution for the syntactic repair P600 (Friederici et al., 2002). These data suggest that syntactic reanalysis and syntactic repair should possibly be viewed as non-identical processes in a processing model.

The functional interpretation of the P600, however, has changed to some degree over the past years. Initially, it was taken to reflect syntactic processes in general (Hagoort et al., 1993), processes of syntactic reanalysis and repair (Friederici et al., 1996), or the difficulty of syntactic integration (Kaan, Harris, Gibson, and Holcomb, 2000). Later studies found the P600 to not only vary as a function of syntactic variables, but also to reflect the interaction of syntactic and semantic anomaly at the sentence level (Gunter, Friederici, and Schriefers, 2000; Kuperberg et al., 2003; Kuperberg, Caplan, Sitnikova, Eddy, and Holcomb, 2006), suggesting that the P600 might reflect late sentence-level integration processes taking into consideration both syntactic and semantic information.

More recently the status of the P600, which was believed to reflect integration processes involving syntactic aspects only, was challenged by additional studies reporting P600 effects for sentence-level semantic violations (Kolk, Chwilla, van Herten, and Oor, 2003; Hoeks, Stowe, and Doedens, 2004; Kim and Osterhout, 2005; Kuperberg, Kreher, Sitnikova, Caplan, and Holcomb, 2007). For example, sentences like *The hearty meal was devouring* led to a P600 (Kim and Osterhout, 2005). This phenomenon, called *semantic reversal anomalies*, has been reported to elicit a P600 in English (Kim and Osterhout, 2005; Kuperberg et al., 2006) and in Dutch (Kolk et al., 2003; Hoeks et al., 2004). As the P600 was traditionally seen to reflect syntactic processes (Hagoort, Brown, and Groothusen, 1993), a *semantic P600* raised a number of questions and resulted in the interpretation of a dominance of semantic processes during online comprehension overriding syntactic processes and thereby leading to a semantic mismatch (Kim and Osterhout, 2005; Kuperberg et al., 2007). In the end three different explanations were put forward for semantic P600 effects: (1) plausibility/semantic attraction between the verb and an argument (Kim and Osterhout, 2005); (2) thematic processing cost (Hoeks et al., 2004); and (3) interaction of thematic and semantic memory (Kuperberg et al., 2007). However, note that all these different interpretations concern thematic role assignment in sentences and may reflect different aspects of it (Bornkessel-Schlesewsky and Schlesewsky, 2008). The most general interpretation of these findings is that the P600 reflects integration of syntactic, thematic, and semantic information provided by the respective subcomponents dealing with syntactic, thematic, and semantic relations. Future research must reveal whether the P600 is a unitary component reflecting one process or whether there is a family of P600 components each reflecting a different subprocess. Unfortunately, the decision on this issue is complicated by the fact that up to now the P600 has not been clearly localized in the human brain.

Localization: MEG and fMRI

When describing the language-brain relationship, one wonders where the processes of integration take place. The information about this is sparse. For thematic role assignment, semantic and syntactic information have to come together. The ERP work investigating thematic role assignment has related the P600 to this process. But we have learned that the P600 is hard to localize. In the functional magnetic resonance imaging work there are some studies investigating thematic role assignment using different paradigms. These suggest that thematic role assignment involves the posterior temporal cortex. Activation in the posterior superior temporal gyrus/superior temporal sulcus has been seen to be modulated by predictability at the sentential level, when the stimulus material involves the processing of the relation between the verb and its arguments—be it in correct sentences when considering a sentence's semantic cloze probability with respect to the verb-argument relation (Obleser and Kotz, 2010), or in sentences which contain a restriction violation between the verb and its arguments (Friederici et al., 2003). When different verb classes and their argument order were investigated, these two factors were found to interact in the posterior superior temporal gyrus/superior temporal sulcus (Bornkessel, Zyssett, von Cramon, and Schlesewsky, 2005). Together, these studies suggest that the left posterior superior temporal gyrus/superior temporal sulcus may be the region in which syntactic information and verb-argument-based information are integrated (Grodzinsky and Friederici, 2006).

The brain basis of the P600 effect is still unresolved. A few studies tried to localize the syntactic P600 and report the middle temporal gyrus and the posterior portion of the temporal cortex as a possible location site (Kwon, Kuriki, Kim, Lee, and Kim, 2005; Service, Helenius, Maury, and Salmelin, 2007). Moreover, there is some indication that the basal ganglia are part of the circuit supporting processes reflected in the syntax-related P600, since patients with lesions in the basal ganglia show reduced P600 amplitudes (Friederici et al., 1999; Frisch, Kotz, von Cramon, and Friederici, 2003). This latter finding of an involvement of the basal ganglia in syntactic processes is, in general, compatible with the model proposed by Ullman (2001), which holds that syntactic processes are supported by a circuit involving the basal ganglia and the frontal cortex, although this model does not distinguish between early and late syntactic processes.

The localization of the P600 in the functional magnetic resonance imaging is difficult, because it is difficult to separate the P600 from earlier language ERP components effects given the poor time resolution of functional magnetic resonance imaging. At present, therefore, the neural basis underlying the P600 effect has not yet been specified in much detail. Because it occurs late in the processing timeline, it is likely that it reflects integration of different information types, including semantic and syntactic processes. Based on the available magnetic resonance imaging data discussed above, I suggest that the posterior superior temporal gyrus/superior temporal sulcus observed to be activated at the sentential level for semantic and syntactic processes is a candidate region to represent the integration process.[10]

Interpretation
The ultimate step in language comprehension as displayed in figure 1.1 is called integration and refers to the interpretation of a linguistic representation in a most general sense. It depends on a number of aspects that are not part of the core language system. These are situational and communicative aspects briefly discussed below, but also an individual's knowledge about the world—be it technical knowledge, cultural knowledge in general, or knowledge about the communication partner's social and family background in particular. This knowledge probably has a distributed representation across different regions in the brain and is therefore not easy to capture.

Summary

This section reviewed the available neurobiological literature on the issue of thematic role assignment. Thematic role assignment takes place in a phase during which the relation between the verb and its arguments is computed to assign the thematic roles in a sentence. Morphosyntactic information (subject-verb agreement, LAN), case information (LAN or N400, depending on the particular language), and lexical selectional restriction information (N400) are taken into consideration to achieve assignment of the relation between the different elements in a sentence. We saw that linguistic theory was backed up by empirical data with respect to a separation between selectional restriction requirements of the verb and requirement concerning verb argument structure. The brain basis responsible for the understanding of "who is doing what to whom" in a sentence comprises a left fronto-temporal network involving the inferior frontal gyrus as well as the middle and superior temporal gyri.

1.8 Processing Prosodic Information

When processing spoken sentences, phonological information in addition to semantic and syntactic information must be processed. Section 1.2 discussed acoustic-phonological processes at the segmental level concerning phonemes and features of these. The acoustic signal, however, also conveys suprasegmental phonological information. These rhythmic and melodic variations in speech are called prosody. Two types of prosodic information are usually distinguished: linguistic prosody and emotional prosody. Emotional prosody is an extralinguistic cue signaling either the speaker's emotional state or emotional aspects of the content conveyed by the speaker. In contrast, linguistic prosody plays a crucial role in the buildup of syntactic and thematic structures. I will review and discuss studies on linguistic prosody and then briefly turn to studies on emotional prosody, as linguistic and emotional prosody are often investigated and discussed together.

Researchers have different views on where in the brain linguistic prosody and emotional prosody are processed. A simple view is that linguistic prosody is processed in the left hemisphere, whereas emotional prosody is processed in the right hemisphere (Van Lancker, 1980). But as we will see, the picture is somewhat more complex.

Processing Linguistic Prosody

Studies on linguistic prosody mostly show an involvement of the two hemispheres, although to a different degree. Therefore, another view holds that pure prosody (pitch information) is predominantly processed in the right hemisphere, but that processes may shift to the left hemisphere the more the phonetic information is in focus, due to either the stimulus material or task demands (Friederici and Alter, 2004). Prosodic information is mainly encoded in the intonational contour, which signals the separation of constituents (syntactic phrases) in a spoken sentence and the accentuation of (thematically) relevant words in a speech stream. By signaling phrase boundaries, prosodic information becomes most relevant for sentence comprehension and the interpretation of "who is doing what to whom." This can be gathered from the example below. A prosodic boundary (i.e., break, indicated by # in the sentences below) is marked by three parameters: a shift in intonation (pitch), a lengthening of the prefinal syllable before the pause, and a pause itself.

(1) *The man said # the woman is stupid.*

(2) *The man # said the woman # is stupid.*

The phrase boundaries in these sentences are crucial for the interpretation as they signal the noun phrase to which the attribute "to be stupid" has to be assigned, either to the woman (1) or to the man (2). As the example shows, prosodic information is relevant for syntactic processes and ultimately for sentence interpretation. There seems to be a close relation between prosody and syntax. Indeed, almost every prosodic boundary is also a syntactic boundary, while the reverse does not hold.

The first neurophysiological correlate for the processing of sentence-level prosodic information was found in a study that recorded the electroencephalography during the processing of German sentences that contained either one intonational phrase boundary or two. At the intonational phrase boundary, the ERPs revealed a centro-parietally distributed positive shift (wave form going more positive) which we called the Closure Positive Shift because the intonational phrase boundary indicates the closure of a phrase (Steinhauer, Alter, and Friederici, 1999) (see figure 1.18). This effect was taken as a marker for processing intonational phrase boundary.

The Closure Positive Shift was replicated in other studies using a different language, namely Dutch (Kerkhofs, Vonk, Schriefers, and Chwilla, 2007; Bögels, Schriefers, Vonk, Chwilla, and Kerkhofs, 2010), Japanese (Wolff, Schlesewsky, Hirotani, and Bornkessel-Schlesewsky, 2008), Chinese (Li and Yang, 2009), and English (Itzhak, Pauker, Drury, Baum, and Steinhauer, 2010). Crucially, it was shown that the Closure Positive Shift is not triggered by the pause at the intonational phrase boundary per se, but that the two other parameters signaling the intonational phrase boundary, namely the pitch change and the lengthening of the syllable prior to the pause, are sufficient to evoke boundary perception in the adult listener. This was evidenced in an experiment in which a Closure Positive Shift was present although the pause at the intonational phrase boundary had been deleted (Steinhauer et al., 1999).

Additional experiments with adults showed that the Closure Positive Shift can also be elicited when only prosodic information of a sentence is delivered (i.e., when segmental information is deleted by a filtering procedure). Under this condition the Closure Positive Shift is lateralized to the right hemisphere (Pannekamp, Toepel, Alter, Hahne, and Friederici, 2005). Moreover, the Closure Positive Shift is reported for sentence reading triggered by the comma indicating the syntactic phrase boundary (Steinhauer and Friederici, 2001;

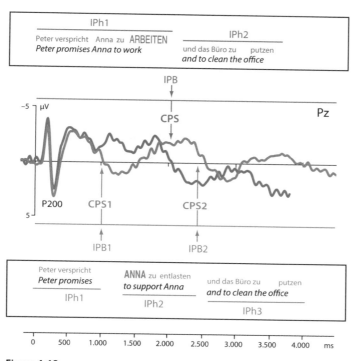

Figure 1.18
Prosody during sentence processing. The closure positivity shift (CPS) is an event-related brain potential (ERP) effect reflecting the processing of an intonational phrase boundary (IPB). Middle: Grand-average ERPs recorded at the Pz electrode. The waveforms for sentences with two intonational phrases (IPh) and one intonational phrase boundary (IPB) (green) and sentences with three IPhs and two IPBs (blue) are superimposed. The onsets of the sentence examples displayed in the top and bottom rows are aligned to the time axis in milliseconds (ms). Both conditions evoke a CPS (positive-going wave form) at their respective IPBs and the resulting CPS indicated by arrows: one CPS in the green condition and two CPS in the blue condition. Adapted from Steinhauer, Alter, and Friederici (1999). Brain potentials indicate immediate use of prosodic cues in natural speech processing. *Nature Neuroscience*, 2 (2): 191–196.

Steinhauer, 2003; Kerkhofs, Vonk, Schriefers, and Chwilla, 2008). Thus, the Closure Positive Shift can be viewed as an ERP component to correlate with prosodic phrasing both when realized openly in the speech stream and when realized covertly in written sentences, and even when the pause in the speech stream is deleted. This suggests that the Closure Positive Shift indeed signals the closure of a phrase rather than being an acoustic phenomenon. Strong support for this view comes from language development. In infants and toddlers, a Closure Positive Shift boundary response cannot be elicited when the pause is deleted, but only when the pause is present (Männel and Friederici, 2009, 2011). However, older children show a Closure Positive Shift as boundary response. Once sufficient syntactic knowledge is acquired, pitch information and syllable lengthening alone can trigger a Closure Positive Shift, just as in adults (Männel and Friederici, 2011).

The brain basis of prosodic processes, in particular the involvement of the two hemispheres, is much debated. And it appears that a number of different factors are relevant here. Initially this issue was investigated behaviorally in patients with cortical lesions in the left hemisphere and the right hemisphere. While some studies came to the conclusion that linguistic prosody is mainly processed in the right hemisphere (Weintraub et al., 1981; Brădvik et al., 1991), others voted for an involvement of both hemispheres as they found that both left hemisphere and right hemisphere patients showed deficits in processing sentence level prosody (Bryan, 1989; Pell and Baum, 1997). Most of these studies used normal speech that contains segmental information, that is, phonemic information allowing the identification of words, and suprasegmental information, that is, the "melody" (intonational contour) of the stimulus. However, when segmental information was filtered out, thereby increasing the reliance on suprasegmental information, right hemisphere patients demonstrated significantly worse performance than left hemisphere patients (Bryan, 1989). These and other studies (e.g., Perkins, Baran, and Gandour, 1996) suggested that the right hemisphere is primarily involved in processing prosodic information. As a result from these findings one can conclude that the less segmental information is available, the more dominant the right hemisphere. More recent studies, moreover, argue for a better control of the lesion site within the hemispheres as a crucial factor, as it was shown that patients with lesions in the temporo-parietal regions performed worse on prosodic tasks than patients with lesions in the frontal or subcortical regions (Rymarczyk and Grabowska, 2007).

Neuroimaging studies provide support for the observation that the amount of segmental and suprasegmental information in the stimulus material matters. Processing of pitch information (intonational contour) is correlated with increased activation in the right hemisphere, but this can be modulated by task demands (Plante, Creusere, and Sabin, 2002). A functional magnetic resonance imaging study that systematically varied the presence/ absence of suprasegmental and segmental information reported a shift in brain activation in the posterior temporal and fronto-opercular cortices to the right hemisphere as a function of the absence of segmental information (Meyer, Alter, Friederici, Lohmann, and von Cramon, 2002; Meyer, Steinhauer, Alter, Friederici, and von Cramon, 2004) (see figure 1.19).

Right prefrontal cortex and right cerebellar activation were also reported for prosodic segmentation during sentence processing (Strelnikov, Vorobyev, Chernigovskaya, and Medvedev, 2006). A study investigating sentences and word lists both with sentence prosody and word list prosody found bilateral activation in the anterior temporal cortex for syntactic and prosodic information, with the left hemisphere being more selective for syntactic information (Humphries et al., 2005). In this study also, clear right-hemisphere dominance was found for prosody. Together, the studies suggest that the right hemisphere is involved in processing intonational (pitch) information during sentence processing, but, in addition, they indicate that the actual lateralization partly depends on task demands

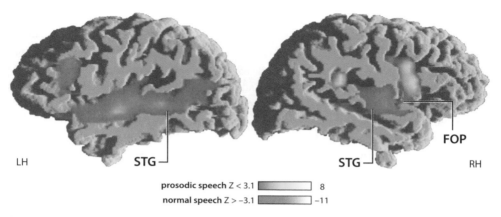

Figure 1.19
Brain activation as a function of prosody. Activation for prosodic speech (no segmental information, only suprasegmental information) compared to normal speech (segmental and suprasegmental information) is color-coded in red-yellow, activation for normal speech compared to prosodic speech is color-coded in green-blue. Adapted from Meyer et al. (2002). FMRI reveals brain regions mediating slow prosodic modulations in spoken sentences. *Human Brain Mapping*, 17 (2): 73–88. © 2002 Wiley-Liss, Inc.

(Plante et al., 2002; Gandour et al., 2004) and on the presence of concurrent segmental information (Bryan, 1989; Friederici and Alter, 2004).

This conclusion holds for non-tonal languages like, for example, Indo-European languages. The hemispheric lateralization of linguistic prosody, moreover, depends on the particular information prosody signals in a given language. In tonal languages like Thai and Mandarin Chinese, pitch patterns are used to distinguish lexical meaning. The syllable *ma* has different meanings in Mandarin Chinese depending on whether it follows a rising or falling pitch contour. Interestingly, when encoding lexical information, pitch is processed in the left hemisphere, similar to lexical information in non-tonal languages (Gandour et al., 2000). From this, it appears that the localization of language processes in the brain is determined by its function (lexical information) and not its form (pitch information). Only when intonation marks suprasegmental sentential prosody is it localized in the right hemisphere. When intonation is lexically relevant, it is processed in the left hemisphere.

Interaction between Linguistic Prosody and Syntax

Prosody and syntax are known to interact during language comprehension as this is indicated by behavioral studies on syntactic ambiguity resolution (Marslen-Wilson, Tyler, Warren, Grenier, and Lee, 1992; Warren, Grabe, and Nolan, 1995). The neuroscientific studies discussed in section 1.5 indicate that syntax is mainly processed in the left hemisphere whereas prosody as such may mainly be proposed in the right hemisphere. Any interaction of syntax and prosody must, therefore, rely on an interaction between the two hemispheres.

The two hemispheres are neuroanatomically connected via the corpus callosum (Hofer and Frahm, 2006; Huang et al., 2005) and would therefore provide the neural structure to support the information transfer between the hemispheres. If the above view about the functional role of the left hemisphere and right hemisphere in language processing is valid, a lesion to the corpus callosum that disallows the information transfer between the two hemispheres should have an effect on any interaction between syntactic (left hemisphere) and prosodic (right hemisphere) information (see figure 1.20).

The interaction of syntax and prosody has been evidenced at the neural level. ERP studies have reported a right anterior negativity for prosodic violations in sentences in which, for example, final-phrase prosodic information was presented at a non-final position thereby leading to an incorrect intonation. These types of prosodic violations were shown to interact with syntactic phrase structure violations suggesting a close interplay between prosodic and syntactic information during an early processing stage (Eckstein and Friederici, 2006). Patients with lesions in the posterior portion of the corpus callosum partly resulting in a disconnection of the left and right hemisphere did not show such syntax-prosody interaction effect, although they exhibited a prosody-independent syntactic effect (Sammler, Kotz, Eckstein, Ott, and Friederici, 2010). These data indicate that the corpus callosum connecting the two hemispheres builds the brain basis for the integration of phrase structure syntactic and prosodic features during auditory speech comprehension.

An interaction of prosodic and syntactic information is also observed when it comes to assigning relations between a verb and its arguments. For example, in the following prosodically correct German sentences as in (1), in which *Anna* is the object of *promise* and (2), in which *Anna* is the object of *support* (the relation between the verb and its object noun phrase is marked by the arrow):

(1) *Peter verspricht Anna # zu arbeiten.*

Peter promises Anna # to work.

(2) *Peter verspricht # Anna zu helfen.*

Peter promises # to support Anna.

Due to German word order, the two sentences appear identical up to the word *zu*, but their syntactic structure is marked differently by intonation in speech (indicated by # marking the intonational phrase boundary). Sentence 1 becomes prosodically incorrect, as in sentence 3, by inserting the intonational phrase boundary after the verb as in sentence 2. In sentence 3, below, the prosodic information signals that *Anna* is the object of the following verb *arbeiten*, but the verb *arbeiten* (*work*) cannot take a direct object.

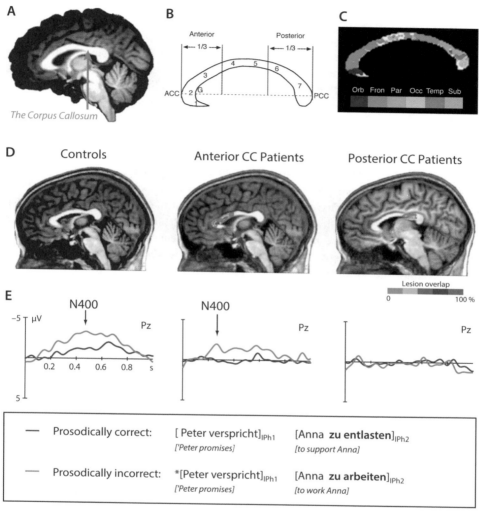

Figure 1.20
The corpus callosum and prosodic processes. (A) Corpus callosum in the human brain. This structure connects the left and right hemisphere. (B) Results of partition of the corpus callosum according to Witelson (1989) from a healthy population. (C) A map determined by the maximum of the histogram representing the fibers crossing from one to the other hemisphere at each pixel in the corpus callosum. Pixels with the maximum fibers projecting to the orbital lobe are color-coded purple, the frontal lobe green, the parietal lobe orange, the occipital lobe yellow, the temporal lobe red, and the subcortical nuclei cyan. Adapted from Huang et al. (2005). DTI tractography-based parcellation of white matter: Application to the mid-sagittal morphology of corpus callosum. *NeuroImage*, 26 (1): 195–205, with permission from Elsevier. (D) Lesion density maps of healthy controls and patients with anterior or posterior lesion contributions in the corpus callosum (CC). For each voxel, the percentage of lesion overlap is depicted. The color scale shows five levels: each bar represents 20% increments. (E) Event-related potentials (grand average) for the prosody mismatch effect at the critical verb in the sentence for the prosodically correct (blue) and incorrect (red) sentence phrasing for the healthy controls and patients. For sentences examples see box at the bottom of the figure. Adapted from Friederici, von Cramon, and Kotz (2007). Role of the corpus callosum in speech comprehension: interfacing syntax and prosody. *Neuron*, 53 (1): 135–145, with permission from Elsevier.

(3) *Peter verspricht # Anna zu arbeiten.

Peter promises # to work Anna.

With the use of such prosodically incorrect sentences, we were able to demonstrate that prosody guides syntactic parsing (Steinhauer, Alter, and Friederici, 1999). This was evidenced by an ERP effect observed at *zu arbeiten* (*work*) in the prosodically incorrect sentence 3. Based on the prosodic information, the parsing system expects a transitive verb (such as *support* as in sentence 2), but it receives an intransitive verb (namely *work* as in sentence 1). This unexpected verb form is a misfit because it violates the expected word form and, moreover, it violates even the expected verb class (i.e., transitive verb). This double violation of the expectation leads to a mismatch effect in the ERP, namely, an N400–P600 pattern. The N400 reflects a reaction to the unexpected verb and the P600 reflects processes of syntactic reanalysis. This functional interpretation of the two ERP components was supported by experiments that, in contrast to the original experiment (Steinhauer et al., 1999), did not use a syntactic judgment task. Without such a grammatical task (and simple passive listening), only an N400 was observed at the critical verb, reflecting simply the unexpectedness of the verb (Friederici, von Cramon, and Kotz, 2007; Bögels et al., 2010).

To test the hypothesis that the prosody-syntax interaction is indeed based on the information exchange between the left and the right hemisphere, sentences 1–3 were presented to patients with lesions in the corpus callosum. With the application of the passive listening paradigm, a prosody-syntax mismatch effect (N400) was observed in healthy controls and in patients with lesions in the anterior corpus callosum (though smaller than in healthy controls), but not in patients with lesions in the posterior corpus callosum (Friederici et al., 2007) (see figure 1.20D, E). This finding provides clear evidence for the view that the interaction of prosody and syntax relies on the interaction between the left hemisphere and the right hemisphere supported by the posterior portion of the corpus callosum through which the temporal cortices of the left and the right hemisphere are connected.

This conclusion is further supported by a functional connectivity analysis of neuroimaging data, in this case magnetoencephalography data investigating the activation coupling between different brain regions (David, Maess, Eckstein, and Friederici, 2011). This analysis revealed functional connectivity between the left and the right hemisphere during processing of syntactic and prosodic information, and moreover, suggested the involvement of a deep subcortical structure—possibly the thalamus or the basal ganglia—during syntax-prosody interaction.

Emotional Prosody

An utterance may encode more than the information simply conveyed by the meaning of the words. When a speaker wants to convey emotional information in addition to what is uttered, he or she has the option to modulate the prosodic parameters of speech. This part of an utterance is called emotional prosody. Parameters that can be changed in speech to signal emotion are intensity and pitch, but also lengthening of speech elements and pausing between these. Emotional prosody carries meaning, such as happy and sad, or fear and anger, and is, therefore, by definition not part of the core language system. This information is encoded in the auditory signal and can in principle interact with the content of an utterance.

The relation between emotional prosody and the brain was first investigated in patients with lesions in either the left or the right hemisphere. It was reported that patients with temporo-parietal lesions in the right hemisphere, but not patients with lesions in the left hemisphere, had deficits in identifying the emotional tone of a speaker producing semantically neutral sentences (Heilman, Scholes, and Watson, 1975; Heilman, Bowers, Speedie, and Coslett, 1984). Other studies could not find such a general difference between patient groups (Van Lancker and Sidtis, 1992; Starkstein, Federoff, Price, Leiguarda, and Robinson, 1994). These studies instead argue that if anything were to differentiate left hemisphere patients and right hemisphere patients, it would be in how they use acoustic cues to judge emotional prosody, which, in turn, would suggest a possible bilateral network for the processing of emotional prosody (Van Lancker and Sidtis, 1992).

Contrastingly, the majority of the functional magnetic resonance imaging studies report a right-lateralized network for emotional prosody involving the superior and middle temporal gyri (Mitchell, Elliott, Barry, Cruttenden, and Woodruff, 2003; Beaucousin et al., 2007) and in addition also right frontal regions when participants were engaged in an emotional identification task (Wildgruber et al., 2005). However, with a different task, namely, a simple recognition task, more bilateral activation was observed, suggesting the influence of the task demands on the active part of the network (Wildgruber et al., 2005). Bilateral activation for emotional prosody, moreover, was reported to be accompanied by subcortical activation in the basal ganglia (Kotz et al., 2003) and in the amygdala (as a general structure relevant for emotion) (Schirmer et al., 2008; Leitman et al., 2010).

There are a number of ERP studies on emotional prosody (see Schirmer and Kotz, 2006; Paulmann and Pell, 2010; Kotz and Paulmann, 2011; Paulmann, Titone, and Pell, 2012), which, due to their high temporal resolution, led to the distinction of a multistage process of early and late stages. The early stages possibly involving subcortical structures such as the basal ganglia represent processes of rapid early emotional appraisal, and later processes involving cortical regions are responsible for emotional meaning comprehension.

One intriguing question is the following: Given that there is emotional-prosodic and emotional-lexical information how do these information types come together during speech comprehension? For example, consider a sentence such as *Yesterday she had her final exam*, which is spoken in a happy or sad intonation, and the sentence is then followed by a visually presented word that had a positive meaning like *success* or a negative meaning like *failure*. This results in four conditions: two conditions in which the emotional sentence intonation and the word match (happy/positive, sad/negative), and two conditions in which they do not match (happy/negative, sad/positive). These conditions were used in two ERP studies. Since we know that a semantic mismatch elicits an N400 in the ERP, it was predicted that the emotional mismatch conditions should lead to N400 effects. The two studies that used this sentence-word material asked a communicatively interesting question: Would male and female participants differ in the way they process emotional-prosodic and emotional-lexical information (Schirmer, Kotz, and Friederici, 2002, 2005)? To get more information about the timing parameters of the processes, the studies varied the time lag (Interstimulus Interval) between the auditorily presented sentence and visually presented word, with a pause of either 200 ms or 750 ms. In the first study (Schirmer et al., 2002) participants were only required to decide whether the visually presented word following the sentence was a real word or not. This study revealed that women showed an N400 effect with a small pause between the sentence and the word, whereas men only showed an N400 with a longer pause. These ERP findings indicated that women make earlier use of the emotional-prosodic information than men.

This raised the question: Do men have a general inability to use emotional-prosodic information, or could one force the men to act like early users under certain conditions? Therefore, in the second study (Schirmer et al., 2005) participants were required to judge not only whether or not the written word was a real word, but, moreover, whether or not the word matched or mismatched the emotional prosody of the preceding sentence. This time, both males and females showed similar N400 effects, suggesting that men are able to process emotional prosody quickly when instructed to take emotional prosody into account. These results show that emotional prosody and emotional meaning of a sentence can interact in principle.

Processing Pitch in Speech and Music
Questions regarding the similarities and dissimilarities of pitch processing apply to both language and music because pitch information not only plays a crucial role in processing prosody in languages, but it is also a crucial parameter in music when processing melodies.[11]

A similarity in the brain basis of pitch processing in music and language is suggested by patient studies. Patients with congenital deficits in music perception (Peretz, 1993) often also show deficits in pitch perception in language (Liu, Patel, Fourcin, and Stewart, 2010; Nan, Sun, and Peretz, 2010; Tillmann et al., 2011). These observations suggested a common brain basis for pitch across domains. This hypothesis was evaluated in a functional magnetic resonance imaging study carefully controlling the familiarity with a tonal language and with music by testing a group of Chinese musicians (Nan and Friederici, 2013). Chinese language phrases and music phrases of similar length ended either correctly (congruously) or incorrectly (incongruously) due to a pitch manipulation. Participants were required to judge congruity of the stimulus heard. A direct comparison between processing Chinese language and processing music revealed larger activation for language than for music in the fronto-temporal language network and in the right superior temporal gyrus. An analysis taking into consideration the activation from both the language condition and the music condition revealed a neural network for pitch processing that involved the left anterior Broca's area and the right superior temporal gyrus. Activation in Broca's area (BA 45) was tightly linked to pitch congruity judgment, whereas right superior temporal gyrus activation was not. These findings promote the idea of right superior temporal gyrus as a general processing region for pitch and anterior Broca's area (BA 45) as a region supporting judgment. The combined data lead to the proposal that pitch information as such may be processed primarily in the superior temporal gyrus in the right hemisphere, but that pitch judgment, pitch-related lexical processes, and pitch-related syntactic processes rather involve the left hemisphere.

The more general conclusion is that pitch information after its initial acoustic analysis is processed according to its function rather than according to its form. If the function is lexical or syntax-related it is processed in the left hemisphere; if it is not lexical or syntax-related, it is processed in the right hemisphere. This view is supported by a study investigating young Chinese children who during early language acquisition have to learn that pitch information is lexically relevant (Xiao, Friederici, Margulies, and Brauer, 2016a). The use of resting-state functional connectivity analyses revealed a shift in brain activation, moving from the right hemisphere by age 3 to the left hemisphere by age 5. These data can be interpreted to indicate a shift from processing pitch as acoustic information in the young children toward processing pitch as lexical information in the older children.

Summary

Pitch information is crucial in linguistic prosody and emotional prosody, but also in processing music melody. In this section I reviewed the neuroscientific studies on processing prosody during speech perception and the processing of music. The processing of prosodic information during speech comprehension crucially involves the right hemisphere, whereas syntax is processed in the left hemisphere. During speech processing the right and the left hemisphere interact online to assign phrase boundaries as the borders of constituents. They do this via the corpus callosum—a white matter structure that connects the two hemispheres. In sum these studies conclude that although pure pitch information may be processed primarily in the right superior temporal gyrus, the localization of higher pitch-related processes in the human brain is dependent on the particular function that pitch conveys: lexical tone is processed in the left hemisphere, linguistic prosody is processed bilaterally with a leftward lateralization, and music is processed bilaterally with a rightward lateralization. Thus it appears that the localization of pitch is determined not by its form, but by its cognitive function.

1.9 Functional Neuroanatomy of Language Comprehension

This chapter started from a cognitive model of auditory language comprehension processes starting from auditory input up to sentence understanding, all of which are represented in figure 1.1. These processes take place in less than a second. This is most remarkable. In the course of the chapter I described the brain basis of these different processes. We have learned that in order to achieve successful comprehension, the language system has to access and process the relevant information within the language network that consists of regions in the left inferior frontal cortex and temporal cortex as well as in the right hemispheric regions. The picture that can be drawn from the studies reviewed here is neuroanatomically quite concise with respect to sensory processes and those cognitive processes that follow fixed rules, namely, the syntactic processes. Semantic processes that involve associative aspects are less narrowly localized. The reason for the latter is partly grounded in the circumstance that purely linguistic semantic processes are hard to separate from semantic memory and semantic associations, which in turn are of large interindividual variance and may even change within the individual over time—depending on personal experiences and the respective associations.

Based on the available data the functional neuroanatomy of auditory sentence comprehension can be described according to both the temporal sequence of the different subprocesses and their localization (see figure 1.21)

Auditory language processing follows a fluctuating time course, as indicated by the different ERP effects in figure 1.21. Acoustic processes and processes of speech sound categorization take place around 100 ms (N100). Initial phrase structure building takes place between 120 and 250 ms (ELAN), while the processing of semantic, thematic, and syntactic relations is performed between 300 and 500 ms (LAN, N400). Integration of different information types takes place around 600 ms (P600). These processes mainly involve the left hemisphere. In addition, prosodic processes engage the right hemisphere and rely on the interaction with the left hemisphere for the processing of prosodic phrase boundaries, which are reflected in the ERP component called Closure Positive Shift.

These processes can be localized in different regions of the left and right hemisphere. The primary auditory cortex and the planum temporale support acoustic-phonetic processes. Initial phrase structure building is subserved by the frontal operculum and the most ventral part of the pars opercularis together with the anterior superior temporal gyrus, with the frontal operculum being held responsible for bringing two elements together and with ventral BA 44 being held responsible for building a hierarchy. Semantic processes and thematic role assignment taking place during a second stage involve the left superior temporal gyrus and middle temporal gyrus and BA 45/47 in the inferior frontal cortex. The processing of complex syntactic hierarchies is supported by BA 44 and the posterior superior temporal gyrus in the left hemisphere. Prosodic processes are mainly located in the right hemisphere involving superior temporal and inferior frontal regions. The interaction

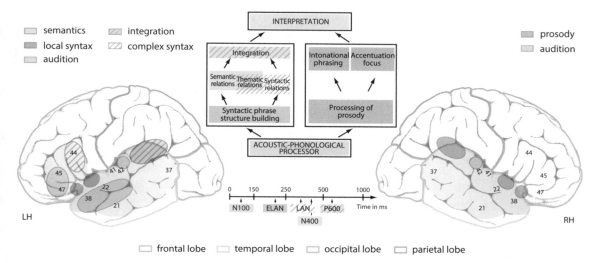

Figure 1.21
Neurocognitive language comprehension model. Functional neuroanatomy of language with its processing phases as defined in the initial neurocognitive language comprehension model (different processing stages are indicated by the color-coded boxes in the model displayed in the center). Language-relevant brain areas are displayed schematically in the left hemisphere (LH) and right hemisphere (RH). Functions of the areas are color-coded and labeled in the legend to the left and the right of the figure. The temporal course of the different subprocesses is displayed at the bottom of the figure as a time line in milliseconds (ms) listing the language-related ERP effects. For details see text. Adapted from Friederici (2011). The brain basis of language processing: From structure to function. *Physiological Reviews*, 91 (4): 1357–1392.

between syntactic information and prosodic information during sentence comprehension is guaranteed by the corpus callosum, which connects the left and the right hemisphere (see figure 1.20).

Thus, there seem to be different functional networks supporting early and late syntactic processes, and an additional network recruited for processing semantic information as well as a network supporting prosodic processes. In chapter 2 we will see that these regions, which constitute a functional network, are bound together by specific fiber tracts revealing separate structural networks.

Summary

This section presents a model of the brain basis of language comprehension, with respect to the neuroanatomy of the brain regions supporting syntactic, semantic, and syntactic processes, as well as the temporal relation and interaction of these different functions as comprehension proceeds.

2

Excursions

In the first chapter I discussed the brain basis of auditory language comprehension. According to a theoretical view (Chomsky, 1995b; Berwick, Friederici, Chomsky, and Bolhuis, 2013) language is assumed to consist of three larger components, a core language system that contains the syntactic rules and the lexicon, and two interface systems. These two interface systems (as I mentioned in the introduction) are the external sensory-motor interface and the internal conceptual-intentional interface. The former interface guarantees perception and production, whereas the latter interface guarantees the relation of the core system to mental concepts and intentions.

In this chapter I will touch on two issues. First, in section 2.1, I will evaluate the relation of language comprehension and production with respect to the theoretically assumed common knowledge base of the language system, consisting of syntactic rules and words. Second, in section 2.2, I will go beyond the core language system and discuss communication including intentional and situational aspects. These aspects are not part of the core computational language system but are important in everyday communication, and therefore deserve mention here.

2.1 Language Comprehension and Production: A Common Knowledge Base of Language

So far we have only discussed language comprehension and its neurobiological bases. However, there is a clear assumption that language comprehension and language production access a common knowledge base consisting of a lexicon and syntactic rules (Chomsky, 1965; Friederici and Levelt, 1988). The model by Friederici and Levelt (1988) was formulated on the basis of this theoretical consideration as well as empirical data available at the time. Here I will first describe the model as depicted in figure 2.1 and then turn to recent empirical data to back up the model.

The model assumes language production (left side of figure 2.1) and language comprehension (right side of figure 2.1) to consist of different subprocesses that are partly sequentially organized. These processes access a central knowledge base, known as the grammar, which consists of the lexicon and syntactic rules (depicted as a circle in the middle of the

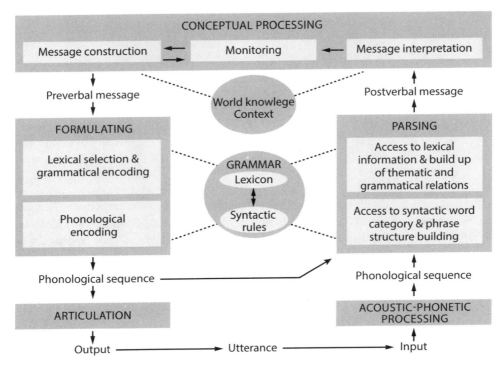

Figure 2.1
Model of language production and comprehension. Central knowledge bases are represented as circles, processes are represented as boxes. For a detailed description see text. Adapted from Friederici and Levelt (1988). Sprache. In *Psychobiologie. Grundlagen des Verhaltens*, ed. Klaus Immelmann, Klaus Scherer, Christian Vogel and Peter Schmoock, 648–671. Stuttgart: Gustav Fischer.

figure). Also common to production and comprehension is the level of conceptual processing at which messages are constructed and at which perceived messages are interpreted. This is done with respect to context and the available world knowledge. Message interpretation and construction are permanently monitored. The intermediate representations of a preverbal message and postverbal message mediate between the conceptual system and the language system (production/comprehension). The subprocesses during comprehension—including the stages of syntactic phrase structure building on the basis of word category information and the subsequent processes of assigning syntactic and thematic relations—have been described in detail in chapter 1. This entire process is often called parsing.

The subprocesses during language production, known as formulating, are described in more detail and substantiated by empirical evidence in this section. The phonological sequences at the bottom of the figure are again an intermediate representation between the language system and the input/output processes. The arrow going from the phonological sequence representation during production to the comprehension system is evidenced by the observation that we as humans are able to stop the production process before articulation in case we realize that it would be better not to utter the constructed message. Once it is articulated, not only our own comprehension system (as indicated by the arrow at the very bottom), but also the comprehension system of others, can perceive the utterance.

Neuroscientific work on language production—at least at the sentence level—is still conducted less frequently than that on comprehension. This stands in clear contrast to research on the mechanisms of speech production with its motoric and articulatoric aspects. I refer readers to a recent book, *Neural Control of Speech* (Guenther, 2016), which describes these processes in much detail. My focus in this section is rather on those processes of language production that precede articulatory speech output, and the relation of these to language comprehension.

I will first discuss a number of studies that led to the model by Friederici and Levelt (1988), and then subsequent studies in support of this model. As studies of language comprehension were presented in detail in chapter 1, I will focus here primarily on studies of language production.

Over the last few decades, researchers have taken different approaches to investigate language production. One method was to study language breakdown after brain lesions, a second way was to systematically investigate speech errors in healthy speakers, and a third approach was to trigger the to-be-formulated message by means of visually or auditorily presented material in experimental settings. I will address these different approaches in turn and show that although the neuroscientific database is limited, they converge to a model of language production that appears to display the reverse order of the three phases/stages that we discussed for language comprehension.

Language Breakdown in Patients

Historically, it was Paul Broca (1865) who, after investigating a patient with severe production deficits and dissecting his brain, claimed that the inferior frontal gyrus was responsible for language production (see details in the introduction of this book). Arnold Pick (1913) described an interesting production deficit after brain lesion, called agrammatic aphasia, as a deficit characterized by an inability to produce grammatical function words (conjunctions, auxiliaries, etc.) and grammatical morphemes (verb inflections and plural morphemes) in a sentence leading to a telegraphic style. The output of these patients was non-fluent with a distortion in the prosodic range of speech. Pick already assumed different stages in the process of language production because Broca's aphasia, a deficit in language production, could be separated from dysarthria, a deficit in articulation. Moreover, he postulated that syntactic and prosodic planning must go hand-in-hand during language production. Work with progressive non-fluent aphasia, that is, patients who suffer from cortical atrophy in prefrontal regions bilaterally and perisylvian regions in the left hemisphere, has demonstrated that speech errors at the phonemic level of language production are not caused by a motor planning impairment (Ash et al., 2010). These data provide evidence for a separation of levels of language planning and motoric articulation. Other types of aphasia (Wernicke's aphasia and so-called conduction aphasia) were characterized by a seemingly fluent language production, but with sound and word substitutions as well as a misordering of these elements in the speech output (Goodglass and Kaplan, 1972). These different types of aphasia already suggested that syntax, lexical processing, and articulation are to be thought of as different subcomponents or processing stages in language production.

Different levels of processing during language production were also evidenced by language impairments caused by stroke. At the phonological level Broca's aphasics with lesions in the frontal cortex, often including motor areas, show production errors that suggest deficits at the level of articulatory planning. They show impairments in the timing relations required for the production of consonants (Blumstein, Cooper, Goodglass, Statlender, and Gottlieb, 1980; Kurowski, Hazen, and Blumstein, 2003). The characteristic deficit of Broca's aphasia, however, lies at the level of grammatical encoding. In their production they often leave out function words and inflectional morphemes, resulting in utterances that appear like a telegram.

Aphasia research was the first to provide neurobiological evidence for a common knowledge base, called grammar in the model presented in figure 2.1. In two seminal papers (Zurif, Caramazza, and Myerson, 1972; Caramazza and Zurif, 1976) researchers demonstrated that Broca's aphasics, characterized by their deficit in syntactic processing during language production, were also deficient in sentence comprehension when sentences could only be understood on the basis of syntactic cues, e.g., *The lion was chased by the tiger*. For such a sentence the correct interpretation relies on the processing of the function words *was* and *by*. The conclusion from these observations was in line with the linguistic theory postulating a common knowledge base of syntax for language production and language comprehension.

Linguistic theory even made more fine-grained assumptions with respect to the two functional subcomponents of grammar: lexicon and syntactic rules (Chomsky, 1965). A crucial test of these would be to show that the same phonological word would be processed differently depending on its grammatical function. For example, the ambiguous word *can* should be processed differently in the phrasal context of *the can* compared to *I can*. In the latter context the word *can* is syntactic function word, whereas in the former context it is lexical item. It was demonstrated that an agrammatic Broca's patient was able to read the word *can* when presented in the *the can* context but not in the *I can* context (Andreewsky and Seron, 1975). A similar function-based processing differentiation was demonstrated in German for the prepositional form *auf*, which can either function as syntactic obligatory preposition carrying no lexical meaning (*Peter hofft auf den Sommer* / Peter hopes for the summer) or as a lexical preposition carrying referential meaning (*Peter steht auf dem Stuhl* / Peter stands on the chair) (Friederici, 1982, 1985). While Broca's aphasics were able to produce and process prepositions in their lexical role, they failed to produce and process syntactically based prepositions.

Production errors made by patients with Wernicke's aphasia, which is caused by lesions in the temporal cortex, are partly located at the lexical level. These patients produce semantic paraphasias and have problems with the correct naming of pictures representing objects. They show deficits in accessing the lexical item in the lexicon not only during production but also during perception when judging and selecting lexical items (Milberg, Blumstein, and Dworetzky, 1987). These findings again are in line with the view of a central knowledge base of words—the lexicon—for language production and comprehension and thereby with the model proposed by Friederici and Levelt (1988).

These findings impressively demonstrate that the brain appreciates and reflects crucial differentiations made in linguistic theory. Moreover, they provide neurological evidence for a common knowledge base of language for both comprehension and production.

Speech Errors and Models of Language Production

A second line of research in language production provided crucial information for model building. This was the research on speech errors, often referred to as *slip of the tongue*. Accepting speech errors as linguistic evidence was a successful way to advance our understanding of the processing stages underlying language production. The speech error corpora of Meringer and Mayer (1895) mark the beginning of this approach. Fromkin (1973) continued to use speech errors as a systematic reflection of how sounds and words of a language are organized in the mind prior to speech output. The analysis of speech errors is still viewed as a means to uncover the workings of a highly complex system that sometimes only allows insights once the system breaks down. Therefore, speech errors are of high explanatory value.

Garrett (1980) developed a model of language production based on a systematic classification of different types of speech errors (anticipations, shifts, exchanges, substitutions, and blends). In particular, exchange errors—*the room to my door* instead of *the door to my room* (Garrett, 1980)—suggest that during production lexical items are put into "syntactic frames." The syntactic frames come with the fixed positions of function words and syntactic word category slots and in such a frame (only) words that belong to the same syntactic category can be exchanged. The cited speech error is only one example of how the analysis of a particular error type can be used to infer the underlying processing steps. Other errors lead to the assumption of other processing stages. The resulting language production model differentiates three processing stages intervening between the conceptual message level and the level of articulation. First, semantic factors select lexical items and their grammatical relations (leading to the functional level of representation). Second, syntactic factors select positional sentence frames with their function words and grammatical morphemes (leading to the positional level of representation). Third, phonetic details of both lexical and grammatical items are specified prior to articulation (at the sound level). In his 1980 paper, Garrett discussed the relation between language production and language comprehension, arguing that both processes must access a common knowledge base of syntactic rules and words. This relation between language production and comprehension was specified in a model by Friederici and Levelt (1988) (see figure 2.1).

Experimental Work

A third approach used in psycholinguistics and in experimental neuroscience is to ask speakers to produce a word or a phrase that describes a given picture. In the context of this book I will only consider those studies on word production that investigate aspects beyond the motor act of articulation. For models of articulation, please refer to Guenther (1995, 2016), Hickok (2012), and Guenther and Hickok (2015). Their work discusses the programming of phoneme and syllable units and their motor output. The neural network supporting the speech output and its timing includes the cerebellum, the thalamus and thalamocortical circuits, the basal ganglia, the supplementary motor area, and the motor cortex and its interplay in time (Kotz and Schwartze, 2015).

Psycholinguistic research including speech error analyses and experimental work has led to solid models of language production (Garrett, 1980; Dell, 1986; Levelt, 1989). These models all agree that there are at least three separate processing levels: meaning, form, and articulation. Some psycholinguistic studies on healthy individuals have tried to investigate the seriality and timing of different processing stages during language production and also found evidence for three different processing phases. These studies used a picture-word-interference paradigm. This paradigm asks a speaker to name an object depicted in a picture, and during the production process the speaker receives an "interfering" word auditorily. The interfering word could be semantically similar (*goat/sheep*), a noun with a similar syntactic gender (article), or a phonologically similar noun. These words are presented at different time points before the target word (i.e., the name of the object) is articulated. Interestingly, interference of these three different conditions is separated in time: first (and furthest away from articulation) a semantic interference is observed; second, the syntactic gender is effective; and finally, immediately prior to articulation, phonological interference can be observed (Levelt, Schriefers, Vorberg, Meyer, Pechmann, and Havinga, 1991; Jescheniak and Levelt, 1994). These results clearly support models assuming a serial and incremental activation of different processing levels during language production, such as those language production models formulated by Garrett (1980) and Levelt (1989).

A brain-based study used motor readiness brain potential during a push-button reaction time paradigm to investigate the temporal sequence of syntactic and phonological processes during language production (van Turennout, Hagoort, and Brown, 1998). The readiness potential indicates the preparation of a motor action prior to its execution and can, therefore, be used as a response-related measure. Participants were requested to name colored pictures of objects in a noun phrase consisting of a morpho-syntactically inflected adjective in Dutch, e.g. *rode tafel* (*red table*). But before producing the noun phrase they were instructed to perform a syntactic-phonological classification task requiring them to push the "syntax" or "phonological" classification button (i.e., to perform a motor action). Syntactic classification involved determining the syntactic gender of the noun, in turn determining the adjective's inflection in Dutch. Phonological classification required the determination of the initial phoneme. The electrophysiological recording was time-locked to the picture onset. The lateralized readiness potential related to the preparation of the motor push-button action revealed a primacy of the "syntactic" response, which was followed in about 40 ms by the "phonological" response. These data provided electrophysiological evidence for the temporal sequence of syntactic preceding phonological processes during language production.

These fine-grained temporal results are evidenced even at the neurotopological level by a study using intracranial electrophysiology (Sahin, Pinker, Cash, Schomer, and Halgren, 2009). With this method, local field potentials are recorded from ensembles of neurons using electrodes implanted over language-related areas including Broca's area. Participants were requested to read words aloud or to grammatically inflect them. Electrodes over Broca's area revealed a sequential activation for lexical processes, grammatical (syntactic) processes, and phonological processes after visual word presentation and before the respective utterance. The observed sequence of processes recorded with the time-sensitive method of electrophysiology is in line with the model of language production as formulated by Levelt, Roelofs, and Meyer (1999), and that postulated in figure 2.1 of Friederici and Levelt (1988).

In addition to these electrophysiological studies revealing the temporal structure for language production, functional brain activation studies tried to localize production processes. Brain activation studies by means of functional magnetic resonance imaging have applied those experimental paradigms used in behavioral studies in language production, such as picture naming (as a single word or in a determiner phrase), reading aloud, verb generation (choose a verb that fits a given noun, e.g., apple: *eat*) and word fluency (generate nouns of a given semantic category (animal: *dog*, *cat*, etc.). These paradigms focus on single-word or phrase production, although they differ in their input to the production system. Note that in these paradigms, the message to be produced is not triggered internally but by external input. Nonetheless, these paradigms may allow us to investigate the process of language production and the neural network supporting this process to a certain extent. In a meta-analysis of 82 activation studies focusing of word production, Indefrey and Levelt (2004) report a set of regions that were reliably found for picture naming and word generation. The network includes the posterior portion of the left inferior frontal gyrus, the left precentral gyrus, the supplementary motor area, the mid and posterior parts of the left superior temporal gyrus and middle temporal gyrus, the left fusiform gyrus, the left anterior insula, the left thalamus, the right mid middle temporal gyrus, and the cerebellum. This is a large network covering all aspects of word production including articulation.

The ultimate brain-based model of language production, however, would want to localize the different aspects of language production in this neural network. These aspects are lexical selection, grammatical (syntactic) encoding, phonological encoding, and articulation. Localizations are partly proposed on the basis of comparing different production studies or on comparing different production conditions in one study.

Lexical selection. Concerning lexical selection, the following conclusions were drawn. As the left middle temporal gyrus was found in picture naming and word generation (Indefrey and Levelt, 2004), but not in studies on word reading (Turkeltaub, Eden, Jones, and Zeffiro, 2002), it was hypothesized that the left middle temporal gyrus could be the neural substrate of lexical selection in word production (Indefrey, 2007). The process of lexical selection was investigated by means of magnetoencephalography, which allows us to measure brain activation with a high temporal resolution (Maess, Friederici, Damian, Meyer, and Levelt, 2002). In this study a semantic category interference paradigm was used to target the lexical selection process. Participants were requested to name pictures of objects in two different blocks: blocks that contained objects of the same semantic category or blocks containing objects of different categories. Behaviorally, naming in same-category blocks is usually slower than naming in different-category blocks due to semantic competition between items in the same-category block. At the neural level, a significant activation difference between same-category and different-category conditions, which was observed in the time window of 150–225 ms after stimulus presentation, was localized in the mid part of the left middle temporal gyrus. Since this time window falls within the time window of lexical selection as indicated by behavioral studies, the left mid part of the middle temporal gyrus is understood to be the neural substrate for lexical selection.

Grammatical encoding. The process of grammatical encoding can be investigated experimentally in healthy people, both at the phrase level and at the sentential level. At the phrase level functional magnetic resonance imaging work points to an involvement of Broca's area (BA 44, BA 45) in syntactic processes during language production as demonstrated in a German functional imaging study on grammatical gender selection when naming a picture in a determiner phrase (Heim, Opitz, and Friederici, 2002). This condition was compared to two other conditions: a control condition (saying *jaja*) and a naming condition (i.e., naming a picture without a determiner). The control condition of articulation alone activated the premotor cortex, the precentral gyrus (BA 6), the anterior insula, and the parietal operculum. Lexical access reflected in naming a picture (without grammatical gender) compared to articulation revealed a network consisting of BA 45, fusiform gyrus, and inferior temporal gyrus. Phrase production (with gender-marked determiner) in contrast to word production (without gender-marked determiner) revealed activation not only in Broca's area but also in the superior temporal gyrus and middle temporal gyrus. These data suggest that a network of areas—which, in addition to Broca's area, involves the middle temporal gyrus where the lexicon is thought to be localized—support determiner phrase production. This is what one might expect, because syntactic gender information of a noun is arbitrary and stored together with the respective item in the lexical entry.

The process of grammatical encoding at the sentential level was targeted in a study using a scene description paradigm (Indefrey et al., 2001). Participants had to describe actions of colored geometrical figures in full sentences (*The red square is pushing the green triangle away*), in single phrases (*red square, green triangle, push, away*) or in single words (*square, red, triangle, green, push, away*). The posterior part of Broca's area (BA 44) and the adjacent BA 6 were activated more strongly in the sentence condition compared to the word list condition. The study was conducted in German, in which the adjectives in phrases and sentences are morphologically marked for syntactic gender agreement with the noun, thus requiring morphosyntactic processes. Given this, the authors take their finding to show that BA 44 and BA 6 are involved in grammatical encoding (Indefrey et al., 2001; see also Indefrey, Hellwig, Herzog, Seitz, and Hagoort, 2004 for a replication of this finding). Other functional magnetic resonance imaging studies on grammatical encoding during production also found activation in BA 44 together with the anterior part of Broca's area (BA 45) when comparing sentence generation to word generation in German (Haller, Radue, Erb, Grodd, and Kircher, 2005) and in English for free sentence generation (Kemeny, Ye, Birn, and Braun, 2005). The additional activation of BA 45 in these studies may be due to increased demands on lexical processes, be it due to lexical access as such or be it due to lexical search and selection in free sentence generation.

Another approach to grammatical encoding during sentence production used a functional magnetic resonance imaging adaptation paradigm (Menenti, Gierhan, Segaert, and Hagoort, 2011; Menenti, Segaert, and Hagoort, 2012). This paradigm is tied to the phenomenon that activity of neural populations—and thereby the functional magnetic resonance imaging signal—decreases once a stimulus is repeated (Krekelberg, Boynton, and van Wezel, 2006). In the language production study Menenti et al., (2012) presented photographs of actions performed by an actor (thereby taking the thematic role of an agent) on a undergoer (thereby taking the thematic role of a patient) and the participants were required to produce a sentence describing the picture. The action was restricted to three verbs (*kiss, help, strangle*) and the agent-noun and patient-noun were either a man or a woman. The study varied the verb (thematic-semantic role: agent, patient), the lexical item (verbs, nouns), and the syntactic structure (active/passive construction). The functional magnetic resonance imaging data reported suggest that thematic-semantic aspects involve the middle temporal gyrus bilaterally but that lexical processes only recruit the left posterior middle temporal gyrus. The activation pattern for the syntactic manipulation involved prefrontal, inferior frontal, temporal, and parietal brain regions.[1] At this point it is interesting to note that a functional magnetic resonance imaging adaptation study on production and comprehension revealed a large overlap of those brain regions demonstrating the adaptation effects in production and in comprehension (Menenti et al., 2011). This can be taken as empirical support for the model presented in figure 2.1 that assumes a common knowledge base in the service of production and comprehension processes.

Phonological encoding and articulation. Since all studies reported Broca's area (BA 45 and/or BA 44) activation for phrase and sentence level production compared to the respective control conditions, we can conclude that this area is involved in grammatical encoding. However, some but not all studies reported activation in the premotor cortex, BA 6. Area BA 6 may not necessarily be part of the core language system, but possibly part of the output system. The role of Broca's area and the adjacent premotor cortex and motor cortex during language production has indeed been the focus of an ongoing discussion. The role of Broca's area during language production has actually been debated since Paul Broca's first publication in 1861. While some researchers proposed that Broca's area accesses *phonological representations* of words whereas the motor regions are responsible for articulations, other researchers see Broca's area to be involved in processing *articulatory representations* (Hickok, 2012). However, in a patient study it was shown that articulatory planning deficits were correlated with brain regions in a discrete region of the left precentral gyrus of the insula, rather than Broca's area (Dronkers, 1996).

A recent study using electrocorticography recording directly from the surface of the cortex which allows to monitor brain activation in the millisecond domain has provided a clear answer to this debate (Flinker et al., 2015). In this study participants were asked to repeat a word they heard. The observed activations sequentially involved the superior temporal gyrus (auditory perception of word), Broca's area (phonological word representation), and the motor cortex (articulation of word). The brain activation in Broca's area (peaking at about 340 ms after the word was heard) clearly preceded activation in the motor cortex during articulation (peaking at about 580 ms) by about 241 ms. Data from this study suggest that during language production Broca's area may generate a phonological/phonetic representation, which is then implemented for output by the motor cortex.

Summary

Historically, first production models were built on the basis of language deficits in patients with brain lesions, and later on the basis of speech errors in healthy people. More recently, attempts have been made to apply neuroscientific methods, such as functional magnetic resonance imaging and electrocorticography, using pictures or perceived words at controlled input to the production system. The available data suggest that language production, apart from brain structures supporting the motor act of speaking, involves Broca's area in addition to temporal regions. In this section we started from the assumption that both language comprehension and language production access a common knowledge base, called grammar, consisting of the lexicon and syntactic rules. The empirical data discussed suggest that the different processing steps assumed for language comprehension reversely mirror those of language production. They, moreover, are in line with the assumption of a common knowledge base.

2.2 Language Comprehension and Communication: Beyond the Core Language System

We have seen already that there are a number of psychological aspects of language use that are not part of the language system, but which are relevant for, and play a role in, daily communication. These include emotional prosody, discussed in section 1.8, as well as aspects of the communicative context and communicative hand gestures accompanying an utterance. In a model of language as proposed by Berwick et al. (2013), these aspects are not part of the core language system. I will touch upon the communicative aspects here because readers may expect to find such a discussion in these pages. But my excursion will be a brief because these aspects are not captured by the neuroanatomical modal of language developed in this book.

Pragmatics during Communication
Contextual knowledge during communication is covered by a research domain called pragmatics. It concerns the way in which context or contextual knowledge contributes to meaning. The understanding in a communicative situation may depend on the communication partner's ability to infer the speaker's intended meaning. To learn more about the field of pragmatics in general, I refer readers to books that cover this research field well (Levinson, 1983; Clark, 1996).

In the context of this book, the question arises how our brains deal with pragmatic information. There are at least two pragmatic domains that have been investigated with respect to their brain representations: one is the recognition of the communicative intention of the speaker and the other is the inference process taking place during discourse understanding.

Interpreting a speaker's message often requires processing the specific communicative context, which means that the listener has to process more than the words and phrases of a given utterance. The interpretation might require inferring the speaker's intention from the contextual situation or other non-linguistic information available. This would also suggest that the neurobiology of this process must extend beyond the core language system. Functional magnetic resonance imaging research indeed revealed an extended network recruiting regions typically involved in reasoning about the mental states of others, such as the medial prefrontal cortex and the temporo-parietal junction area, as well as the anterior temporal lobes bilaterally (Frith and Frith, 2003; Saxe and Kanwisher, 2003).

Some studies have investigated the brain basis of pragmatics in patients with cortical lesions. Many of these, such as the study by Ferstl, Guthke, and von Cramon (2002), investigated patients with brain lesions in the left hemisphere and right hemisphere. These authors found that patients with lesions in the left frontal brain regions, but not patients with lesions in the left temporal or right frontal region, had a deficit in drawing inferences from a text. It is, in particular, the dorsomedial prefrontal cortex that appears to support inferencing (Siebörger, Ferstl, and von Cramon, 2007) often together with the posterior cingulate cortex (for a review, see Ferstl, 2015).

One of the few functional magnetic resonance imaging studies in this domain focused on processing indirect replies rather than direct replies in a dialog situation. A large network was found to be active: it involved the dorsomedial prefrontal cortex, right temporo-parietal junction area, and the insula as well as the right medial temporal and the bilateral inferior frontal gyrus (Basnakova, Weber, Petersson, van Berkum, and Hagoort, 2014). These different regions in the network are said to support different aspects of the inference process. The dorsomedial prefrontal cortex, the right temporo-parietal junction, and the insula may be be recruited during mentalizing processes and theory-of-mind mentalizing (Saxe and Kanwisher, 2003). The left anterior insula, in particular, which is known to be involved in social affective processes and empathy, may have a role in this context (Singer and Lamm, 2009). All these processes might come into play when processing in a pragmatically complex situation, but they are certainly not part of the language system.

Thus it appears that processes drawing on contextual knowledge or information that can only be inferred indirectly from the text involve a large neural network that goes well beyond the core language network.

Gestures during Communication

In a direct interpersonal communication, the role of expressive hand gestures can be very important. And indeed speakers usually move their hands when they talk. These hand movements or gestures may have different functions: they may be automatic hand movements with no direct communicative function (for example, when speaking on the telephone), but they can also convey additional meaning during face-to-face communication.

Different types of co-speech gestures can be deployed during the act of communication: (1) so-called hand signs, like the "OK" sign, are meaningful hand postures with a clear message in a given culture (Nakamura et al., 2004); (2) iconic gestures are hand movements that represent object attributes, spatial relations, or actions; (3) deictic or pointing gestures serve as indicators to objects, events, or spaces and are used in mother-child interaction as a joined attention cue impacting language learning; (4) beats are rhythmic gestures that indicate the significance of an accompanying word or phrase. These different gesture types are used in different situations during online processing to enhance comprehension.

Iconic gestures are shown to play a supportive role in first-language acquisition (Morford and Goldin-Meadow, 1992; Capone and McGregor, 2004; Iverson and Goldin-Meadow, 2005) and in second-language learning in adults (Kelly, McDevitt, and Esch, 2009). A functional magnetic resonance imaging study on word learning in adults revealed that words learned together with iconic gestures could be learned and memorized better than those accompanied by meaningless gestures (Macedonia, Müller, and Friederici, 2011). A comparison of the brain activation for the differentially learned words found activity in the premotor cortices for words encoded with iconic gestures. Words encoded with meaningless gestures activated a network associated with cognitive control. Moreover, other studies showed that iconic gestures enhance native speech comprehension directly. They can be used to disambiguate speech (Holle and Gunter, 2007) and are beneficial in countering the effects of difficult communication conditions independent of whether the difficulties are due to external (babble noise) or internal (hearing impairment) factors (Obermeier, Dolk, and Gunter, 2012).

The neural basis of co-speech gestures has been described by Dick and Broce (2015) based on the dual-stream model of language processing, as proposed by Hickok and Poeppel (2007). Note that this model deviates from the model presented in this book, in that these authors assume only one dorsal stream, namely, the connection from the posterior temporal cortex to the premotor cortex, which subserves a sensory-motor function. This model is considered as a framework for the neurobiology of co-speech gestures (Dick and Broce, 2015).

A number of studies used ERPs to investigate co-speech gestures and their interplay with language during speech comprehension. In some of these studies in which iconic gestures and speech were presented together, the gesture was either semantically congruent or incongruent with the speech. Incongruency led to an N400 component known from semantic incongruency in language-only studies (Kelly, Kravitz, and Hopkins, 2004; Holle and Gunter, 2007; Özyürek, Willems, Kita, and Hagoort, 2007; Wu and Coulson, 2007). The N400 findings in these studies suggest an interaction between gesture and speech at the semantic level during online comprehension.

There are a number of recent imaging studies on the processing of gesture and language, mostly touching the semantic level of sentence processing. Imaging studies on the processing of semantic information in gesture and speech, similar to the electrophysiological studies, manipulated the semantic relation between gesture and speech. In general, these studies report a neural network similar to that known from language-only studies. This includes the inferior frontal gyrus, mostly the pars triangularis, the posterior superior temporal sulcus, and the middle temporal gyrus (Willems, Özyürek, and Hagoort, 2007, 2009; Holle, Gunter, Rüschemeyer, Hennenlotter, and Iacoboni, 2008; Holle, Obleser, Rüschemeyer, and Gunter, 2010; Dick, Mok, Beharelle, Goldin-Meadow, and Small, 2014).

Functionally, this raises a most interesting question concerning gesture and speech: Where in the brain does the integration between gesture and speech occur? A couple of functional magnetic resonance imaging studies address this issue directly. Some researchers suggest the posterior temporal cortex as the region where the integration of iconic gesture and speech takes place (Holle et al., 2008, 2010; Green et al., 2009). Other researchers view the left inferior frontal gyrus as the region to support the integration of gesture and speech (Willems et al., 2007, 2009). The latter two studies compared brain activation of speech with incongruent iconic gestures to speech with iconic gestures. In contrast, the former studies, which found their integration effects in the posterior temporal cortex, compared speech with gestures of different degrees of meaningfulness (dominant and subdominant meaning) to speech with meaningless, that is, grooming hand movements (Holle et al., 2008). In a similar vein, the findings from Gunter, Kroczek, Holle, and Friederici (2013) suggest a major role of the posterior temporal cortex and not of the inferior frontal gyrus as a region where integration of gestures and speech takes place.

Meaningful gestures led to increased activation in the premotor cortex and in the parietal cortex when meaningful gestures were compared to grooming hand movements (Holle et al., 2008). An involvement of the premotor cortex was also found in two other studies on the processing of co-speech gestures, thereby providing supportive evidence (Willems et al., 2007; Josse, Joseph, Bertasi, and Giraud, 2012). However, other studies did not find these activations (Green et al., 2009; Willems et al., 2009), thus leaving a number of open questions. Further research must be carried out to show the exact role of the premotor cortex and the parietal cortex during the processing of meaningful gesture and speech during comprehension.

Whereas some gestures indicate a meaning by itself, in beat gestures the speaker tries to put a certain element of the sentence in focus. Such gestures were highlighted in an electrophysiological study in which the impact of a beat gesture on the processing of syntactically ambiguous German sentences was investigated (Holle et al., 2012). The ambiguous sentences in the study allowed for either a simple subject-first or a more complex, less-preferred, object-first structure. The ERP component P600, usually observed for the more complex object-first structure, was reduced by the beat gesture—indicating an enhanced processing of the complex sentence in the presence of a beat gesture. Note, however, that the P600 reflects a late processing stage during which different types of information are integrated. This result was extended by a functional magnetic resonance imaging version of the experiment (Gunter et al., 2013). Here, clear main effects of syntactic complexity (complex compared to simple syntax) were found in the left inferior frontal gyrus in the pars opercularis (BA 44), the left posterior middle temporal gyrus, the left pre-supplementary motor area, the left precentral gyrus, and bilateral insulae. All these areas, except the left inferior frontal gyrus, showed a significant interaction with beat gesture in such a way that a beat on the first noun phrase facilitates the processing of the easy subject-first sentence structure and inhibits the complex object-first structure, and vice versa for a beat on the second noun phrase. These findings indicate that gesture and speech can interact when trying to overcome preferences in syntax structure. Interestingly, however, no such interaction takes place in BA 44, the only area for which a main effect of syntax is found.

The available data indicate that meaningful gestures can enhance speech comprehension. At the neurophysiological level the integration of meaningful gesture and speech appears to interplay between the posterior temporal cortex and the inferior frontal gyrus. Beat gestures—when relevant for the assignment in syntactic structure—also interact with speech in prefrontal and temporal regions, but not in BA 44. This region known to support syntactic structure building reveals a syntax-only effect. This latter result is of particular interest, since it demonstrates that syntax, as a major part of the core language system, is independent of communicative gestures.

Summary

There are important aspects to be considered for communication that are beyond the core language system we reviewed in this section. These are contextual knowledge, known as pragmatics, as well as communicative hand gestures, which may interact with language during communication. At the neuroscientific level a number of brain regions beyond those involved in language, such as the dorsomedial prefrontal cortex and the temporo-parietal junction, have been identified to support aspects of social communication. The interplay between meaningful gestures with language is thought to take place in the posterior temporal cortex at the junction of the parietal cortex. Interestingly, BA 44 as the main syntactic processing region remains unaffected by communicative gestures. At the moment the available data on the communication aspects of language in general are sparse, but a recent neuroanatomical model of social communication (Catani and Bambini, 2014) may serve as a road map for future research.

II

Language processing does not take place in a single brain region, but in several brain regions that constitute a larger language network. Chapter 1 described the different regions within the network and their particular function. During language processing, however, these regions have to cooperate. And the question is: How do they do it?

In this part of the book I will describe the neural basis that may make cooperation and exchange of information within the language network possible. Transmitting information from one brain region to another region should be based on a neurophysiological principle that can bind the relevant regions, preferably if possible at the structural and functional level.

The available data suggest that the language-related regions are part of a larger network within which information is exchanged. The different language-relevant parts are connected by a large number of white matter fiber tracts responsible for the transmission of information from one region to another. In chapter 1 we have seen that the language-relevant brain areas are distributed over the left hemisphere involving the frontal and temporal lobe and even partly involve the right hemisphere. These brain areas are connected by particular long-range white matter fiber tracts. These fiber tracts together with the language-relevant brain regions constitute the larger structural language network. Within this network different subparts either connect brain areas processing syntactic information or brain areas processing semantic information. These structural connections will be discussed in chapter 3, at the end of which I propose a neuroanatomical pathway model of language.

In order to guarantee language processing these areas also have to constitute a functional network. This means that neuronal ensembles creating the neural activation in different brain regions must find a way to communicate with each other. Neurons and neuronal ensembles send electrical signals through the white matter fiber tracts, but the communication from one neuron to the next is based on neurotransmitters guaranteeing the transmission of a nerve impulse. We will see that at this microscopic neurotransmitter level it can be demonstrated that the cortical language regions constitute a network even at this level. The regions within the larger language are receptorarchitectonically more similar than those outside the language the network. At a macroscopic level we can observe the cooperation between brain regions by means of functional connectivity and oscillatory activity reflecting the simultaneous activation of different brain regions. These functional aspects will be discussed in chapter 4, at the end of which I propose a model describing the functional neural language circuit.

At present it is an open question to what extend functional connectivity and structural connectivity map onto each other. This is because functional connectivity only tells us whether two brain regions are activated simultaneously. This could be mediated by a direct structural connection (i.e., a direct fiber tract between these two regions), or by an indirect way mediated by subcortical structure such as the thalamus or the basal ganglia. Future research will have to solve this open question.

3

The Structural Language Network

An adequate description of the neural basis of language processing must consider the entire network both with respect to its structural white matter connections and the functional connectivities between the different brain regions as the information has to be sent between different language-related regions distributed across the temporal and frontal cortex. Possible data highways are the white matter fiber bundles that connect different brain regions. These fiber bundles can connect quite distant brain regions located in the frontal and temporal cortex and guarantee the information transfer between them. Thus these white matter structures are considered to be the backbone of language processing, and as such deserve a closer look. The white-matter fiber bundles are composed of millions of axons that are surrounded by myelin, which is essential for fast transmission of electrical impulses (Wake, Lee, and Fields, 2011). Previously, scientists could only analyze the fiber bundles postmortem. With the advent of new imaging techniques and analytic methods we are now able to investigate the brain's white matter in vivo. Fractional anisotropy is a value often used in diffusion imaging to quantify the diffusion process. It is taken to reflect fiber density, axonal diameter, and myelination in white matter. In this chapter we will look at the anatomy of the white matter fiber bundles that connect the language-relevant regions.[1]

3.1 The Neuroanatomical Pathways of Language

The relevance of white matter fiber bundles for language perception and production was first recognized in the late 19th century when they were proposed to form possible connections between the different language centers (Wernicke, 1874). In his model, Wernicke proposed a speech production center (based on the work of Broca, 1861), a sensory language center, and a connection between the centers supporting their interaction. Broca's area in the inferior frontal cortex and Wernicke's area in the superior temporal cortex as the classical language areas are subparts of the network that supports language functions at different levels (Vigneau et al., 2006; Hickok and Poeppel, 2007; Friederici, 2011).

The connections between the classic language regions that have been identified over the past decades are multifold. Initially, two broad functional processing streams connecting temporal and frontal areas were proposed (Hickok and Poeppel, 2000). Without basing their model on structural magnetic resonance imaging evidence of the fiber tracts connecting the language-related brain regions, Hickok and Poeppel (2004) discussed a functional "dorsal stream" as being responsible for sensory-to-motor mapping in speech processing and a functional "ventral stream" as supporting sound-to-meaning mapping. Based on neuroanatomical studies in non-human primates and functional studies in human and non-human primates, Rauschecker and Scott (2009) proposed a model that assumes a dorsal stream going forward from the temporal cortex to the premotor cortex, a ventral stream going forward from the temporal cortex to Broca's area and from there going backward via a dorsal stream to the parietal and temporal cortex. The authors take these processing streams to underlie human speech processing during perception and production.

In recent years, new imaging techniques, such as diffusion-weighted magnetic resonance imaging, also called diffusion tensor imaging, allow white matter fiber bundles to be tracked in vivo in the human brain (Mori and Zijl, 2002). Diffusion tensor imaging can provide information about the internal fibrous structure based on the measure of water diffusion. Since water will diffuse more rapidly in the direction aligned with the internal structure, the principal direction of the diffusion tensor can be used to infer white matter connectivity in the brain. Often diffusion tensor imaging is used to identify the different fiber tracts in the human brain (Behrens et al., 2003; Catani and Thiebaut de Schotten, 2008; Berthier, Lambon Ralph, Pujol, and Green, 2012).

Diffusion tensor imaging is a method that can only tell us something about the white matter structure; but it cannot inform us directly about the particular white matter pathway's function (Friederici, 2009a, 2009b). When it comes to allocate function to particular fiber bundles several approaches are used. A first approach is primarily structure-based and independent of function: it identifies tracts and their target brain regions. Then the knowledge about these regions' function is used to allocate the functional role of the tracts, for example, language, are discussed (Catani, Jones, and Ffytche, 2005; Makris and Pandya, 2009). This approach only allows very indirect conclusions about a particular fiber tract's function. A second approach is called a function-based approach. It combines functional magnetic resonance imaging and diffusion magnetic resonance imaging, and thereby allows specification of a given fiber tract's function, indirectly, in a two-step approach. In a first step, particular brain regions relevant for a specific language task are identified in a functional experiment by functional magnetic resonance imaging, and then in a second step these regions are used as seed regions from which a tractogram is calculated (Friederici, Bahlmann, Heim, Schubotz, and Anwander, 2006; Saur et al., 2008, 2010). The resulting fiber tract is taken to transmit information relevant for the particular function investigated by the functional magnetic resonance imaging, although it is not always certain whether the tracing originates only from the seed region or also from adjacent parts.

Within this function-based tractography approach, two methodologies can be applied: probabilistic and deterministic tracking. The probabilistic approach takes one seed point, which is neurofunctionally defined as the starting point of tractography (Friederici, Bahlmann, et al., 2006). The deterministic approach takes two regions that are activated simultaneously as a function of a particular task and calculates the fiber tract between these (Saur et al., 2008). The most direct test of a particular fiber tract's function, however, is to investigate language performance in patients with lesions in the white matter fiber tracts (Galantucci et al., 2011; Papoutsi, Stamatakis, Griffiths, Marslen-Wilson, and Tyler, 2011; Turken and Dronkers, 2011; Wilson et al., 2011), or to correlate language performance with the degree of myelination of particular fiber tracts in the developing brain (Brauer, Anwander, and Friederici, 2011; Skeide, Brauer, and Friederici, 2016). Based on these approaches, different white matter pathways that are relevant for auditory language processing have been identified. Most generally there are different long-range fiber bundles connecting the language-relevant areas in the frontal cortex and in the temporal cortex dorsally and ventrally. Each of the pathways consists of more than one fiber bundle terminating in different cortical regions.

Before turning to a functional description of the fiber tracts based on the function of the respective termination regions, which I provide in section 2.2, I offer here a purely structural description. This neuroanatomical description will primarily focus on the left hemisphere but also consider the connections in the right hemisphere. Anatomical connections within each hemisphere are relevant, as both hemispheres are involved in auditory language processing. Within the hemisphere a distinction has been made between dorsally and ventrally located fiber tracts. For a schematic overview of these fiber tracts, refer to figure 0.1.

The distinction between a dorsal and a ventral way of connecting the temporal cortex and the frontal cortex is based on the neuroanatomic differentiation in the monkey between the "where" stream, which goes from the auditory cortex dorsally to the dorsolateral prefrontal cortex, and the ventrally located "what" stream, originally defined by Rauschecker and Tian (2000).

Neuroanatomically, the picture is somewhat more complex. The dorsal pathway appears to consist of more than one white matter fiber bundle (Catani et al., 2005; for recent reviews, see Gierhan, 2013; Catani and Bambini, 2014). This work has identified a major dorsal fiber bundle connecting Broca's area with the posterior temporal cortex. The fiber bundle that connects the language areas (Broca's area and Wernicke's area) has long been described as relevant for language processing. It is classically labeled as the arcuate fasciculus (AF). Catani et al. (2005) tracked the arcuate fasciculus using diffusion magnetic resonance imaging and suggested two dorsal connections, one connecting Broca's region and Wernicke's region directly (long segment), and one connecting these regions indirectly via the inferior parietal cortex (posterior segment and anterior segment). The figure in their

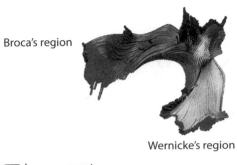

- long segment
- posterior segment
- anterior segment

Figure 3.1
Dorsal fiber connections. Tractography reconstruction of the dorsally located fiber bundles. Broca's and Wernicke's regions are connected through direct and indirect pathways in the average brain. The pathway targeting Broca's area directly (long segment shown in purple) runs medially and corresponds to classical descriptions of the arcuate fasciculus. The indirect pathway runs laterally and is composed of an anterior segment (blue) connecting the inferior parietal cortex and Broca's region and a posterior segment (yellow) connecting the parietal cortex and Wernicke's region. Adapted from Catani, Jones, and Ffytche (2005). Perisylvian language networks of the human brain. *Annals of Neurology*, 57 (1): 8–16, with permission from John Wiley & Sons, Inc.

publication suggests the anterior segment terminates in the premotor cortex rather than in Broca's area (figure 3.1).

This early work provided some indication of there being two major dorsal connections, a direct one and an indirect one mediated by the inferior parietal cortex. The neuroanatomic description and the labeling of the dorsally located fiber bundles in the literature are unfortunately somewhat heterogeneous and have even changed over the past few decades. I will present these here to introduce the reader to the different terminology used in different studies. Today, the fiber bundles that connect the inferior parietal lobe to the premotor cortex are most commonly named the superior longitudinal fasciculus (SLF) (Petrides and Pandya, 1984; Makris et al., 2005; Schmahmann et al., 2007, Frey, Campbell, Pike, and Petrides, 2008; Thiebaut de Schotten, Dell'Acqua, Valabreque, and Catani, 2012). The arcuate fasciculus, which connects the prefrontal cortex to the posterior superior temporal gyrus dorsally, partly runs closely in parallel with the superior longitudinal fasciculus from prefrontal to parietal regions, but not in its posterior portion, curving into the posterior temporal cortex. But since the arcuate fasciculus and the superior longitudinal fasciculus do run partly in parallel between the prefrontal and parietal cortex, some researchers refer to this fiber tract as the arcuate fasciculus/superior longitudinal fasciculus (AF/SLF) or superior longitudinal fasciculus/arcuate fasciculus (SLF/AF).

There are two major dorsally located fiber bundles that are relevant for speech and language, respectively. These two fiber bundles can be differentiated by their termination

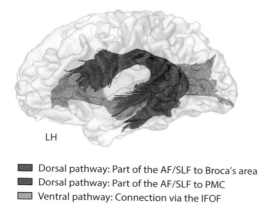

■ Dorsal pathway: Part of the AF/SLF to Broca's area
■ Dorsal pathway: Part of the AF/SLF to PMC
□ Ventral pathway: Connection via the IFOF

Figure 3.2
Dorsal and ventral fiber connections. Structural connectivity between language-relevant areas. (A) Fiber pathways as revealed by probabilistic fiber tracking of diffusion tensor imaging data for adults with seed in Broca's area and seed in the precentral gyrus/premotor cortex. Two dorsal pathways are present in adults—one connecting the temporal cortex via the arcuate fasciculus (AF) and the superior longitudinal fasciculus (SLF) to the inferior frontal gyrus, i.e., Broca's area (purple), and one connecting the temporal cortex via the AF and SLF to the precentral gyrus, i.e., premotor cortex (PMC) (blue). The ventral pathway connects the ventral inferior frontal gyrus via the extreme capsule fiber system and the inferior fronto-occipital fasciculus (IFOF) to the temporal cortex (orange). Adapted from Perani et al. (2011). The neural language networks at birth. *Proceedings of the National Academy of Sciences of the United States of America*, 108 (38): 16056–16061.

regions in the frontal cortex and by the particular language subfunctions of these termination regions. One dorsal fiber bundle connects the temporal cortex with the premotor cortex through the mediation of the inferior parietal lobe, whereas the other bundle connects the temporal cortex to Broca's area, in particular BA 44 (Perani et al., 2011) (figure 3.2). The fiber bundle that terminates in the premotor cortex appears to be functionally relevant for acoustic-motor mapping, whereas the one that terminates in BA 44 appears to support the processing of syntactically complex sentences—as discussed in more detail below.

The language regions in the frontal cortex and in the temporal cortex are connected not only by dorsally located pathways but also by at least two ventrally located pathways. First, a fiber system usually referred to as the extreme capsule fiber system (ECFS), but also named the inferior-fronto-occipital fasciculus (IFOF), as it connects inferior frontal regions (BA 45 and BA 47) with the temporal and occipital cortex. Second, there is the uncinate fasciculus, which connects the frontal operculum to the anterior temporal cortex. As these two fasciculi run closely parallel their respective function is hard to disentangle, except with respect to their termination points.

Thus, neuroanatomically there are at least two dorsal and two ventral fiber tracts connecting the temporal to the inferior frontal cortex. The fiber tracts can be separated with respect to their target regions in the frontal cortex. These different fiber tracts appear to support different language functions, as we will see below.[2]

Language-Relevant Fiber Tracts

When information flows through fiber tracts, its function cannot be registered directly in the tracts. The allocation of functions to different fiber tracts must be based on the functions of the brain areas that those tracts connect, as the following examples will show. A first study using a combined functional and structural magnetic resonance imaging approach analyzed both the dorsal and ventral fiber bundles connecting those prefrontal and temporal regions that are found to be activated in response to a particular language function (Friederici, Bahlmann, et al., 2006). The functional magnetic resonance imaging experiment applied an artificial grammar paradigm with rule-based syllable sequences after either adjacent or hierarchical nonadjacent dependency rules. Although the former activated the left frontal operculum, the latter additionally activated the posterior portion of Broca's area (BA 44) (compare figure 1.6). The diffusion magnetic resonance imaging data of probabilistic tracking from these two regions, located in close vicinity, revealed two distinct fiber tracts. First, a dorsal pathway connecting Broca's area to the posterior superior temporal gyrus/middle temporal gyrus was found when seeding in Broca's area (BA 44); given this area's activation in the processing of nonadjacent, embedded hierarchical dependencies (see figure 3.3), this pathway was interpreted to support the processing of complex syntactically structured sequences. Second, fiber tracking with a seed in the frontal operculum that was activated for the processing of adjacent dependencies revealed a ventral pathway to the temporal cortex; therefore, this pathway was viewed to support the processing of adjacent dependencies.

Subsequently, a second study (Saur et al., 2008) using a combined functional magnetic resonance imaging/diffusion magnetic resonance imaging approach in language investigated the comprehension of simple sentences (*The pilot is flying the plane*) and the oral repetition of single words and pseudowords. The functional data show activation in the premotor cortex and in the temporal cortex for the oral repetition of a heard utterance, whereas sentence comprehension activated regions in the anterior frontal and the temporal cortex. Using a deterministic tracking approach, a dorsal and a ventral connection were identified. The authors interpret their data to show that the dorsal pathway supports sensory-to-motor mapping necessary for sensorimotor integration during speech processing, and that the ventral pathway subserves sound-to-meaning mapping necessary for comprehension.

This apparent contradiction between the two studies in their functional interpretation with respect to the dorsal pathway, however, can be explained on the basis of novel diffusion magnetic resonance imaging data suggesting two dorsal fiber tracts defined on the basis of their different termination points in the adult brain. One dorsal fiber tract connects the posterior temporal cortex with the posterior portion of Broca's area, whereas a second dorsal fiber tract connects the posterior temporal cortex and the premotor cortex, not reaching into Broca's area (Perani et al., 2011).

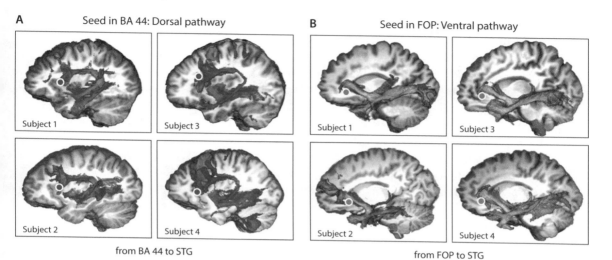

Figure 3.3
Fiber connections for different grammar types. Structural connectivity for two brain regions in the inferior frontal cortex: Broca's area (BA 44) and frontal operculum (FOP) as seed taken from a functional experiment. Three-dimensional rendering of the distribution of the connectivity values of two start regions with all voxels in the brain volume (purple, tractograms from Broca's area (BA 44); orange, tractograms from FOP). (A) Four representative subjects of the group processing a phrase structure grammar with their individual activation maxima in Broca's area (BA 44) in the critical contrast incorrect vs. correct sequences ($P > 0.005$). For all subjects, the tractography detected connections from Broca's area (BA 44) to the posterior and middle portion of the superior temporal region via the fasciculus longitudinalis superior. (B) Four representative subjects of the group processing a finite-state grammar with their individual activation maxima in the FOP in the critical contrast incorrect vs. correct sequences ($P > 0.005$). The individual peaks of the functional activation were taken as starting points for the tractography. For all subjects, connections to the anterior temporal lobe via the fasciculus uncinatus were detected. BA, Brodmann's area; STG, superior temporal gyrus. Adapted from Friederici, Bahlmann, et al. (2006). The brain differentiates human and non-human grammars: Functional localization and structural connectivity. *Proceedings of the National Academy of Sciences of the USA*, 103 (7): 2458–2463. Copyright (2006) National Academy of Sciences, U.S.A.

The dorsal fiber tracts. The specific function of the dorsal fiber tract connecting the posterior temporal cortex to the posterior part of Broca's area (BA 44), the arcuate fasciculus, is still under debate. The proposed function is that it is particularly involved in processing syntactically complex sequences and complex sentences (Friederici, 2011). This proposal is supported by a number of studies using different approaches to identify the function of this fiber tract in language. The first approach is a function-based diffusion magnetic resonance imaging study using probabilistic fiber tracking with a seed in BA 44 as the activation peak for processing syntactically complex sequences. Since BA 44 is structurally connected by a dorsal fiber tract to the posterior temporal cortex, this tract was taken to support and be crucially involved in processing complex syntax (Friederici, Bahlmann, et al., 2006). Second, a study with non-fluent primary progressive aphasic patients, who had suffered from lesions in this fiber bundle and were found to be specifically deficient in processing syntactically complex sentences, provides additional evidence for this view (Wilson et al., 2011). Third, two developmental studies indicate that the maturation of the arcuate fasciculus is directly related to the behavioral performance in processing syntactically complex sentences (Brauer et al., 2011; Skeide et al., 2016).

The function of the second dorsal pathway connecting the posterior temporal cortex with the premotor cortex is to support auditory-to-motor mapping. This fiber tract is necessary when it comes to repeating what somebody else has said, as in a so-called repetition task (Saur et al., 2008). It should also be most relevant for infants throughout their babbling phase, during which they try to imitate speech sounds.

The ventral fiber tracts. The ventral pathway connecting the frontal and temporal cortex ventrally was originally viewed as a unifunctional tract subserving sound-to-meaning mapping (Hickok and Poeppel, 2004, 2007; Rauschecker and Scott, 2009), but later different functions were assigned to subparts of the ventral pathway, specifically the IFOF (Saur et al., 2008; Saur et al., 2010; Turken and Dronkers, 2011; Sarubbo, De Benedictis, Maldonado, Basso, and Duffau, 2013) and the uncinate fasciculus. For a discussion of the ventral pathway see Weiller, Musso, Rijntjes, and Saur, 2009; Weiller, Bormann, Saur, Musso, and Rijntjes, 2011.

The ventral pathway—with frontal terminations in BA 45 / BA 47, and the orbitofrontal cortex and temporal terminations in the middle temporal gyrus, the superior temporal gyrus, and the temporo-parietal cortex—is seen as the major pathway supporting comprehension of semantically simple, high-constraint sentences (Saur et al., 2008). As this pathway (IFOF) targets BA 45 and BA 47 in the inferior frontal gyrus, a region known to support semantic processes, this pathway is widely taken to subserve semantic processing.

The specific function of the ventrally located uncinate fasciculus, which connects the frontal operculum and orbitofrontal cortex to the anterior superior temporal gyrus (Anwander, Tittgemeyer, von Cramon, Friederici, and Knösche, 2007; Hua et al., 2009; Thiebaut de Schotten et al., 2012), is still being debated. Early work proposed that the uncinate fasciculus is relevant for language processing in general (Wise, 2003; Grossman et al., 2004; Catani and Mesulam, 2008; Matsuo et al., 2008). However, Friederici, Bahlmann, and colleagues (2006), using function-based diffusion magnetic resonance imaging, suggested that the uncinate fasciculus connecting the ventral part of the inferior frontal gyrus and the anterior superior temporal gyrus might also be syntax-relevant at a most basic level as it supports basic combinatorics when combining two elements.

Both the proposal of allocating semantic processes to the ventral pathway and the proposal of allocating local combinatorial processes to the ventral pathway may be valid because, functionally, the anterior superior temporal sulcus has been found to be relevant for the intelligibility of speech in general (Scott, Blank, Rosen, and Wise, 2000; Crinion,

Lambon Ralph, Warburton, Howard, and Wise, 2003; Narain et al., 2003; Obleser, Zimmermann, Van Meter, and Rauschecker, 2007; Friederici, Kotz, Scott, and Obleser, 2010; Obleser and Kotz, 2010) and the anterior superior temporal gyrus as well as the anterior temporal lobe have been reported to be activated as a function of semantic and syntactic aspects (for a review of the functional studies, see Mazoyer et al., 1993; Stowe et al., 1998; Friederici, Meyer, and von Cramon, 2000; Friederici, Rüschemeyer, Hahne, and Fiebach, 2003; Vandenberghe, Nobre, and Price, 2002; Humphries, Love, Swinney, and Hickok, 2005; Humphries, Binder, Medler, and Liebenthal, 2006; Snijders et al., 2009; Friederici, 2011). Future research will have to reveal whether both semantic and syntactic combinatorial processes that are elementary for language involve the anterior temporal cortex, possibly with different partnering regions.[3]

In general, combinatorial processes appear to be supported by a ventral pathway for simple syntactic structures (Griffiths, Marslen-Wilson, Stamatakis, and Tyler, 2013), whereas the dorsal pathway (arcuate fasciculus/superior longitudinal fasciculus) is needed to process syntactically complex sentences (Brauer et al., 2011; Wilson et al., 2011).

Communication-Relevant Fiber Tracts

Catani and Bambini (2014) propose additional structural networks relevant for communication and social cognition. The authors, however, also take the arcuate fasciculus linking those regions devoted to the formal aspects of language as the central connection in their model that in total assumes five different structural networks: two are relevant for language (Network 3 and 4) and three are relevant for communication (Network 1, 2 and 5). Network 1, relevant for informative actions, consists of a subset of fibers of the arcuate fasciculus that links Broca's area to the parietal cortex. Network 2, relevant for communicative intentions, connects Broca's area to the dorsomedial frontal cortex via the so-called frontal aslant tract. Network 3 represents the network for lexical and semantic processes: it involves the anterior temporal lobe connected via the ventrally located uncinate fasciculus and the ventrally running inferior fronto-occipital fasciculus connecting anterior Broca's area to Wernicke's area in the temporal lobe. Network 4 represents the fronto-temporal network for syntactic processes: it consists of Broca's area directly connected via the arcuate fasciculus to Wernicke's area. The language networks 3 and 4 are largely in line with the proposal made by Friederici (2011, 2012a). Network 5 is the network for pragmatic integration. This network consists of a part of the arcuate fasciculus that connects the parietal cortex to Wernicke's area and especially the angular gyrus.

Summary

This section discussed the white matter fiber tracts connecting the language-relevant regions in the frontal and temporal cortices. These can be classified into two dorsal and two ventral pathways. Each pathway consists of more than one major fiber tract. Within the ventral pathway one tract connects BA 45/47 with the superior temporal gyrus and middle temporal gyrus and the other tract connects the orbitofrontal cortex including the frontal operculum with the anterior temporal cortex. The former ventral tract (as we will see in chapter 4) supports semantic processes, while the latter most likely supports combinatorial processes. Within the dorsal pathway, one tract connects the posterior temporal cortex to the premotor cortex and another one connects it to the posterior portion of Broca's area (BA 44). While the former dorsal pathway is taken to support sensory-to-motor mapping, the latter appears to be relevant for the performance of complex syntactic tasks.

3.2 Pathways in the Right Hemisphere and Cross-Hemispheric Pathways

During auditory sentence comprehension, prosodic information—that is, the rhythmic and melodic variation in speech—conveys relevant information at different levels: it marks syntactic phrase boundaries, but it also conveys information about the speakers' emotions and intentions. These different aspects of prosody have been discussed under the terms *linguistic prosody* and *emotional prosody*, respectively. To what extent these prosody types are processed in the left and/or right hemisphere is a long-standing debate (Van Lancker, 1980; Pell and Baum, 1997; Friederici and Alter, 2004; Schirmer and Kotz, 2006; Wildgruber, Ackermann, Kreifelts, and Ethofer, 2006; see also section 1.8).

A most general view holds that emotional prosody is mainly processed in the right hemisphere and that linguistic prosody involves both hemispheres, but that left-hemispheric lateralization for linguistic prosody can be observed which is modulated by the type of stimulus material (normal speech vs. hummed "speech," i.e., melody only) and by the type of task (explicit judgment vs. passive listening) (Friederici and Alter, 2004; Kreitewolf, Friederici, and von Kriegstein, 2014; Frühholz, Gschwind, and Grandjean, 2015). That means that the more linguistically complete the stimulus material and the more the focus is on processing of segmental information (speech sounds and words), the more we see a lateralization of linguistic prosody to the left hemisphere. In consonance with this view, it was also found that emotional prosody is not restricted to the right hemisphere, but can elicit bilateral fronto-temporal activation when emotion is conveyed not only by prosody alone but also by the semantic content of a sentence (Frühholz et al., 2015).

Independent of the relative involvement of the right hemisphere in processing prosody, two things are clear from the literature. First, temporal and frontal regions in the right hemisphere are involved in auditory language comprehension. Second, both hemispheres interact during auditory language comprehension. These brain functional findings call for a structural description of the pathways within the right hemisphere and the pathway connecting the right and the left hemisphere.

Pathways in the right hemisphere. Similar to the dorsal and ventral pathways in the left hemisphere, a dual-stream model and its underlying structural connections between the temporal and frontal regions have been proposed for the right hemisphere (Frühholz et al., 2015; Sammler, Grosbras, Anwander, Bestelmeyer, and Belin, 2015). Neuroanatomical studies using structural imaging have investigated the arcuate fasciculus, the white matter tract that connects the temporal cortex with the frontal cortex both in the left and right hemisphere, in particular with respect to a possible asymmetry (Catani et al., 2007; Vernooij et al., 2007; Thiebaut de Schotten et al., 2011; Fernandez-Miranda et al., 2015). The studies often report stronger dorsal fiber tracts in the left hemisphere compared to the right hemisphere. It has been argued that this is a result of a "pre-programmed" development of the arcuate fasciculus (Chi, Dooling, and Gilles, 1977), which is probably due to a reduced fiber density in the right hemisphere (Galaburda, LeMay, Kemper, and Geschwind, 1978). For the right hemisphere not only a dorsal pathway but also a ventral pathway has been identified in the context of prosody processing. Based on functional regions involved in emotional prosody and in analogy to the known fiber pathways in the left hemisphere, a novel study reconstructed fiber tracts in the right hemisphere. Using a probabilistic approach, seeds were placed in the inferior frontal gyrus, the frontal operculum, the anterior superior temporal gyrus, and the posterior superior temporal gyrus. The most posterior superior temporal gyrus in the right hemisphere was found to be dorsally connected to the right frontal operculum (Frühholz et al., 2015, and see also Glasser and Rilling, 2008; Ethofer et al., 2012). The fundus of the posterior superior temporal sulcus was connected to the right inferior frontal gyrus, equally strong via ventral and dorsal pathways. The ventral pathway consisted of fiber bundles similar to those constituting the ventral pathway in the left hemisphere, the inferior-fronto-occipital fasciculus and the extreme capsule (Frühholz et al., 2015).

Starting from a functional study investigating linguistic prosody, indicating a statement versus a question based on pitch information, two strong dorsal fiber tracts connecting the superior temporal sulcus to the frontal cortex were found—one that connected the right anterior superior temporal sulcus to the right premotor cortex and one that connected the right posterior superior temporal sulcus via the arcuate fasciculus/superior longitudinal fasciculus to the right inferior frontal gyrus (Sammler et al., 2015) (see figure 3.4). Note that the discrimination between statement and question appears to also recruit a ventral pathway along the superior temporal lobe, which, however, does not target the inferior frontal gyrus. When considering the pathways between the temporal and the frontal cortex in the right hemisphere, there are two fiber tracts: one pathway targeting the inferior frontal gyrus, and a second one targeting the premotor cortex. These two target regions are strongly interconnected via a short-range fiber connection. The function of the pathway to premotor cortex was investigated indirectly by inhibitory stimulation of the right premotor cortex as a key region for prosodic processes. Inhibitory stimulation of this region leads to a decrease in participants' performance in prosodic categorization (Sammler et al., 2015). The dorsal pathway targeting premotor cortex is taken to represent an action-perception

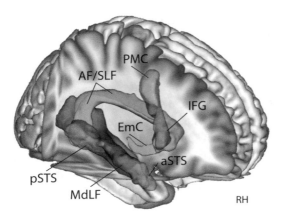

Figure 3.4
Fiber tract pathways in the right hemisphere. Structural connectivity data for the right hemisphere (RH). Group overlay of fiber tracts connecting the functional clusters: pSTS with aSTS (red), PMC (yellow), and IFG (green). Only voxels with fibers in more than 50% of participants are highlighted. Fiber tracts were slightly smoothed (1 mm full width at half maximum) for visualization. AF, arcuate fasciculus; SLF, superior longitudinal fasciculus; MdLF, middle longitudinal fasciculus, aSTS, anterior superior temporal sulcus; pSTS, posterior superior temporal sulcus; PMC, premotor cortex; IFG, inferior frontal gyrus; EMC, extreme capsule. Adapted from Sammler et al. (2015). Dorsal and ventral pathways for prosody. *Current Biology,* 25 (23): 3079–3085, with permission from Elsevier.

network interlinking auditory and motor areas. The other dorsal pathway, which terminates in BA 44/45 in the inferior frontal gyrus, is seen as a network in which the "IFG contributes task-dependent cognitive resources via the dorsal pathways to parse and explicitly label the dynamic prosodic contour" (Sammler et al., 2015, p. 3083). This latter interpretation clearly needs further empirical evidence. For the time being, however, similar to the left hemisphere, ventral and dorsal pathways that connect the inferior frontal to the temporal cortex can also be identified in the right hemisphere.

Cross-hemispheric pathways. Linguistic prosody has been shown to interact with syntactic information during speech comprehension. Thus, there must be an information transfer between the right hemisphere mainly processing suprasegmental information and the left hemisphere processing syntactic information. The structure that guarantees this information transfer between the two hemispheres is the so-called corpus callosum.

The fibers that connect the left and right hemisphere run through the corpus callosum in a fine-grained, ordered way; fibers with orbital and frontal termination points run through the anterior third of the corpus callosum, whereas fibers with parietal, occipital, and temporal termination points cross through the posterior third of the corpus callosum. The middle part is taken by fibers of subcortical termination points and partly by those with frontal termination points (Huang et al., 2005 [also see figure 1.20C]; Hofer and Frahm, 2006; Park et al., 2008).

Anterior and posterior parts of the corpus callosum have been discussed to be involved in language-relevant processes. A lesion in the anterior portion of the corpus callosum has been reported to impact the processing of emotional and linguistic prosody (Klouda, Robin, Graff-Radford, and Copper, 1988). Other studies using imaging and lesion approaches indicated that the posterior part of the corpus callosum is relevant for the interhemispheric transfer of auditory information (Rumsey et al., 1996; Pollmann, Maertens, von Cramon, Lepsien, and Hughdahl, 2002) and for the development of verbal abilities (Nosarti et al., 2004). A combined ERP/lesion study investigated linguistic prosody processing during sentence comprehension in patients with lesions in the posterior and the anterior part of the corpus callosum (Friederici, von Cramon, and Kotz, 2007). The posterior part of the corpus callosum was found to be crucial for the interhemispheric interplay between syntax and linguistic prosody at the phrase level. A similar finding was reported in such patients for the interplay between morphosyntactic information and prosodic intonation (Sammler, Kotz, Eckstein, Ott, and Friederici, 2010).

Summary
The white matter fiber tracts connecting the language-relevant regions in the frontal and temporal cortices in the left hemisphere can be classified into two dorsal and two ventral pathways. Each pathway consists of more than one major fiber tract. Within the ventral pathway one tract connects BA 45/47 with the superior temporal gyrus and middle temporal gyrus, and the other tract connects the orbitofrontal cortex including the frontal operculum with the anterior temporal cortex. The former ventral tract supports semantic processes, while the function most likely supports combinatorial processes. Within the dorsal pathway, one tract connects the posterior temporal cortex to the premotor cortex and another one connects it to the posterior portion of Broca's area (BA 44). While the former dorsal fiber tract is taken to support sensory-to-motor mapping, the latter appears to be relevant for the performance of complex syntactic tasks. The right hemisphere shows ventral and dorsal pathways connecting temporal and frontal regions, although with a lower density of fiber bundles in the right compared to the left hemisphere. Similar to the left hemisphere, there are several fiber tracts in the right hemisphere, one targeting the premotor cortex and one targeting BA 44/45 in the inferior frontal gyrus. The corpus callosum is the structure that connects the two hemispheres. Within the corpus callosum, specifically the posterior third, lie the fibers that connect to the temporal lobes of the two hemispheres, which in turn are crucial for auditory language processing.

3.3 The Neuroanatomical Pathway Model of Language: Syntactic and Semantic Networks

The language comprehension model described in section 1.1 proposed a weak syntax-first model, which assumes that the processing system initially builds up a local phrase structure on the basis of the available word category information (Friederici, 2002, 2011). Semantic and higher-order syntactic relations are only processed after this, unless the context is syntactically and semantically highly predictive. The neuroanatomical pathway model proposed here assumes that the two different stages of syntactic processing (Friederici, 2002) are represented in two different syntactic networks (Friederici, 2011), and that an additional, separate network supports semantic processes. The white matter fiber bundles that are part of these networks will be described in this section.[4]

Syntactic Networks
The first processing step of local structure building is functionally based on the grammatical knowledge of the target language. This concerns the basic knowledge about the structure of adjacent dependencies, as in local phrases, and there are only a few in each language, such as the determiner phrase (*the ship*) and prepositional phrase (*on ships*). This knowledge must be acquired during language acquisition and its use becomes automatic as learning proceeds. In the adult brain, this process is highly automatic and the detection of violations in local phrases involves the frontal operculum and the anterior superior temporal gyrus (Friederici et al., 2003). If the process is less automatized, as in second-language processing (Rüschemeyer, Fiebach, Kempe, and Friederici, 2005) and during development (Brauer and Friederici, 2007), then BA 44 is also recruited for violation detection. This is interesting because these two regions located adjacent to each other differ in their phylogeny, with the frontal operculum being phylogenetically older than BA 44 (Amunts and Zilles, 2012; Sanides, 1962). Thus, it appears that a phylogenetically older cortex can deal with the more simple processes of detecting violations in adjacent dependencies rather than the more complex process of building structural hierarchies. The frontal operculum and the anterior superior temporal gyrus are connected via a ventral fiber tract, that is, the uncinate fasciculus. The function of the respective regions could be defined as follows: In the adult brain, the anterior superior temporal gyrus that receives its input from the auditory cortex represents templates of local phrases (determiner phrase, prepositional phrase), against which the incoming information is checked (Bornkessel and Schlesewsky, 2006). From here, the information is transferred via the uncinate fasciculus to the frontal operculum, which in turn transmits this information to the most ventral part of BA 44 for further processing. This ventral network is involved in the most basic combinational processes, with the frontal operculum subserving the combinatorics of elements independent of hierarchy and the adjacent most ventral anterior portion of BA 44 supporting hierarchical phrase structure building (Zaccarella and Friederici, 2015a).

A second syntactic network deals with the processing of hierarchical dependencies in syntactically complex sentences. The term *complexity* is used to cover different sentence-level phenomena, including sentences with non-canonical word order (Röder, Stock, Neville, Bien, and Rösler, 2002; Grewe et al., 2005; Friederici, Fiebach, Schlesewsky, Bornkessel, and von Cramon, 2006; Meyer, Obleser, Anwander, and Friederici, 2012), sentences with varying degrees of embedding (Makuuchi, Bahlmann, Anwander, and Friederici, 2009), sentences with varying degrees of syntactically merged elements (Ohta, Fukui, and Sakai,

2013), and the interplay of these sentence structures with working memory. These studies indicate that across the different languages such as English, German, Hebrew, and Japanese, the factor of syntactic hierarchy operationalized as the reordering in non-canonical sentences or processing of embedded structures is localized in Broca's area, mostly in its posterior portion (BA 44). All these studies show that an increase in the level of hierarchy as defined in a syntactic tree leads to an increase in activation in BA 44.

A second region also reported to be activated as a function of syntactic complexity and of verb-argument resolution is the posterior superior temporal gyrus/superior temporal sulcus (Ben-Shachar, Palti, and Grodzinsky, 2004; Kinno, Kawamura, Shioda, and Sakai, 2008; Friederici, Makuuchi, and Bahlmann, 2009; Newman, Ikuta, and Burns, 2010; Santi and Grodzinsky, 2010). This region has also been activated when the semantic relation between a verb and its argument cannot be resolved (Friederici et al., 2003; Obleser and Kotz, 2010). Moreover, it was found that the factors of verb class and argument order interact in this region (Bornkessel, Zysset, Friederici, von Cramon, and Schlesewsky, 2005). Therefore, it appears that the posterior superior temporal gyrus/superior temporal sulcus is a region in which syntactic information and semantic verb-argument information are integrated (Grodzinsky and Friederici, 2006). Thus, the posterior Broca's area (BA 44) together with the posterior superior temporal gyrus/superior temporal sulcus constitute the second syntactic network, which is responsible for processing syntactically complex sentences. Within this dorsal syntactic network, BA 44 supports the buildup of hierarchical structures, whereas the posterior superior temporal gyrus/superior temporal sulcus subserves the integration of semantic and syntactic information in complex sentences.

From these conclusions, one can identify two syntax-relevant networks, a ventral network and a dorsal syntactic network, each responsible for a different aspect during syntactic processing. But the question arises whether there is further independent support for the view of two syntax-relevant networks from either patient or developmental studies. Unfortunately, patient studies do not allow us to distinguish between BA 44 and the frontal operculum, because both regions are adjacent and both lie in the supply region of the middle artery. However, lesions in the inferior frontal gyrus involving these two regions are reported to mostly result in syntactic processing deficits (for a review, see Grodzinsky, 2000), although the variation in group selection, design, and methodology leads to some diversity between the outcomes of different studies.

Recent functional magnetic resonance imaging and diffusion magnetic resonance imaging studies on patients revealed some interesting results concerning the dorsal and ventral syntax-relevant networks. A study by Griffiths and colleagues (2013) reported that patients with lesions in the left hemisphere involving either parts of the ventral network or the dorsal network showed some deficit in syntactic processing. Although this study did not systematically vary the syntactic complexity of their sentences, the results generally support the idea that both networks are involved in syntactic processes. The study by Wilson and colleagues (2011) investigating non-fluent progressive aphasics, however, indicated that degeneration of the dorsal fiber tract connecting the temporal cortex and posterior Broca's area lead, in particular, to deficits in the processing of syntactically complex sentences. This finding is clearly in line with the current interpretation of the dorsal fiber tract targeting Broca's area as a pathway supporting the processing of complex syntax. Further support for the view that this pathway subserves the processing of syntactically complex sentences stems from developmental studies on language. These report that children, at an age when they are still deficient in processing non-canonical sentences, demonstrate a dorsal fiber tract targeting BA 44 that is not yet fully myelinized (Brauer, Anwander, and Friederici, 2011), and, moreover, that the degree of myelination predicts behavior in processing syntactically complex non-canonical sentences (Skeide et al., 2016).

Concerning the ventral syntax-relevant system, one must admit that reports on the relation between syntactic abilities and lesions in the temporal cortex are quite sparse because lesions in the temporal lobe are primarily related to semantic deficits. However, interestingly, those patients with temporal lesions that extend and include the anterior portion of the temporal lobe show syntactic comprehension deficits (Dronkers, Wilkins, Van Valin, Redfern, and Jaeger, 2004). Additionally, a correlational analysis of the white matter integrity in stroke patients and their behavioral performance indicates that syntax is processed both by the dorsal and ventral system (Rolheiser, Stamatakis, and Tyler, 2011). This latter finding can be explained under the view that both systems are relevant for syntax, but with different subfunctions. The ventral system may be involved in combinatorial aspects independent of syntactic hierarchy, whereas the dorsal system is crucially involved in hierarchy building (see Zaccarella and Friederici, 2015a).

Thus quite a number of studies agree with the view that there are two syntax-relevant networks, a dorsal one and a ventral one. Moreover, some studies specifically demonstrate that the dorsal syntactic system is necessary for processing hierarchically structured sentences (Brauer et al., 2011; Skeide et al., 2016; Wilson et al., 2011).

Semantic Networks
The ventral stream has long been taken to support semantic processes (Hickok and Poeppel, 2004; Saur et al., 2008). The ventral stream can be divided into two pathways: one involves the uncinate fasciculus and the other involves the ECFS or IFOF. The uncinate fasciculus reaches the anterior temporal lobe, whereas the ECFS or IFOF additionally also reach more posterior portions of the temporal lobe. The functional allocation of these pathways is a matter of ongoing debate. Many researchers see the uncinate fasciculus as being involved in language processes (Parker et al., 2005; Friederici, Bahlmann, et al., 2006; Catani and Mesulam, 2008; Duffau, 2008; Duffau, Gatignol, Moritz-Gasser, and Mandonnet, 2009); however, its particular function is a matter of discussion. The functional magnetic resonance imaging study by Friederici, Bahlmann, and colleagues (2006) suggests that the uncinate fasciculus supports the processing of adjacent dependencies during perception of structured sequences. The IFOF has been reported to support semantic processes in many

studies either using an approach combining functional magnetic resonance imaging and diffusion magnetic resonance imaging or combining diffusion magnetic resonance imaging and behavioral measures (Saur et al., 2008; Wilson et al., 2010; Turken and Dronkers, 2011; Wilson et al., 2011). This fiber system runs from the ventral portion of the inferior frontal gyrus along the temporal cortex to the occipital cortex. This way, those inferior frontal brain areas that are reported to be involved in semantic processes, such as BA 45 and BA 47 (Bookheimer, 2002), are connected to the temporal cortex, which is known to support semantic processes including aspects of semantic memory (Patterson, Nestor, and Rogers, 2007). BA 45/47, as one of the regions in the semantic network, is activated particularly when lexical semantic processes are under strategic control; that is, when participants are required to perform some kind of semantic relatedness or plausibility judgment (Fiez, 1997; Thompson-Schill, D'Esposito, Aguirre, and Farah, 1997; Dapretto and Bookheimer, 1999; Kuperberg et al., 2000; Newman et al., 2010). Another region considered to be part of the semantic network is the anterior temporal lobe, since degeneration of this brain region leads to semantic deficits at single-word level and is discussed as a general semantic "hub" (Hodges, Patterson, Oxbury, and Funnell, 1992; Patterson et al., 2007; Lambon Ralph and Patterson, 2008).

Functional magnetic resonance imaging studies investigating sentence-level semantic networks used different types of paradigms and revealed different findings. Studies varying the semantic plausibility found activation in BA 45/47 in the inferior frontal gyrus (Newman et al., 2010), whereas studies investigating semantic predictability reported activation in the supramarginal gyrus and the angular gyrus in the posterior temporo-parietal region (Obleser and Kotz, 2010). At present, it is not entirely clear how to model these sentence-level semantic processes. The regions have been seen to be involved in semantic prediction as well as in combinational semantics (Molinaro, Paz-Alonso, Duñabeitia, and Carreiras, 2015; Price, Bonner, Peele, and Grossman, 2015). What is clear, however, is that the anterior temporal lobe, the inferior frontal cortex, and the posterior temporo-parietal cortex including the angular gyrus are involved in sentence-level semantic processes. Nevertheless, their interplay in the service of language understanding remains to be specified. The ventral pathway connecting the inferior frontal gyrus and the superior temporal gyrus/middle temporal gyrus plays a crucial part in semantic processes, but it is also conceived that the dorsal connection is involved whenever predictive processes are in dispute.

Summary

Sentence processing is based on different neuroanatomically defined networks: two syntax-relevant networks and at least one semantic network. The semantic network involves the anterior temporal lobe, the anterior inferior frontal cortex, and the posterior temporo-parietal region. The former two regions are connected by a ventrally located pathway. Syntactic processes are based on two syntax-relevant networks: a ventral and a dorsal

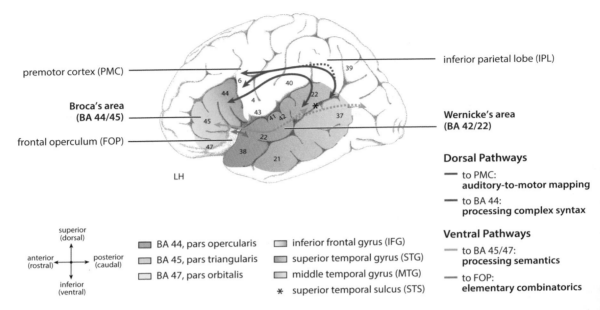

Figure 3.5
Functional and structural language network. Language-relevant brain regions and fiber tracts (schematic and condensed view of the left hemisphere (LH)). The dorsal pathway connecting dorsal premotor (PMC) with posterior STG/STS involves the superior longitudinal fasciculus (SLF); the dorsal pathway connecting BA 44 with the posterior STG involves the arcuate fasciculus (AF). The ventral pathway connecting BA 45 and BA 47, with the posterior STG/MTG, and the occipital cortex, involves the inferior fronto-occipital fasciculus (IFOF) (also called the Extreme Capsule Fiber System (ECFS); the ventral pathway connecting the FOP with the anterior STG involves the uncinate fasciculus. These different fiber tracts support different language functions indicated in the legend at the right side of the figure. STG, superior temporal gyrus; STS, superior temporal sulcus; MTG, middle temporal gyrus. From Friederici & Gierhan (2013). The language network. *Current Opinion in Neurobiology*, 23 (2): 250–254, with permission from Elsevier.

network. The ventral network supports the combining of adjacent dependencies, and thereby provides the basis for later phrase structure building processes. It involves the anterior superior temporal gyrus and the frontal operculum connected via a ventrally located pathway adjacent to the ventral portion of BA 44. The dorsal network supports the processing of syntactic hierarchies and involves the posterior portion of Broca's area (BA 44) and the posterior superior temporal gyrus connected via a dorsally located pathway (see figure 3.5).

4

The Functional Language Network

In the previous chapters I described the neural language network of functionally relevant brain regions (gray matter) and its structural connections (white matter). It became clear that online language processing requires information transfer via the long-range white matter fiber pathways that connect the language-relevant brain regions within each hemisphere and between hemispheres. In this chapter I will discuss the available data related to the functional neural network of language, although our knowledge about how and what type of information is transferred is very rudimentary.

The basic neurophysiological principle is that electrical pulses are transmitted via white matter fiber bundles in the human brain. Today we can look at the functional language network at different levels. At the macrolevel we can analyze the level where brain activation within the language network can be measured. This can be done either in the form of functional connectivity that reflects the join activation of different well-localized brain regions within the network, or in form of oscillatory activation that reflects synchronous activation across brain regions, which can be localized coarsely. At the microlevel the functional language network can also be identified at the level of neurotransmitters that make up the basis of information transfer at the neuronal level. The latter measurement at this point in time, however, can only be analyzed in ex vivo brains.

How information content is encoded and decoded in the sending and receiving brain areas is still an open issue—not only with respect to language, but also with respect to the neurophysiology of information processing in general. A possible though speculative view is that encoding and decoding requires similarity at the neuronal level in the encoding and decoding regions. I will first describe the language network at the neurotransmitter level and then discuss the available data at the level of functional connectivity and oscillatory activity.

4.1 The Neuroreceptorarchitectonic Basis of the Language Network

We know that at the neuronal level the transmission of information from one neuron to the next is based on neurotransmitters. Neurotransmitters play a crucial role at the synaptic level when electric action potentials are sent via axons from one neuron to the next neuron, which receives these via post-synaptic dendrites. Neurotransmitters and their receptors are the key molecules of neural function. Different brain regions are characterized by different densities of the neurotransmitter receptor–binding types (GABA, dopamine, muscarine, etc.). These different densities, which indicate the specific receptorarchitectonic structure of a brain region, are called *fingerprints* (Zilles, Palomero-Gallagher, and Schleicher, 2004). The receptor fingerprints represent the molecular structure of the local information processing in each cortical region. It is conceivable that even distributed brain areas within a functional network have similar molecular characteristics (i.e., receptor fingerprints). Below we will see that this is indeed the case.

In cooperation with Katrin Amunts' group (Amunts et al., 2010) we conducted a neuro-receptorarchitectonic analysis of the inferior frontal cortex. This analysis provided crucial information for the parcellation of the inferior frontal cortex, showing that BA 45 was subdivided into two portions, a more anterior area (45a) bordering BA 47 and a more posterior area (45p) bordering BA 44. Moreover, BA 44 was receptorarchitectonically subdivided into a dorsal (44d) covering the superior part of BA 44 and a ventral (44v) covering the inferior part of BA 44. These subdivisions were discussed to be of particular functional importance as different language experiments have allocated different functions to these different subregions. BA 44 has been shown to be crucially involved in syntactic processes with its dorsal part (44d) supporting phonological-memory-related processes in syntax and its ventral part (44v) being crucial for hierarchical processing. BA 45 has been discussed as an important area in semantic processing. These semantic aspects appear to recruit the anterior part of BA 45 at the boarder of the more anterior located BA 47, thus 45a. The more posterior part—45p bordering BA 44—has rather been reported to be involved in some syntactic studies.

In a receptorarchitectonic study I conducted with colleagues (Zilles, Bacha-Trams, Palomero-Gallagher, Amunts, and Friederici, 2014), we found that the language network is characterized and segregated from other brain regions on its molecular basis. In this study we investigated the hypothesis that the brain regions of the larger language network are characterized by a large similarity in their receptorarchitectonic fingerprints—which would be an indirect indication of their functional neurophysiological connectedness through neuronal potentials. This hypothesis was tested in a study that analyzed the receptorarchitecture of human postmortem brains (such analyses can only be conducted ex vivo). In the study 15 different neurotransmitter receptor–binding sites were measured in vitro by means of quantitative receptor autoradiography. Two analyses were performed. The first one covered the entire language network including all regions known to be involved in language perception and production (all regions indicated by red labels in figure 4.1A).

The second, more specific analysis only included those regions that were shown to be part of the neural functional network for the human ability to process syntactically complex sentences, as previously identified in a functional magnetic resonance imaging study on

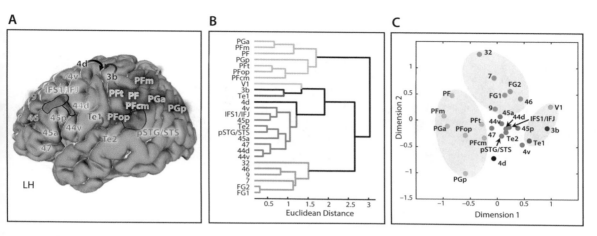

Figure 4.1
Receptorarchitectonic clusters of brain regions in the left hemisphere. (A) Localization of receptorarchitectonically examined cortical regions projected on the lateral surface of the single-subject MNI template brain (Evans et al., 2012). Areas color-coded in red are areas that have been found to be involved in different aspects of speech and language processing. These are 44d (dorsal BA 44); 44v (ventral BA 44); 45a (anterior BA 45); 45p (posterior BA 45); 47 (BA 47); IFS1/IFJ (areas in the inferior frontal sulcus (IFS) and at the junction (IFJ) between the inferior frontal and precentral sulci); pSTG/STS (areas of the posterior superior temporal gyrus and sulcus); Te1 (primary auditory cortex, BA 41); Te2 (auditory belt area, BA 42). Dark blue, dark green, yellow, and black encode the primary somatosensory, auditory, and visual cortices, and the hand representation region of the motor cortex, respectively. Light blue encodes IPL areas, whereas light green represents prefrontal, superior parietal, cingulate, and extrastriate fusiform areas. BA: Brodmann areas (Brodmann, 1909). (B) Hierarchical cluster tree and multidimensional scaling of receptor fingerprints in 26 cortical brain regions. Hierarchical cluster tree of receptor distribution patterns in the left hemisphere. (C) Multidimensional scaling resulting in a 2D display of the 15-dimensional receptor feature vectors of the receptor fingerprints of 26 cortical regions measured in the left hemisphere. Adapted from Zilles et al. (2014). Common molecular basis of the sentence comprehension network revealed by neurotransmitter receptor fingerprints. *Cortex*, 63: 79–89, with permission from Elsevier.

processing center-embedded sentences (Makuuchi, Bahlmann, Anwander, and Friederici, 2009). This latter study revealed activation in the posterior portion of BA 44, the inferior frontal sulcus, and in the posterior superior temporal gyrus/superior temporal sulcus (indicated as extended regions color-coded in red in figure 4.1A). For each region a receptorarchitectonic fingerprint was constructed (for more details, see Zilles et al., 2014). A striking receptorarchitectonic similarity was observed across the regions of the left hemispheric language network, but not across other non-language regions within the same hemisphere. Increased receptorarchitectonic similarity was specifically found for regions belonging to the syntactic network (i.e., BA 44; dorsal 44d and ventral 44v); the inferior frontal sulcus; the posterior temporal gyrus and sulcus, posterior superior temporal gyrus/superior temporal sulcus; figure 4.1B and C (language regions of activation indicated in red). This is the first empirical evidence for a common molecular structure of those regions that constitute the neural language network.

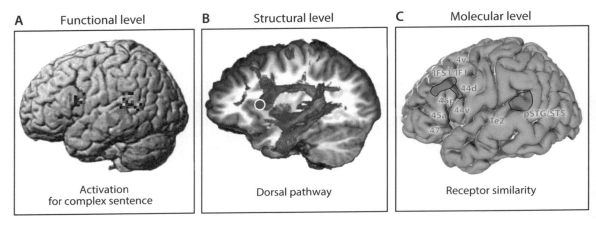

Figure 4.2
The language network in the left hemisphere. The language network as defined at (A) the functional level, (B) the structural level, and (C) the molecular level. (A) Adapted with permission from Wolters Kluwer from Friederici, Makuuchi, and Bahlmann (2009). The role of the posterior superior temporal cortex in sentence comprehension. *NeuroReport*, 20 (6): 563–568. (B) From Friederici, Bahlmann, et al. (2006). The brain differentiates human and non-human grammars: Functional localization and structural connectivity. *Proceedings of the National Academy of Sciences of the USA*, 103 (7): 2458–2463. Copyright (2006) National Academy of Sciences, U.S.A. (C) Adapted from Zilles et al. (2014). Common molecular basis of the sentence comprehension network revealed by neurotransmitter receptor fingerprints. *Cortex*, 63: 79–89, with permission from Elsevier.

Most interestingly, a comparable receptorarchitectonic analysis of the homolog areas in the right hemisphere did not reveal such a similarity between the respective regions, indicating that the left hemisphere is unique in this respect (see supplementary information in Zilles et al., 2014). These findings indicate that only areas that are part of the language network in the left hemisphere show a similar multireceptor organization. These data suggest that this neurotransmitter feature pattern is an important prerequisite for the computational processes underlying language functions, and maybe a general principle for effective communication within a functional neural network.

For the domain of syntax processing, the fronto-temporal network can now be described at the functional, structural, and molecular level (figure 4.2A, B, and C). The left-hemispheric language network for processing complex sentences functionally involves BA 44 and the posterior superior temporal gyrus/superior temporal sulcus. These two regions are (A) functionally involved in processing complex sentences, (B) structurally connected by long-range white matter fiber tracts via the dorsal pathway, and (C) receptorarchitectonically most similar, providing the basis for an effective cooperation between them.

Summary
This section looked at the neural basis of information transfer, namely at the neurotransmitters that are crucially involved in the transmission of information from one neuron to the

next. We have seen that the larger language network is characterized and segregated from other brain regions by its network internal similarity. The neurotransmitter similarity is even larger for subnetworks as demonstrated for the syntactic network.

4.2 Functional Connectivity and Cortical Oscillations

An exhaustive brain-based model of language processing needs to specify not only the language-related brain regions, their structural connectedness, and their receptorarchitectonic properties, but also the functional interplay of the network components. Different types of approaches allow us to describe this functional interplay. The cortical dynamics underlying language functions is restricted to methods that can be applied to humans. From biophysiological work at the cellular level comes the suggestion that the cortex operates by means of an interaction between feed-forward and feedback information, not only at the level of larger neural networks, but even at the level of a single neuron (Larkum, 2013). Neurophysiological methods allow us to describe the information exchange between neurons and neuronal ensembles in different brain areas. The dynamics in the language network can be explored by functional connectivity and cortical oscillations. While functional connectivity analyses can in principle be conducted for both functional magnetic resonance imaging as well as electroencephalography or magnetoencephalography data, the analysis of cortical oscillations, in recent years, has emerged as a more direct electrophysiological marker of the functional relation within neural networks.

Functional Connectivity

Brain dynamics is often assessed by *functional connectivity* (see the introduction of this book). Methods that assess functional connectivity quantify the synchronicity of neuronal charge/discharge alternations in local and remote brain networks, using direct (electroencephalography or magnetoencephalography) and indirect (functional magnetic resonance imaging) data sources. Functional connectivity refers to maps of synchronic brain oscillations in brain regions. These maps indicate which brain regions work together without providing information about the direction of the data flow between the regions (Friston et al., 1997; Fox and Raichle, 2007). This method has been used to analyze the functional connectivity between brain regions. *Effective connectivity* also allows us to determine the information flow between defined regions in the neural language network. Effective connectivity refers to statistical-mathematical approaches to assess the direction of the data flow from one region to another in the neural network. The complex mathematical approaches to quantifying effective connectivity are detailed elsewhere (see Friston, Harrison, and Penny, 2003).

There are two ways of using functional magnetic resonance imaging data as the basis for functional connectivity analysis. One approach uses *resting-state* functional magnetic resonance imaging data gathered when participants are "at rest," that is to say, not involved in a task. However, in order to make these data relevant for the issue of the functional language network, these data are usually correlated with behavioral language processing data gathered independently. The second approach uses *task-related* functional magnetic resonance imaging data, for example from language studies, but separates out the condition-related activation and only uses the remaining data for analysis. Both approaches when combined with behavioral language data can provide valuable data concerning the functional connectivity between different brain regions in the language network. We will discuss these two approaches in turn.

Resting-state functional connectivity. This approach is based on the discovery that the low frequency (< 0.1 Hz) BOLD signal acquired in a task-free resting condition indicates coherent activities within the brain's major cognitive networks (Biswal, Yetkin, Haughton, and Hyde, 1995; Fox and Raichle, 2007). These spontaneous low-frequency fluctuations are considered to reflect unconstrained cognitive processes during resting state (Buckner, Andrews-Hanna, and Schacter, 2008).

The relation between the functional connectivity of the language network and the functional magnetic resonance imaging resting state can be made indirectly by setting the regions known to be functionally involved in language processing as regions of interest for the functional connectivity analyses; it can then be assessed when other brain regions display a similar activation time course, in turn indicating functional connectivity. Resting-state functional connectivity can, moreover, be correlated with behavioral language tasks conducted independently in the same participants in order to determine its relevance for language.

There are several papers reporting task-free resting-state functional connectivity with seeds in Broca's area and Wernicke's area (Xiang, Fonteijn, Norris, and Hagoort, 2010; Tomasi and Volkow, 2012; Bernal, Ardila, and Rosselli, 2015). The data revealed networks that involve a functional connectivity between Broca's area and Wernicke's area, including prefrontal and parietal regions as well as subcortical regions such as parts of the basal ganglia and the thalamus (Tomasi and Volkow, 2012; Bernal et al., 2015). A study that pooled data from 84 experiments that had reported Broca's area activation in language tasks revealed functional connectivity of Broca's area with the left frontal operculum, left posterior temporal region, the parietal lobe, as well as the supplementary motor area (Bernal et al., 2015)—that is, those component regions of the language network which I have discussed already as supporting language processing. Furthermore, in another resting-state study it was shown that three subregions of the larger Broca's regions, namely, pars opercularis (BA 44), pars triangularis (BA 45), and pars orbitalis (BA 47), connect to three different subregions in the temporal cortex (i.e., posterior superior temporal gyrus and middle temporal gyrus, inferior posterior middle temporal gyrus, posterior inferior temporal gyrus) (Xiang et al., 2010). These data as well as additional data suggest an intrinsic connectivity between the frontal and temporal language areas.

Another way to gain task-independent low frequency fluctuation data is based on the fact that even during task-dependent functional magnetic resonance imaging, only about 20 percent of the activation is explained by the specific task whereas about 80 percent of the low frequency fluctuation is unrelated (Lohmann et al., 2010). Using this latter part of the functional data across different language studies indicated a so-called default language network in the left perisylvian cortex. Seeds in left Broca's area also revealed a functional connectivity with the left posterior superior temporal gyrus.

These data raise an intriguing question: To what extent is the neural language network the same across different languages? One study analyzed 970 subjects from 22 research sites covering languages such as English, German, Dutch, and Chinese (Tomasi and Volkow, 2012). With seeds in Broca's area and Wernicke's area, an extended network including prefrontal, temporal, and parietal cortical regions, as well as subcortical structures (basal ganglia and subthalamic nucleus) was found (Tomasi and Volkow, 2012). The analyses showed a striking similarity of the resting-state functional connectivity across different scanning sites including subjects with different native language backgrounds, suggesting a certain generality of the intrinsic neural network.

Language-dependent functional connectivity. In contrast to a large number of resting-state functional connectivity studies, only a few studies focus on functional connectivity based on task-dependent language functions.

One way to investigate the functional connectivity among predefined functional brain regions is called Psychophysiological Interaction analysis (Friston et al., 1997). This analysis is an exploratory multiregression analysis to specify the interaction between regions of interest as revealed by functional magnetic resonance imaging. For sentence comprehension, Psychophysiological Interaction has shown that two regions of interest in the inferior frontal cortex, one that varied as a function of syntactic complexity (embedded sentences; i.e., BA 44) and one that varied as a function of the distance between dependent elements (subject-noun and verb; i.e., the inferior frontal sulcus) cooperate to achieve sentence comprehension (Makuuchi et al., 2009).

The effective connectivity method used to determine the direction of the information flow in the activated areas revealed by functional magnetic resonance imaging that has frequently been used in cognitive neuroscience is Dynamic Causal Modeling (Friston et al., 2003). Until now it has been applied to some functional magnetic resonance imaging language data, both at the word processing level (Heim, Friederici, Schiller, Rueschemeyer, and Amunts, 2009) and the sentence processing level (den Ouden et al., 2012; Makuuchi and Friederici, 2013). The studies on sentence processing are of particular interest here.

The Dynamic Causal Modeling approach tests various models that are formulated as neurophysiologically plausible against each other. A first Dynamic Causal Modeling study used data from an auditory processing experiment that had identified four activation clusters for processing syntactically complex sentences in four different regions: inferior frontal gyrus (BA 45), premotor cortex, posterior superior temporal sulcus, and anterior middle temporal gyrus (den Ouden et al., 2012). All tests using the theoretical Dynamic Causal Modeling approach assumed bidirectional intrinsic connectivity between these four regions. The prevailing model was the model with inferior frontal gyrus as the input, where syntactic complexity modulated the flow of information from the inferior frontal gyrus to the posterior superior temporal sulcus, reflecting the importance of this connection for processing complex syntactic sentences.

Another Dynamic Causal Modeling study analyzed data from a reading experiment that also varied the syntactic complexity of the sentences (Makuuchi and Friederici, 2013). In this study four activation clusters were identified: BA 44 and the inferior frontal sulcus (in the frontal cortex), the inferior parietal cortex, and the temporal cortex—here the middle temporal gyrus. As this was a reading study, the visual word form area in the fusiform gyrus was taken as the starting point, with the different models varying in their connections between the regions starting from the fusiform gyrus (see figure 4.3A). The prevailing model indicated bottom-up information flow from the fusiform gyrus via the inferior parietal sulcus (IPS) (known as the phonological working memory system) to the inferior frontal sulcus (known as the syntactic working memory system). And most interestingly the information from the inferior frontal gyrus flows back to the posterior temporal cortex, via two processing streams, as a direct functional connection from BA 44 (known to

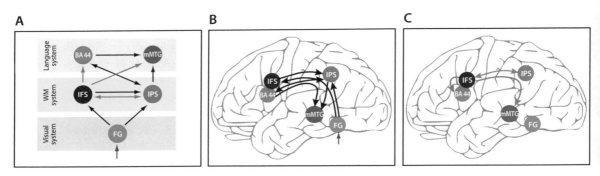

Figure 4.3
Functional connectivity for sentence processing. Significant endogenous connections and significant modulatory effects on connections by the increased working memory load. (A) Hierarchical diagram for the functionally segregated regions (WM, core language, and visual systems) and their connections. Pars opercularis (BA 44), inferior frontal sulcus, mid portion of middle temporal gyrus (mMTG), inferior parietal sulcus (IPS), and fusiform gyrus (FG). Statistically significant endogenous connections and significant modulation of connections are shown by arrows. Black arrows indicate endogenous connections. Red arrows indicate significantly increased connection strengths by the factor distance between relevant words. The vertical gray arrow to the FG represents the input to the visual system. (B) Significant endogenous connections plotted on a schematic brain. (C) Significantly modulated connections by the factor distance between relevant words plotted on a schematic brain. For more details see text. Adapted from Makuuchi and Friederici (2013). Hierarchical functional connectivity between the core language system and the working memory system. *Cortex*, 49 (9): 2416–2423, with permission from Elsevier.

process the hierarchical syntactic structure) to the middle temporal gyrus and as an indirect connection from inferior frontal sulcus (IFS) via the inferior parietal cortex to the middle temporal gyrus (see figure 4.3B and C). Thus it appears that during the comprehension of syntactically complex sentences, working memory systems in the inferior frontal sulcus and in the inferior parietal sulcus come into play. This study together with the study by den Ouden et al. (2012) provide the first modelling evidence that information flows from the inferior frontal gyrus back to the posterior temporal cortex via the dorsal pathway. These findings are compatible with the model of Rauschecker and Scott (2009) assuming a backward mapping from the inferior frontal gyrus to the temporal cortex.

The ultimate description of information flow, however, should not be based solely on Dynamic Causal Modeling (Lohmann, Erfurth, Muller, and Turner, 2012), as only strong neurophysiological priors allow realistic modeling—and priors in many instances may be weak. As a result, different approaches have been used to describe the dynamics of the information transfer within a functionally predefined language network. So far, none of them qualifies as the gold standard. Functional connectivity is investigated mostly at the macroscopic level using functional magnetic resonance imaging data or electrophysiological data indicating the correlated activation of two or more brain regions. Studies that used these methods were able to show that Broca's area and the posterior temporal cortex work together.

What one would really like to know is how these language-related brain regions work together. There is one intriguing study that provides a direct answer with respect to the directionality of the information flow between these brain regions. This study reported cortico-cortical evoked brain potentials (Matsumoto, Nair, LaPresto, Najm, Bingaman, Shibasaki, and Lüders, 2004). Electrocorticograms were taken from epilepsy patients who underwent invasive monitoring with subdural electrodes for epilepsy surgery. Brain potentials were recorded from 3–21 electrodes per patient. Patients were not performing any task. Stimulation of Broca's area elicited evoked potentials in the middle and posterior part of the superior temporal gyrus, the middle temporal gyrus, and the supramarginal gyrus. Stimulation of the posterior language area elicited brain potentials in Broca's area, however, with less defined responses in this region. This study was the first to demonstrate this inter-areal connectivity in the human language system in vivo using electro-cortical evoked potentials. The authors take the arcuate fasciculus as the possible fiber to mediate the activation to proceed from Broca's area to the posterior temporal region (see figure 4.4). This conclusion is in line with our model presented later in figure 4.5.

Neural Oscillations

Neural oscillations as reconstructed from electroencephalography and magnetoencephalography data are a more direct way of approaching the neural dynamics within the language network, although often with a less restricted localization. Neural ensembles are believed to operate through coherent (i.e., synchronous) oscillations in a frequency range

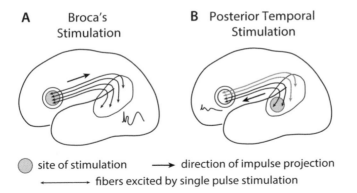

Figure 4.4
Cortico-cortical evoked potentials. Schematic illustration of presumed connections between the anterior and posterior language areas. The gray circle denotes the site of stimulation. Arrows indicate the direction of impulse projection evoked by single-pulse stimulation. The excited fibers are shown as thick black lines, and those not excited as thin gray lines. Reprinted from Matsumoto et al. (2004). Functional connectivity in the human language system: a cortico-cortical evoked potential study. *Brain*, 127: 2316–2330, by permission of Oxford University Press.

of 30–80 Hz (gamma band), whereas it is believed that coherence in spatially extended ensembles of oscillations is reflected in lower frequency bands (beta, 15–30 Hz; theta, 4–8 Hz; delta, 0.5–4 Hz) (for a review, see Buzsáki, Logothethis, and Singer, 2013). Different frequency bands are not domain-specific, but rather reflect general processes. Without being able to provide a clear classification of the processes underlying the different frequency range, one can most generally assume the following relations between frequency band and mental processes. Gamma is thought to be important for binding information and for learning. Beta is thought to be involved in conscious thinking. Alpha is taken to bridge between the conscious and subconscious mind. Theta has been associated with working memory processes. Delta, the slowest wave, has been observed with neural oscillations that can be localized to regions of the language network using electroencephalography and magnetoencephalography source-localization techniques (Oostenveld, Fries, Maris, and Schoffelen, 2011; Bastos and Schoffelen, 2016).

The size of functional networks defined by synchronous activity reflected in oscillations covers many different scales. Small-scale, local networks are usually identified by measuring the power of local oscillatory signals, known as local field potentials, when recorded with invasive techniques or in animals or as sensor signals when recorded with electroencephalography or magnetoencephalography techniques. Large-scale networks—usually extending across several, sometimes distant, cortical areas—are identified by determining the synchronicity of oscillatory signals picked up from different electrodes or sources positioned across the scalp. Using coherence measures, large-scale networks have been identified for numerous cognitive functions including language functions.

However, studies that focus on the neural dynamics of language processing remain rare, in particular those that localize the oscillatory activity. These studies have focused on different aspects of language processing: phonological processing, word processing, and sentence processing. The available studies provide first, but not sufficient, data sets for an adequate model. In the domain of speech processing it has been shown that the interhemispheric transfer between primary and secondary areas of the auditory cortex during conscious syllable processing is based on synchronous gamma oscillations (Steinmann et al., 2014). Gamma activity in the human auditory cortex was found to track the speech envelope (Kubanek, Brunner, Gunduz, Poeppel, and Schalk, 2013). The speech envelope refers to the spectral components of sounds in the sound waveform of speech. The related gamma effect was primarily located in the non-primary auditory cortex, but the data suggest that higher-order processing regions such as Broca's area and the superior temporal gyrus are also involved, but to a lesser degree. Neural oscillatory activity as obtained by invasive electrocorticography suggests left lateralized processing syllables (Morillon et al., 2010). For word processing, differences in gamma power were reported for content words versus function words (Bastiaansen, van der Linden, ter Keurs, Dijkstra, and Hagoort, 2005) and for nouns versus verbs (Pulvermüller, Lutzenberger, and Preissl, 1999).

Oscillations in the theta range have been reported in large-scale networks involving inferior frontal and temporo-parietal regions for processing pseudowords versus words (Strauss, Kotz, Scharinger, and Obleser, 2014) and for word-word priming (Mellem, Friedman, and Medvedev, 2013). Theta range oscillations have also been observed in studies investigating the retrieval of word from memory. It is conceivable that pseudoword processing as in Strauss et al. (2014) reflects an intensive (non-successful) search in the lexicon and that semantically unrelated words, as in the priming study of Mellem et al. (2013), require more search in the lexicon than a related word.

Successful retrieval of words from memory has been associated with enhanced coherence in the beta band between anterior and posterior brain regions (Sarnthein, Petsche, Rappelsberger, Shaw, and von Stein, 1998; Weiss and Rappelsberger, 2000). Such findings agree well with the notion that memory-related processes (Siapas, Lubenov, and Wilson, 2005; Sigurdsson, Stark, Karayiorgou, Gogos, and Gordon, 2010; Fell et al., 2011) and very extensive networks comprising widely distributed cortical areas are typically coordinated in the theta frequency band and in the beta range frequency band (Sarnthein et al., 1998; Sauseng, Klimesch, Schabus, and Doppelmayr, 2005; Summerfield and Mangels, 2005). Lewis and colleagues recently suggested that beta-gamma oscillations in language, in particular, are related to predictive coding because beta effects correlated with the N400 in the ERP (Lewis and Bastiaansen, 2015; Lewis, Schoffelen, Schriefers, and Bastiaansen, 2016).

A number of studies have investigated the cortical dynamics during sentence processing in general (Haarmann, Cameron, and Ruchkin, 2002; Haarmann and Cameron, 2005;

Bastiaansen and Hagoort, 2006; Bastiaansen, Magyari, and Hagoort, 2010). A few studies have investigated the processing of syntactic structure using frequency analyses. One of these was able to show that neural oscillations in the delta band track the processing of abstract linguistic structures at different levels of hierarchy, such as phrases and sentences (Ding, Melloni, Zhang, Tian, and Poeppel, 2015). The data suggest that neural activation reflects the online mental construction of hierarchical linguistic structures, namely phrases and sentences. Additionally performed electrocorticography measures, which allowed a good localization of the respective processes, revealed the middle and posterior superior temporal gyrus bilaterally and the left inferior frontal gyrus to support phrase and sentence processing. This localization result complements functional magnetic resonance imaging study on phrase and sentence processing data discussed in chapter 1, indicating that BA 44, in particular, is responsible for hierarchy building (Zaccarella, Meyer, Makuuchi, and Friederici, 2015).

Another study investigated the processing of syntactically complex sentences (Meyer, Obleser, and Friederici, 2013) by systematically varying the verbal working memory demands required for processing. Increasing memory demands were found to increase alpha power over the left parietal cortex. This brain region had been identified in a prior functional magnetic resonance imaging study to support verbal working memory (Meyer, Obleser, Anwander, and Friederici, 2012; Meyer et al., 2013). An additional study varied the part of the sentence structure (embedded/non-embedded) from which a pronoun antecedent had to be retrieved in order to understand a sentence. Retrieval, constrained by syntactic structure, was reflected in a theta power increase located in the left frontal, left parietal, and bilateral inferior temporal cortices. Coherence analyses suggested a synchronicity of these regions during syntactic processing (Meyer, Grigutsch, Schmuck, Gaston, and Friederici, 2015).

The dominance of syntactic knowledge was demonstrated in a study on the interplay of prosodic phrasing and syntactic phrasing during auditory sentence comprehension (Meyer, Henry, Gaston, Schmuck, and Friederici, 2016). The study used ambiguous sentences, allowing for two alternative syntactic structures (and thereby meanings), one of which was preferred over the other as shown behaviorally. The ultimate interpretation of the ambiguous sentence was dependent on whether an identical word was followed by a prosodic phrase boundary or not. Analyzing delta-band oscillations, it was found that acoustics-based prosodic grouping was overridden by the internal syntactic preference for grouping words into syntactic phrases. This clearly indicates that top-down a syntactic bias can override acoustic cues during speech comprehension via delta-band oscillations.

When discussing oscillation-based studies an obvious issue concerns the relationship between different frequency bands and the underlying processes. It would be premature to provide an ultimate view on the relation between frequency band and process type during language processing. But the available data are in line with the view that local neural ensembles supporting syllable and word processing are related to gamma oscillations.

Retrieval of words from the lexicon or memory is associated with the beta and theta range. At the sentential level alpha, theta, and delta waves come into play depending on the processing demands. Memory demands are reflected by alpha and theta band oscillations, whereas delta band oscillations reflect top-down processes.

At this point the question arises about how the different processes, which during speech comprehension take place in parallel but in different time scales and frequency ranges, are coordinated and integrated. The time-varying "meta-ensembles," which represent the processing of syllables, words, and sentences that take place in different subsystems of the brain, would require a hierarchical nesting of ensembles. A solution to this problem may be offered by a recent report demonstrating the nesting of oscillatory pattern for phonemic and syllabic processes, reflected in a theta-gamma nesting (Giraud and Poeppel, 2012). These data provide a general view of how the dynamics within and between neural ensembles might be realized in the brain.

How to Encode Information for Transmission?

From the data reviewed here we learned that the functional connectivity and the synchronous oscillations for regions in the inferior frontal and temporal lobe are well documented. It appears that neural ensembles in the different brain regions exchange information. However, it is unknown how different types of information—be it semantic or syntactic—are coded.

We do know that electrical potentials are transmitted via the fiber tracts from one region to the next, but we do not know how the communicating regions encode and decode the information content. Here we can only speculate. Neurophysiological research has provided detailed results concerning the propagation of the electrical potential from one neuron to the next and along nerve fibers, but it is an open question how "content" is transferred. One hypothesis is that content—be it semantic meaning or syntactic structure—is represented by a number of neurons within a neuronal ensemble within which each neuron represents a particular feature. For example, for the word *bird* the respective semantic feature neurons could be *animate, non-human, has wings*, and so on. For transmission, the neurons within the neuronal ensemble of the decoding region activate simultaneously, and electrical potentials are propagated to the receiving regions in which neurons representing the same features are activated. Decoding of the "content" requires the activation of at least a number of the same features that represent the "content" in the encoding region. In the decoding regions these neurons have to be activated simultaneously in order to achieve a successful transfer.

The principle described here for language is borrowed from the principle of mirror neurons empirically evidenced in monkeys for the perception-action domain (Fabbri-Destro and Rizzolatti, 1996; Fogassi et al., 2005). It was found that when observing an action not only neurons in the visual system activate, but also neurons in the motor cortex activate although the monkey only views the action without performing it. Such mirror neurons have not been identified with respect to particular features of a given, visually presented action; one could speculate that the mirror ensembles of neurons representing the content or parts of it may be a general principle of effective communication between different brain regions in a network.

Here I propose for language that there are what I call *mirror neural ensembles* through which two distinct brain regions communicate with each other. To achieve successful communication at the neural level there must be a partial, but sufficient activation of neurons representing the same feature in the encoding and decoding region. This partial overlap of feature activation guarantees communication even under the biologically plausible condition that not all neurons representing a given concept are activated at the same time. A non-sufficient overlap of features can, however, lead to errors in understanding, which indeed sometimes occurs.

The underlying principle here is that communication between brain regions is possible by means of common coding by mirror neuronal ensembles as formulated above. To test this speculative idea, however, one would need advanced methods of electrophysiological recording and the possibility to identify respective neurons and mirror neurons in the involved brain regions. Such methods are not yet available, but may be in the future.

Summary

Functional connectivity analyses provide information about how different brain regions work together. They allow us to make statements about which regions work together, and moreover, about the direction of the information flow between these. Neural oscillations in addition allow us to specify the temporal domains in which local and larger neural circuits are active. In this section we reviewed studies suggesting a functional cooperation between the inferior frontal gyrus and the posterior temporal cortex. The available data are compatible with the structural connectivity data revealing strong white matter connections between these regions.

4.3 The Neural Language Circuit

In part I of the book I described the neural language network with respect to the different subprocesses that take place, from auditory perception up to sentence comprehension. The temporal time course of these subprocesses was identified and each subprocess was localized in the temporo-frontal network. In part II, I first presented the structural connections between the language-relevant brain regions and discussed the different ventral and dorsal fiber tracts with respect to their possible language-related function. My discussion of the available studies concerning the information flow between the language-relevant regions followed, taking into account these white matter fiber tracts. In this section, I offer a model of the cortical language circuit as a whole, taking into consideration relevant findings from the three chapters. The language circuit starts with bottom-up, input-driven processes and ends with top-down, controlled processes (see figure 4.5, below).[1] The bottom-up, input-driven processes proceed from the auditory cortex to the anterior superior temporal cortex and from there to the prefrontal cortex, whereas top-down, controlled, and predictive processes go from the prefrontal cortex back to the temporal cortex are proposed to constitute the cortical language circuit.

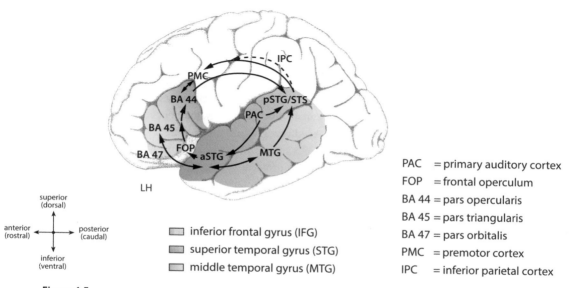

Figure 4.5
Dynamics within the language network. The cortical language circuit (schematic view of the left hemisphere). The major gyri involved in language processing are color-coded. In the frontal cortex, four language-related regions are labeled: three cytoarchitectonically defined Brodmann (Brodmann, 1909) areas (BA 47, 45, 44), the premotor cortex (PMC) and the ventrally located frontal operculum (FOP). In the temporal and parietal cortex the following regions are labeled: the primary auditory cortex (PAC), the anterior (a) and posterior (p) portions of the superior temporal gyrus (STG) and sulcus (STS), the middle temporal gyrus (MTG) and the inferior parietal cortex (IPC). The solid black lines schematically indicate the direct pathways between these regions. The broken black line indicates an indirect connection between the pSTG/STS and the PMC mediated by the IPC. The arrows indicate the assumed major direction of the information flow between these regions. During auditory sentence comprehension, information flow starts from PAC and proceeds from there to the anterior STG and via ventral connections to the frontal cortex. Back-projections from BA 45 to anterior STG and MTG via ventral connections are assumed to support top-down processes in the semantic domain, and the dorsal back-projection from BA 44 to posterior STG/STS to subserve top-down processes relevant for the assignment of grammatical relations. The dorsal pathway going from PAC via pSTG/STS to the PMC is assumed to support auditory-to-motor mapping. Furthermore, within the temporal cortex, anterior and posterior regions are connected via the inferior and middle longitudinal fasciculi, branches of which may allow information flow from and to the mid-MTG. Adapted from Friederici (2012). The cortical language circuit: From auditory perception to sentence comprehension. *Trends in Cognitive Sciences*, 16: 262–268, with permission from Elsevier.

Bottom-Up Processes in the Temporal Cortex: From Auditory Perception to Words and Phrases

The initial stage in the auditory language comprehension process is auditory perception. The acoustic-phonological analysis and the processing of phonemes are performed in the left middle portion of the superior temporal gyrus (Obleser, Zimmermann, Van Meter, and Rauschecker, 2007; Leaver and Rauschecker, 2010), lateral to Heschl's gyrus, which houses the primary auditory cortex. The processing of auditorily presented words is located in a region anterior to Heschl's gyrus in the left superior temporal gyrus. This region has been considered as the area where the processing of auditory word forms (i.e., a word's phonological form) takes place (Binder, Frost, Hammeke, Bellgowan, Kaufman, et al., 2000; Cohen, Jobert, Le Bihan, and Dehaene, 2004; DeWitt and Rauschecker, 2012). Neurophysiological evidence suggests that the recognition of a word form's lexical status (word vs. pseudoword) is ultrarapid (50–80 ms) (MacGregor, Pulvermüller, van Casteren, and Shtyrov, 2012) and so is the initial response to a word syntactic category error (40–90 ms) (Herrmann, Maess, Hahne, Schröder, and Friederici, 2011), which may be due to the recognition of a particular morphological word form. Magnetoencephalography source analyses reported that the very early word recognition effect is supported by left perisylvian sources and the right temporal lobe (MacGregor et al., 2012).

Once the phonological word form is identified, its syntactic information and semantic information must be retrieved in a subsequent step. Words belong to a particular word class and specific categories. The information about a word's syntactic category allows the initial construction of syntactic phrases. Based on neurophysiological data, these phrase structure building processes have been localized in the anterior superior temporal cortex approximately 120–150 ms after word category information is available as an early automatic syntactic process (Friederici, Meyer, and von Cramon, 2000; Shtyrov, Pulvermüller, Näätänen, and Ilmoniemi, 2003; Herrmann, Maess, Hahne, et al., 2011). The involvement of the anterior superior temporal gyrus has been reported in functional magnetic resonance imaging studies using syntactic violation paradigms (Friederici, Rüschemeyer, Hahne, and Fiebach, 2003), as well as a natural listening paradigm (Brennan et al., 2012). It has been proposed that, in the adult brain, these processes can be fast, since templates of different phrase structures (e.g., determiner phrase, prepositional phrase) represented in the anterior superior temporal gyrus/superior temporal sulcus are available automatically once the phrasal head (e.g., determiner, preposition) is encountered (Bornkessel and Schlesewsky, 2006). The observed posterior-to-anterior gradient from primary auditory cortex to anterior superior temporal gyrus/superior temporal sulcus going from phonemes to words and to phrases finds support in a recent meta-analysis (DeWitt and Rauschecker, 2012).

The processing of semantic information of single words is covered in a wide range of literature (Démonet, Thierry, and Cardebat, 2005; Patterson, Nestor, and Rogers, 2007). Here only lexical-semantic access and those integration processes as necessary for sentence comprehension will be considered. Lexical-semantic access occurs fast, that is, approximately 110–170 ms after the word recognition point, whereas the well-known N400 effect (350–400ms) is assumed to reflect controlled processes (Brown and Hagoort, 1993; Chwilla, Brown, and Hagoort, 1995; MacGregor et al., 2012). In sentential context, these lexical-semantic context effects, elicited by low cloze probability compared to high cloze probability words, are usually reported between 350 and 400 ms, starting at 200 ms (Binder, Desai, Graves, and Conant, 2009). In functional magnetic resonance imaging studies, lexical-semantic processes have mainly been observed in the middle temporal gyrus, although they do not seem to be confined to this region: they also include the association cortices in the left and right hemisphere (Binder et al., 2009). Semantic processes at the sentential level are more difficult to localize. They seem to involve the anterior temporal lobe, as well as the posterior temporal cortex and angular gyrus (Obleser, Wise, Dresner, and Scott, 2007; Lau, Phillips, and Poeppel, 2008). The particular function of the anterior and posterior brain regions in semantic processes is still a matter of debate (Patterson et al., 2007).

Since the anterior temporal lobe appears to be involved in processing syntactic information and semantic information at least at the sentential level, the function of this neuroanatomical region has been discussed as reflecting general combinatorial processes (Hickok and Poeppel, 2004, 2007; Humphries, Binder, Medler, and Liebenthal, 2006). These processes are a necessary step for later phrase structure building as well as in semantic processing. Thus, both syntactically relevant information and semantic information may be transferred from the anterior superior temporal gyrus to the inferior frontal cortex for higher-order computations.

From Temporal to Frontal Cortex: Toward Higher-Order Computation

Both syntactic and semantic processes involve the inferior frontal cortex, which can be subdivided cyto- and receptorarchitectonically into different subparts (Amunts, Schleicher, Morosan, Palomero-Gallagher, and Zilles, 1999; Amunts et al., 2010). Within the inferior frontal cortex the frontal operculum and pars opercularis (BA 44) appear to subserve syntactic processes, and the pars triangularis (BA 45) and pars orbitalis (BA 47) seem to support semantic processes. For further language processing, the information thus has to be transferred from the temporal cortex to the inferior frontal cortex where the next processing steps take place.

Concerning syntax, the system now has to deal with higher-order structural aspects in order to establish the grammatical relations between the different phrases, which are delivered by the anterior superior temporal gyrus, the frontal operculum, and the most ventral portion of BA 44. More recently, it has been demonstrated that the frontal operculum is involved in assembling words independent of any syntactic structure and that the most ventral portion of BA 44 comes into play when building syntactic hierarchies is required, even at the lowest level of hierarchy (Zaccarella and Friederici, 2015a). In the case of more complex hierarchies, that is, for sentences with a non-canonical surface structure (e.g., object-first sentence), reordering of phrasal arguments in the hierarchy must additionally be achieved. This process is supported by Broca's area in the inferior frontal gyrus (for a review, see Friederici (2011)). The studies on syntactic complexity reviewed in chapter 1 of this book indicated activation in BA 44 and in the posterior portion of BA 45. It appears that reordering of clearly marked phrases mainly involves the pars opercularis (BA 44), whereas the (re)computation of arguments that are moved from subordinate sentence parts recruits the posterior portion of BA 45 bordering BA 44.

Regarding sentential semantic aspects, the processing system now has to deal with the semantic and thematic fit between the different argument noun phrases and the verb. Semantic aspects in general activate more anterior portions of the inferior frontal gyrus, namely BA 47 and the anterior portion of BA 45, particularly when lexical processes are under strategic control (Rodd, Davis, and Johnsrude, 2005; Newman, Ikuta, and Burns, 2010) or when sentential semantic context is examined (Obleser, Wise, et al., 2007; Newman et al., 2010; Price, 2010).

In order to achieve these higher-order syntactic and semantic processes in the inferior frontal gyrus, the information on the basis of which these computations takes place must be transferred from the anterior temporal cortex to the inferior frontal cortex via structural connections (Hickok and Poeppel, 2004, 2007; Friederici, Bahlmann, Heim, Schubotz, and Anwander, 2006; Anwander, Tittgemeyer, von Cramon, Friederici, and Knösche, 2007; Friederici, 2009a; Weiller, Bormann, Saur, Musso, and Rijntjes, 2011). Two ventral fiber tracts connect the temporal and the frontal cortex: the uncinate fasciculus, which connects the more medio-ventrally located frontal operculum with the anterior temporal cortex and temporal pole, and the IFOF connecting the more laterally located BA 45 and BA 47 with the temporal and occipital cortex (Anwander et al., 2007; Weiller et al., 2011). Semantic information appears to be transferred from the temporal cortex to the anterior portion of the inferior frontal gyrus via the ventral pathway through the IFOF to BA 47 and BA 45, as indicated indirectly by combined functional magnetic resonance imaging and diffusion weighted magnetic resonance imaging (diffusion magnetic resonance imaging) studies, and most directly by patient studies (for reviews, see Hickok and Poeppel, 2004; Weiller et al., 2011). Syntactic information, however, seems to be transferred from the anterior superior temporal gyrus/superior temporal sulcus to the frontal operculum also via a ventral connection, as indicated by a combined functional magnetic resonance imaging/ diffusion magnetic resonance imaging study (Friederici et al., 2006a), and from there to BA 44, where higher-order syntactic computations take place (Grodzinsky, 2000; Friederici, Fiebach, Schlesewsky, Bornkessel, and von Cramon, 2006; Weiller, Musso, Rijntjes, and Saur, 2009). Based on these findings, the temporo-frontal network of syntactic processing can be modeled to initially involve the anterior superior temporal gyrus and the posterior portion of Broca's area (BA 44) mediated ventrally by the frontal operculum. The ventral system of semantic processing is assumed to involve the middle temporal gyrus, the anterior temporal lobe, and the anterior portion of Broca's area (BA 45).

There has been some discussion whether the entire region of the inferior frontal gyrus spanning from BA 47 to BA 44, a unification space, enables integration (Hagoort, 2005). Interestingly, empirical findings suggest that interaction and integration of semantic and syntactic information recruit not only the inferior frontal gyrus but also the posterior temporal cortex (Kuperberg et al., 2000; Friederici, Makuuchi, and Bahlmann, 2009; Snijders et al., 2009). Thus it remains an open issue whether each of the regions support integration processes and if so, whether they do this in concert in a highly interactive manner or not.

Top-Down Processes: From Inferior Frontal Cortex Back to Temporal Cortex
Here it is assumed that the posterior temporal cortex supports final semantic/syntactic integration in phrases and sentences (Bornkessel, Zysset, Friederici, von Cramon, and Schlesewsky, 2005; Newman et al., 2010; Friederici, 2011). This region must, therefore, receive input from BA 44 as the core syntax region (Kuperberg et al., 2000; Friederici, 2011) and from semantic regions (Vigneau et al., 2006), that is, either BA 45 or BA 47 (Upadhyay et al., 2008; Newman et al., 2010), the angular gyrus (Humphries et al., 2006, 2007), and the middle temporal gyrus (Vandenberghe, Nobre, and Price, 2002). Such an information flow is possible via existing structural pathways (see chapter 3).

During syntactic processing, the reported activations in BA 44 and the posterior temporal cortex in the comprehension of syntactically complex sentences (Bornkessel et al., 2005; Friederici et al., 2009) raise the question of how these regions are functionally related. It has been argued that BA 44 plays a particular role in creating argument hierarchies as a sentence is computed (Bornkessel and Schlesewsky, 2006; Friederici, Fiebach et al., 2006). A long-standing debate questions whether the processing of syntactically complex sentences requires support from working memory, in particular phonological working memory system located in the parietal cortex. Now, there is evidence that the syntactic (re)ordering of arguments in a sentence is located in BA 44, and that processing phonological working memory during sentence processing is located in the parietal cortex (Meyer et al., 2012). The necessary structural connection between these regions is provided by parts of the superior longitudinal fasciculus and the arcuate fasciculus (superior longitudinal fasciculus/arcuate fasciculus) (Makris et al., 2005; Meyer et al., 2012). An important function of the posterior inferior frontal gyrus might be to deliver syntactic and semantic predictions about the incoming information in a sentence to the temporal cortex, namely the posterior superior temporal gyrus/superior temporal sulcus.

Activation of the angular gyrus is reported for the sentence when semantic predictability of a word given the prior sentence context is high (Obleser, Wise, et al., 2007; Obleser and Kotz, 2010). The posterior superior temporal gyrus/superior temporal sulcus is activated when expectancy between a verb and its direct object-argument is low, thus when integration of a word is difficult (Obleser and Kotz, 2010). The posterior superior temporal gyrus/superior temporal sulcus, moreover, is often reported to covary with BA 44 (Obleser and Kotz, 2010), whereas the angular gyrus often covaries with the left anterior portion of the inferior frontal gyrus (BA 47) and with the left lateral and medial superior frontal gyri (Obleser, Wise, et al., 2007; Uddin et al., 2010). Moreover, at the level of word combinatorics, angular gyrus activation is reported together with the inferior frontal gyrus (BA 45/47) and the anterior temporal lobe (Bemis and Pylkkänen, 2013; Molinaro, Paz-Alonso, Duñabeitia, and Carreiras, 2015). Functionally this could mean that the semantic processes beyond the word level may partly rely on a semantic network that is based on the ventral pathway and on parts of the dorsal pathway (Obleser, Wise, et al., 2007). In contrast, syntax-based expectations of verb-argument relation may rather involve mainly the dorsal pathway (Obleser and Kotz, 2010). The finding that verb-argument processing is related to activation in the posterior superior temporal gyrus/superior temporal sulcus is in line with earlier studies (Friederici et al., 2003; Bornkessel et al., 2005). Up to now the literature, however, does not provide strong evidence as to whether the activation in these posterior regions is a result of information transfer within the temporal cortex or a result of information also provided by the inferior frontal gyrus. Structurally, it is possible that information can be transferred from the primary auditory cortex to both the anterior superior temporal gyrus and the posterior superior temporal gyrus, as there are two short-range fiber tracts within the superior temporal cortex: one going from the auditory cortex to the anterior superior temporal gyrus and one going from the auditory cortex to the posterior superior temporal gyrus. These tracts have been shown to be functionally relevant in auditory processing (Upadhyay et al., 2008).

The function of the information transfer from primary auditory cortex to the posterior superior temporal gyrus may be seen in the integration of bottom-up and top-down processes. Such top-down predictions on the basis of preceding syntactic information would not concern a particular word but rather a particular class of words (Friederici and Frisch, 2000). These predictions might be transferred top-down via the dorsal pathway connecting the posterior inferior frontal gyrus (BA 44) to the posterior temporal integration cortex, through the superior longitudinal fasciculus/arcuate fasciculus (Rodd et al., 2005; Friederici, Bahlmann, et al., 2006) either via a direct dorsal connection or an indirect dorsal connection mediated by the parietal cortex (Catani, Jones, and Ffytche, 2005).

Dynamic causal modeling of sentence processing has been helpful in determining whether the assumed information flow is valid. One Dynamic Causal Modeling study using data from an auditory sentence processing experiment (den Ouden et al., 2012) suggested a model with the inferior frontal gyrus as the input, where syntactic complexity modulated the flow of information from the inferior frontal gyrus to posterior superior temporal sulcus. Another Dynamic Causal Modeling study (Makuuchi and Friederici, 2013) employing data from a reading experiment varying the syntactic complexity of sentences suggested top-down information flow from the inferior frontal gyrus to the posterior cortex, as a direct functional connection from BA 44 and as an indirect connection from inferior frontal sulcus via the inferior parietal cortex.

The present model remains open with respect to how exactly semantic information is delivered to the posterior temporal cortex for integration. At least two processing streams are possible. First, if the assumption that the function of the anterior inferior frontal gyrus is to mediate top-down controlled semantic retrieval of lexical representations located in the middle temporal gyrus (Lau et al., 2008) is valid, then semantic information could be transferred from BA 47/45 via the ventral pathway through the ECFS to the posterior temporal cortex (Saur et al., 2008), with additional information collected from the lexical-semantic system in the middle temporal gyrus (Tyler and Marslen-Wilson, 2008) by the middle longitudinal fasciculus (Saur et al., 2008). Second, it is also possible that semantic information is processed in BA 47/45 and integrated with syntactic information from BA 44/45 in the inferior frontal gyrus (Hagoort, 2005). If this is the case, then it could be transferred from there—via the superior longitudinal fasciculus/arcuate fasciculus—to the angular gyrus and the posterior temporal cortex.

Summary

The language circuit modeled here is conceptualized as a dynamic temporo-frontal network with initial input-driven information processed bottom-up from the auditory cortex to the frontal cortex along the ventral pathway, with semantic information reaching the anterior inferior frontal gyrus, and syntactic information reaching the posterior inferior frontal gyrus. The anterior inferior frontal gyrus is assumed to mediate top-down controlled lexical-semantic access to the middle temporal gyrus and semantic predictions to the posterior temporal cortex via the parietal cortex. The posterior inferior frontal gyrus is assumed to support hierarchization of phrases and argument and to possibly mediate verb-argument related predictions via the dorsal pathway to the posterior temporal cortex, where integration of syntactic and semantic information takes place.

III

In the first two parts of the book I focused on the mature brain when describing the language network. In the third part I will put my focus on the relation between language acquisition and the brain. In the discussion of language acquisition that either takes place early in life for our native language or later in life for a foreign language, we will see that the age-of-acquisition is an important issue.

This has already been discussed since Eric Lenneberg (1967), who was the first to claim that there is a critical period for language acquisition. The critical time window during native language acquisition is still being debated and so is the critical time window for second language acquisition.

In the following two chapters we will see that language acquisition proceeds according to a fixed program independent of the modality of the language input, be it auditory (as in a typical language acquisition condition) or visual (as in a sign language condition). This holds as long as sign language is provided as input early. The acquisition of a second language proceeds native-like only when the age-of-acquisition is early.[1] I will review in chapter 5 the relevant data and discuss the issue of critical and sensitive periods of language acquisition. In chapter 6, which is devoted to the ontogeny of the language network, I will describe the relation between language development and brain maturation from birth to childhood and, at the end, propose a model of the ontogeny of the language network covering these data.

5

The Brain's Critical Period for Language Acquisition

Whether a critical period for language learning exists (or not) is one issue that interests many people who want to learn a foreign language. Learning a foreign language, as we know, is difficult later in life. Ongoing debates question why second language learning appears to be easy early in life but much more difficult as we age. We may reasonably assume this is due to neurobiological constraints that determine the time window of the brain's plasticity. In this context discussions often take place about a so-called critical period for language acquisition early on in life and, if there is one, when this period ends.

The first three years of life are indeed thought to represent a critical or sensitive period in which humans are especially open to learning a language. After this critical period, language acquisition is taken to be more difficult. Arguments in favor of a critical period for language acquisition are the relative ease of language learning in infants and children compared to adults learning a new language.

Our knowledge concerning the neurobiological underpinning of this behavioral fact in the language domain is sparse. For the sensory domain such critical periods have long been demonstrated in animal studies (Hubel and Wiesel, 1962). Only recently have animal studies at the molecular level begun to unravel the biological mechanisms that control the beginning and the closure of critical periods at the cellular level (Hensch, 2005). For human cognitive domains such as language, the biological mechanisms underlying sensitive periods remain a mystery. At present we can only describe these periods at the macrolevel.

In the language domain, Lenneberg (1967) was the first to propose a maturational model of language acquisition, suggesting a critical period for language acquisition that ends around the onset of puberty. Today researchers see the closure of the critical period, especially for syntax acquisition, either at or no later than age 6, whereas some claim that the critical period of native-like syntax acquisition is even earlier, around age 3 (Weber-Fox and Neville, 1996; Meisel, 2011). Neurophysiological and behavioral evidence shows that individuals who learned their second language within the first three years cannot be distinguished from native learners later in life—in their performance of semantic or syntactic tasks (Weber-Fox and Neville, 1996; figure 5.1). Those who learned their second language

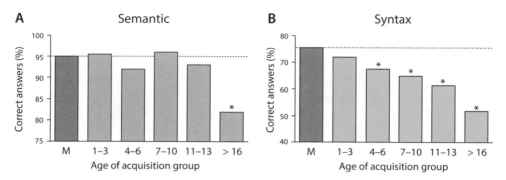

Figure 5.1
Behavioral language performance as a function of age-of-acquisition. Behavioral performance (% correct answers) on (A) semantic and (B) syntactic tasks for monolinguals (M) and for bilinguals with different ages-of-acquisition for the second language, in this case English (* indicates statistically significant difference from the monolingual group). Adapted from Rösler (2011). Psychophysiologie der Kognition: Eine Einführung in die Kognitive Neurowissenschaft. Heidelberg: Spektrum Akademischer Verlag; and from Weber-Fox and Neville (1996). Maturational constraints on functional specializations for language processing: ERP and behavioral evidence in bilingual speakers. *Journal of Cognitive Neuroscience*, 8 (3): 231–256. © 1996 by the Massachusetts Institute of Technology.

after puberty showed no significant difference to native learners later in life in semantic tasks, but they did perform worse in syntactic tasks. Individuals acquiring the second language after age 3 performed significantly worse in all the syntactic tasks (Meisel, 2011). These findings are in line with the assumption of critical, most sensitive time periods for language learning: (a) in the period from birth to 3 years of age language learning for a second language is native-like for both syntax and semantics, (b) after the age of 3 years acquisition is native-like for semantics, but already non-native-like for syntax, and (c) after puberty a second language does not seem to be acquired in a native-like manner. These findings suggest that the human brain is most responsive to language input during critical periods in the first years of life, be it one or two languages, and whether in an auditory language or a sign language.

5.1 Neurophysiology of Second Language Learning

The research on second language learning is relevant for the scientific debate on critical periods of language learning in general and, moreover, for educational decisions of when and how second language learning and language teaching is most effective. The solution of this issue is still open, as the available studies are mostly conducted in bilingual adults.

A general problem present in the research on bilingualism in adults is the fact that it is almost impossible to recover the individual history of language learning, although such a history may be crucial for the language proficiency and its neural representation later on in life. For a solid scientific analysis it would be important to know under what conditions,

The Brain's Critical Period for Language Acquisition

and when during development, the second language was learned, how intensively the second language was used, and what happened to the first years of life with respect to the language input. Therefore, although researchers have controlled for one or the other of these important factors, it was not always possible for them to document all factors of interest. Most studies on bilingualism focused either on the factor of *age of acquisition*, on the factor of *proficiency*, or both. Other studies instead investigated learning of an entirely novel language at adult age in order to be able to control the input directly. However, as discussed above, language learning in the first years of life may be different from language learning in adulthood. Therefore, studies on adult language learning have taken both approaches and provided interesting results on how the developing and adult brain accomplishes language learning.

Both electrophysiological and functional magnetic resonance imaging studies have investigated the brain basis of second language processing. These addressed when and under what conditions the brain is ready to learn a foreign language.[1]

The Time Course: ERP Studies

Electrophysiological studies conducted in bilingual adults have mostly used language materials that have led to clear ERP effects in monolinguals. These are the N400 for semantic violations and the ELAN–P600 effect for word category violations as well as the LAN–P600 for morphosyntactic violations (see chapter 1).

A pioneering study by Weber-Fox and Neville (1996) approached the question of a sensitive period in language acquisition by comparing the ERPs of adult bilinguals who had begun to acquire their second language (English) at different ages. ERPs recorded during sentence processing revealed a strong link between age of acquisition and syntactic processes: the processing of syntactic violations in sentences is affected if children begin their second language after age 3, while semantic violations are less affected by the age of acquisition. These electrophysiological findings are in line with those of behavioral studies in suggesting different critical periods for learning syntactic and semantic aspects of language.

Similar findings were reported in other ERP studies using different languages. In Japanese late learners of German, a native-like N400 was evident in response to semantically violated sentences, while components typically seen in response to syntactic phrase structure violations (i.e., the ELAN and the P600 component) were missing in late learners (Hahne and Friederici, 2001). Similarly, Russian late learners of German showed an N400 (although slightly delayed and diminished) in response to semantic violations and a delayed P600 in response to syntactic phrase structure violations, but no ELAN component (Hahne, 2001). Since the ELAN has been shown to reflect early automatic processes and the P600 is taken to reflect late controlled processes (Hahne and Friederici, 1999), the missing ELAN indicates less automatic syntactic processes in late learners. This conclusion is supported and extended by an ERP study with participants who differed in the age of acquisition of English, but were equal in their language proficiency (Pakulak and Neville, 2011). Late learners, in contrast to native English speakers, did not show a LAN, but only a P600. These findings suggest that the acquisition of the neural representation of syntactic processes falls within a sensitive period, independent of a later proficiency of language use.

This view, however, was challenged by a study demonstrating in principle that native-like ERP pattern can be elicited even when learning a novel language in adulthood. In this study adults were trained with an artificial phrase structure grammar which obeyed natural grammar rules (Friederici, Steinhauer, and Pfeifer, 2002). This grammar called BRO-CANTO, however, had only very few elements (words) in each syntactic category (noun, verb, determiner, adjective, adverb), in order to ease learning. Very high proficient (95% correct) users of this artificial language showed an early anterior negativity and a P600 to rule violations, whereas untrained users did not. This finding demonstrates that ERP patterns—at least under extremely high proficiency—can look similar but not identical to those observed in natives. In this study the early anterior negativity was not left lateralized, which raises the possibility that the underlying processes, although similar to native ones, may not be identical to these.

Investigating syntactic violations in high and low proficient German-Italian bilinguals, an early anterior negativity and P600 was found in highly proficient German-Italian bilinguals who learned the second language late (Rossi, Gugler, Friederici, and Hahne, 2006). Again in this study the early anterior negativity was not left lateralized. On the one hand, these results may challenge the view of an early and short critical period for the acquisition of syntactic processes. On the other hand, one has to bear in mind that the latter two studies showed a bilateral early anterior negativity and not a left lateralized ELAN/LAN often observed in native speakers, thereby suggesting similar but not identical syntactic processes in proficient bilinguals and in natives.

The data from the different second language learning ERP studies suggest that the processing of semantic violations as reflected in the N400 is less vulnerable to the factors age of acquisition and proficiency, whereas the processing of syntactic violations and its ERP reflection depends on various factors such as age of acquisition and level of proficiency.

Other ERP studies used the method to monitor foreign language learning in adults. One study reported a rapid plasticity of the brain at least for semantic aspects as compared to syntactic aspects (McLaughlin, Osterhout, and Kim, 2004). The authors observed a fast emergence of the N400 component in English learners of French. After only 14 hours of classroom instruction, the brain distinguished between words and pseudowords of the new language revealed by an N400 effect while a behavioral learning effect was first seen after 63 hours of instruction. Osterhout, McLaughlin, Pitkanen, Frenck-Mestre, and Molinaro (2006) propose that the development of morphosyntactic ERP components in the foreign language depends on the similarity between the native and the second language as well as the phonological realization of the relevant morphological structure. Their ERP data show that initially an N400-like response is evoked by morphosyntactic violations; yet, with increasing proficiency, this component is replaced by a P600. This ERP pattern might be due to the fact that at the beginning of the learning process the morphosyntactic structures are stored lexically, which are memorized as units rather than decomposed, and that rule-based processes only occur later during learning (Osterhout et al., 2006).

A principle difference between word learning (semantics) and rule learning (syntax) has been evidenced by an ERP study (De Diego Balaguer, Toro, Rodriguez-Fornells, and Bachoud-Levi, 2007) and functional magnetic resonance imaging studies (see below). In the ERP study, participants had to learn an artificial grammar. Changes in the ERP pattern were strongly correlated with the discovery of the structural rules underlying the trisyllabic words presented to the participants. Once the rules were learned, violations of these led to an early anterior negativity followed by a late positivity (P600), similar to the pattern observed by Friederici et al. (2002). Clear word-based learning effects were reflected in a modulation of the N400 amplitude (De Diego Balaguer et al., 2007).

The view that—at least in adults—initial language learning is word-based and should be reflected in an N400-like ERP component is supported by studies in which adults had to learn syntactic dependencies in a novel language (Mueller, Oberecker, and Friederici, 2009; Friederici, Mueller, Sehm, and Ragert, 2013). These studies revealed that second language learners display an N400 component for syntactic violations after short learning periods, but develop a native-like N400–P600 pattern after a longer continuous learning period (Citron, Oberecker, Friederici, and Mueller, 2011). This view, moreover, receives some support from a learning study using functional magnetic resonance imaging, which shows that during the initial learning phase an item-based learning brain system (the hippocampus) is involved, whereas later the syntax-related brain system (Broca's area) comes into play (Opitz and Friederici, 2007, see below).

Localization: fMRI Studies

In the context of second language learning, one particular question often springs to mind: Is the second language represented in the brain in the same way as the first language? When investigating second language acquisition experimentally in adults, again two major factors have to be considered: *age of acquisition* and *proficiency*. This is because language performance and brain activation are clearly a result of the interplay between these two factors. In their studies some but not all researchers varied these two factors separately.

The first functional magnetic resonance imaging studies on second language processing focused on whether we use the same or different brain systems when processing different languages. This study (Kim, Relkin, Lee, and Hirsch, 1997) found separate brain regions of the two languages during sentence generation in the left inferior frontal cortex if the second language was learned late, while the overlapping regions were reported for early bilinguals. However, other studies showed similar brain activation patterns in early and late bilinguals when controlling for proficiency, thereby highlighting the importance of the factors *proficiency* and *age of acquisition* when looking at the brain organization of second language learning (e.g., Chee, Tan, and Thiel, 1999; Perani et al., 1998).

Over the next few years a number of neuroimaging studies compared the brain activation for bilinguals and monolinguals. These studies did not systematically vary the age of acquisition or proficiency factors. One functional magnetic resonance imaging study monitored brain activation patterns in response to semantic and syntactically violating sentences in contrast to non-violating sentences in Russian-German late bilinguals and compared these to monolingual native Germans (Rüschemeyer, Fiebach, Kempe, and Friederici, 2005). While for the semantic condition, no activation difference, but an activation overlap was found between groups, an increased activation for bilinguals compared to monolinguals in the left inferior frontal gyrus was observed for the syntactic condition, suggesting a stronger recruitment of this region in bilinguals. Another study presenting correct sentences with different syntactic complexity also found significantly greater activation in the left inferior frontal gyrus (Broca's area) for Spanish-English bilinguals than for English monolinguals (Kovelman, Baker, and Petitto, 2008). These studies suggest that bilinguals show more activation than monolinguals in particular in the left inferior gyrus when processing aspects of syntax.

A more recent study, however, did not directly confirm this conclusion with respect to the activation in the left inferior frontal gyrus. This study used two different tasks—a semantic judgment and a grammaticality judgment task—in a comprehension experiment with highly proficient Catalan-Spanish bilinguals and Spanish monolinguals (Román et al., 2015). During both tasks bilinguals displayed more brain activity than monolinguals in the left temporal cortex, but no activation in the left inferior frontal gyrus. The difference with respect to the inferior frontal gyrus findings between these two studies, however, may be explained by the age of acquisition factor, as indicated by Jasinska and Petitto's study (2013). These authors found more inferior frontal gyrus activation for Spanish-English late bilinguals (second language acquired after age 4) compared to early bilinguals (second language acquired at birth) and monolinguals. Thus it appears that the brain activation of those bilinguals who learned the second language early during life is more similar to that of monolinguals than the brain activation of those who learned the second language later in life.

In addition to the age of acquisition factor, it may be of relevance how typologically "close" the two different languages are in the investigated bilinguals. Catalan and Spanish, the languages investigated in the study by Román et al., (2015), have more overlap in the lexical domain than English and Spanish, which may have led to more interference between the languages in the former case. Controlled processes may thus be needed to avoid interference with the other language. This could explain why highly proficient Catalan-Spanish bilinguals demonstrate more activation in the inferior frontal gyrus than English-Spanish bilinguals when compared to monolinguals.

Proficiency is another crucial factor in studies on bilingualism, as it is quite rare that a person is more proficient in his or her second language than in the first language. For all other cases, most studies report a negative relationship of proficiency and brain activation; that is to say, the higher the level of proficiency (i.e., the more native-like), the lower the brain activation to process the cognitively more demanding second language. This was shown in fronto-temporal brain regions for a number of different tasks, for word generation tasks, and also for semantic or prosodic matching tasks, as well as sentence comprehension tasks (for review, see Abutalebi, 2008). The negative relation between proficiency level and brain activation or alternatively the positive relation between task demands and brain activation is known from other linguistic tasks: the more difficult or complex a task, the greater the functional magnetic resonance imaging signals.

A critical point is that some bilingual studies did not control equally for language proficiency and the age of acquisition. This means that often a highly proficient first language was compared to a less proficient second language that was also acquired later. A well-designed functional magnetic resonance imaging study tried to disentangle the influence of age of acquisition and proficiency level (Wartenburger et al., 2003). This study found that the functional magnetic resonance imaging activation during lexical-semantic processing is dependent on the proficiency level: greater activation in the left inferior and right middle frontal cortex was observed for low versus high proficient bilinguals who learnt the second language late. Syntactic processing, in contrast, is more dependent on the age of acquisition. For the syntactic condition, greater activation in bilateral inferior frontal cortex was found in late highly proficient versus early highly proficient bilinguals as measured in Italian-German bilinguals during a grammatical judgment task. The group that learned the second language late showed larger activation differences for the second language than for the first language. The group that learned the second language early did not show such an activation difference. Although both factors, age of acquisition and proficiency, are relevant for the brain activation of a second language, it can be concluded from this study that the age of acquisition (early/late) is more important for a neural similarity of the first and second language than the level of proficiency (high/low).

Similarly, when processing irregular versus regular syntactic morphology at the word level, Broca's region in the left inferior frontal cortex is recruited more strongly by late bilinguals as compared to early bilinguals with comparable proficiency, indicating again that in late bilinguals neural resources are higher for syntactic processing than in early bilinguals (Hernandez, Hofmann, and Kotz, 2007). Greater activation in left inferior frontal regions was also seen in late highly proficient bilinguals when processing non-canonical sentences in their second versus their first language, while equally proficient early bilinguals showed overlapping brain activation patterns. Moreover, it was demonstrated that the observed activation patterns for second language processing depend on the similarity of the syntactic features in the first and the second language as well as on the sequence in which the two languages were acquired (Saur et al., 2009; see also Kovelman et al., 2008). One study (Bloch et al., 2009) even investigated the impact of age of acquisition of the second language on learning a third language. It seems that the age of acquisition of the second language is a crucial factor even in this investigation. While subjects who learned their second language early showed low variability in the activation pattern during language production, those who learned their second language later showed a high variability in the language network involved. These data suggest a certain separation of the languages in late learners and a shared network if the second language is learned early.

Another study investigated the impact of the age of acquisition on the intrinsic functional connectivity as measured during resting-state functional magnetic resonance imaging in two groups (Berken, Chai, Chen, Gracco, and Klein, 2016). One group had acquired French and English simultaneously from birth, whereas the other group had learned the two languages sequentially, learning one of the languages first and the other language after age 5. When seeding in the left inferior frontal gyrus, a stronger functional connectivity was found for the simultaneous bilinguals between the left inferior frontal gyrus and the right inferior frontal gyrus, as well as between the left inferior frontal gyrus and some brain areas involved in language control including the dorsolateral prefrontal cortex, inferior parietal lobe, and the cerebellum. In sequential bilinguals this functional connectivity significantly correlated with the age of second language acquisition; the earlier the second language was acquired, the stronger the connection. This functional connectivity underscores the importance of the left inferior frontal gyrus with its intrinsic functional connectivities for native language learning, and, moreover, the importance of learning a second language early.

With respect to the phonological level, an interesting functional magnetic resonance imaging study suggests that some aspects of the phonological representation of language learned at an early age are maintained over time even without continued input (Pierce, Klein, Chen, Delcenserie, and Genessee, 2014). Children aged 9–17 years who were adopted from China as infants by the age of 13 months, but after that were exclusively exposed to French, were asked to perform a Chinese lexical tone task. The brain activation pattern was very similar to those of early Chinese-French bilinguals and differed from those of French monolinguals. Both adopted individuals, who after the age of 13 months never heard the Chinese language again, showed that some brain activation when performing a lexical tone task as those that had learned Chinese and French early. Both adopted individuals and bilinguals recruited the left superior temporal gyrus, whereas French monolinguals activated the right superior temporal gyrus. The involvement of the left hemisphere for the processing of lexical tone in tone languages such as Mandarin Chinese and Thai for natives in contrast to non-natives has long been reported in the literature (Gandour, Wong, Hsieh, Weinzapfel, Van Lancker, and Hutchins, 2000; Klein, Zatorre, Milner, and Zhao, 2001; Gandour et al., 2002; Wong, Parsons, Martinez, and Diehl, 2004). The similarity between the adoptees and the early bilinguals indicates that the early acquired phonology of lexical tone and its representation is maintained in the brain even when this information is no longer consciously available (Pierce et al., 2014).

There is an interesting bilingual study of two languages that differ with respect to whether the parameter of tonal information is lexically relevant or not, here Chinese as a tonal language and English as a non-tonal language (Ge et al., 2015). Measuring functional magnetic resonance imaging and conducting a functional connectivity analysis by means of Dynamic Causal Modeling, the authors found that although speech processing in both languages was supported by a common fronto-temporal network, differences were present in their functional connectivity patterns with a higher interaction between the right and left anterior temporal gyrus cortex for Chinese as a tone language compared to English.

Another approach to language learning is to monitor language learning for a completely novel language that is constructed according to particular syntactic or word-based aspects. One functional magnetic resonance imaging study used an artificial grammar obeying rules of a natural grammar (possible grammar) (similar to BROCANTO used in the ERP study by Friederici et al., 2002 discussed above). In this study we found different activation patterns for rule-based learning and word-based learning (Opitz and Friederici, 2004). After an initial training phase, participants were confronted with two types of violation, here changes: a rule-change and a word-change. Brain activity was scanned during the learning of these novel (changed) forms. Word-change led to a clear modulation of the left anterior hippocampus, a system known to be crucial for item learning. Rule-change led to an increased activation in the left ventral premotor cortex slightly posterior to BA 44. Rule-dependent activation in this area is in line with other studies that reported similar activation patterns for the acquisition of possible compared to impossible grammar rules (Tettamanti et al., 2002), for local rule violations in a natural language (Friederici, Rüschemeyer, Hahne, and Fiebach, 2003), and for processes at the first level of hierarchy building in processing natural language and mathematical formulae (Makuuchi, Bahlmann, and Friederici, 2012). The study by Opitz and Friederici (2004) suggests that during second language learning there is an initial phase of word-based learning that is followed by a phase of rule-based learning.

In a first study on pure syntax learning, Tettamanti and colleagues (2001) asked a simple question: Is there a distinction between learning grammatical (possible) and non-grammatical (impossible) rules, that is, rules that never occur in a natural language? Italian monolingual adults were requested to learn the novel possible and impossible rules in sequences constructed of Italian words. A possible rule, for example, was the following: "The article immediately follows the noun to which it refers." This is not a rule in Italian, but it is a possible rule in other languages. An impossible rule, however—"The article immediately follows the second word in the sentence"—was a rule that does not occur in any natural language. Possible rules specifically activated a left hemispheric network with Broca's area (BA 44), showing the crucial difference in activation between the two rule types. In a second study Italian monolinguals had to learn a new language that consisted of Japanese words and possible as well as impossible rules (Musso et al., 2003). Again, the learning of possible rules revealed activation in Broca's area, but not the learning of impossible rules.

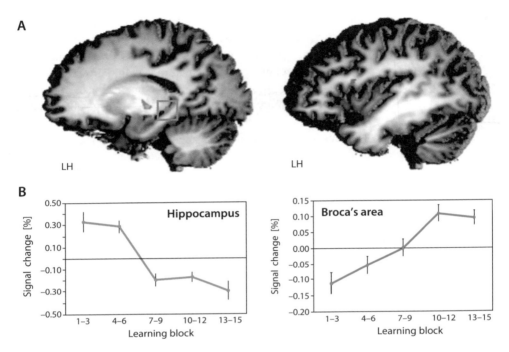

Figure 5.2
Brain activation during artificial grammar learning. The grammar to be learned called BROCANTO conforms to rules of a natural grammar. (A) Location of brain activation in the hippocampus (green) and Broca's area (red). (B) Signal change in the two brain regions plotted as a function of learning over time (from learning across the learning blocks from 1 to 15). Activation in the hippocampus decreases and activation in Broca's area increases as the grammar is learned. Adapted from Opitz and Friederici (2003). Interactions of the hippocampal system and the prefrontal cortex in learning language-like rules. *NeuroImage*, 19 (4): 1730–1737, with permission from Elsevier.

Another functional magnetic resonance imaging study monitored the process of language learning, for a possible language BROCANTO, while the learner is in the scanner (Opitz and Friederici, 2003). In this study we monitored brain activity during the learning process itself which lasted approximately one hour. While the initial learning process showed a high level of activation in the hippocampus and low activation in Broca's area, the later learning stage was characterized by an increase in activation in Broca's area and a decrease in hippocampal activation (see figure 5.2). These data suggest that once an initial phase of hippocampal-based item learning has passed, the frontal system known to support procedural processes comes into play. The exact time course of this transition for natural language learning remains to be seen. Neurophysiological data suggest that this may take months (Osterhout et al., 2006) or years (Hahne, 2001; Rossi et al., 2006).

Summary

All the studies on second language learning reviewed here suggest that a unitary neural system with largely overlapping brain regions is involved when processing more than one language. The activation strength of this network for second language processing depends on the age of acquisition, the proficiency level, and similarities between the languages. Neurophysiological and brain imaging evidence suggests that the age of acquisition dominantly affects syntax processing in the bilingual brain. The earlier a second language is learned, the more similar the neural language networks for the two languages, and this is most crucial for syntax processing. The available data suggest that language learning may be constrained by critical periods for the learning of particular aspects of language.

5.2 Critical and Sensitive Periods of Learning: Facts and Speculations

Critical periods can clearly be observed in various perceptual sensory systems, and during these periods a system is open and most responsive to the input of the environment. It is thought that the system and its neurobiological basis are dependent on the maturation of the neural circuits responsible for the particular perceptual or cognitive domain (Werker and Hensch, 2015).

The term *critical period* refers to a strict time window during which input is essential for normal functional development (Wiesel, 1982), whereas the term *sensitive period* refers to a relatively limited time window during which input is most effective—as demonstrated for cognitive functions such as language (Neville and Bavelier, 1998). There is preliminary evidence that such critical or sensitive periods are associated with functional and structural changes in the brain, and at the network level of connected brain regions. To what extent direct biological brain maturation and environmental experience are mutually dependent during a child's cognitive development can be particularly well illustrated by the example of language. At the beginning, the human language processing system is completely open, and it must be, because every child learns the language(s) into which he or she is born. Children can learn a second language without any problem until age 3. If they start learning the second language after age 3, however, they will not learn the syntax of the second language in a native-like manner (Meisel, 2011). In this case the neural processing mechanisms are different (Weber-Fox and Neville, 1996). Even if proficiency of the second language reaches a high level, the neural basis deviates from that of the native language. Semantic aspects can be learned quite efficiently even after age 3, but the language learning ability deteriorates after puberty.

Interestingly, puberty is precisely the point during development at which in humans the brain maturation within its large structures seems to have come to a relative end. The gray matter—that is, a major component of the central nervous system—continuously decreases from birth and remains relatively constant after age 10. The white matter—at least the long-range fiber bundles that connect brain areas—increases continuously from birth and then increases at a slower rate after age 10 over the next few years. Initial studies show that both the decrease in gray matter (Fengler, Meyer, and Friederici, 2016) and the increase in the functionality of the fiber tracts by means of fractional anisotropy highly correlate and define the sensitive period with the development of language skills (especially syntax) (Skeide, Brauer, and Friederici, 2016). It is still unclear how this decrease in gray matter and an increase in white matter in the brain are related, and whether this is a fundamental principle of plasticity and learning. It is known, however, that in humans the brain maturation determines the optimal acquisition of a language in its sensitive phase.

In the domain of first language learning it has been shown that the maturation of the white matter fiber bundles connecting language-relevant brain regions in the inferior frontal gyrus and the posterior superior temporal cortex predicts language development (Skeide et al., 2016). In particular, the late maturing fiber tract connecting BA 44 and the posterior temporal cortex predicts comprehension performance on syntactically complex sentences. Moreover, it appears that the functional brain activation for syntax processing only gradually shifts toward the adult syntax region BA 44, reaching this location only after age 10 (Skeide et al., 2014). Partial correlation analyses indicate that both activation in BA 44 together with the posterior superior temporal gyrus / superior temporal sulcus and the white matter maturity (i.e., fiber density, axonal diameter, and myelination) predict the correctness of sentence comprehension, but that only the myelin strength additionally predicts the speed of sentence comprehension (Skeide et al., 2016). These data are discussed in more detail in chapter 6. Given that myelin surrounding the white matter fiber tracts is responsible for the propagation of the electrical signal, it is not surprising that the strength of the myelin is also correlated with the speed of sentence comprehension.

The effects of myelination are summarized in a recent article on myelin imaging (Turner, 2015). It has been shown that myelination of axons accelerates the conduction speed of action potentials by an order of magnitude (Waxman, 1980), and thereby leads to a decrease of energy cost of transmission (Harris and Attwell, 2012). Myelin increase also has effects on the neuronal structure: it reduces synaptic plasticity by inhibiting neurite growth (Ng, Cartel, Roder, Roach, and Lozano, 1996; McGee, Yang, Fischer, Daw, and Strittmatter, 2005). It has been speculated (Klaus Nave, personal communication) that during neuronal development the increase of myelin with age has two functions: to enhance propagation in the established network, and to inhibit sprouting of neurons in the target region. Under this view, learning as such should work best in the absence of myelin, whereas the establishment of what has been learned would require myelin. Thus, if the myelin is low as in the immature brain, the brain should be open and plastic for novel input and learning should be easy. If, in contrast, myelin is high, as in the adult system for most of the fiber bundles, then learning new things should be more difficult. This has been observed in general, and for language learning in particular. In the case of language, this means that once one language is learned and the white matter is in place and structurally fixed, learning a new language, which requires different processes, becomes more difficult.

Moreover, the direct relation between function and white matter structure during learning allows an additional prediction: if the growth of the white matter is dependent on the particular processes taking place during learning, we would expect the adult white matter structures of those who have learned different languages to be different. In the next section we will see that this is indeed the case.

Summary

There appears to be a close relation between the developmental trajectory of white matter maturation and behavioral language skills. Learning as discussed here may depend on the interaction between gray matter and the myelin strength of the white matter. Future research will have to show whether this idea holds.

5.3 Universality of the Neural Language Network

Sign language is the perfect testing ground to evaluate language as a representational and neural system independent of speech as the input or output system. Ferdinand de Saussure (1857–1913) made the original fundamental distinction between language and speech using three French terms—*language*, *langue*, and *parole*. According to de Saussure, *language* refers to human language, *langue* refers to a specific language such as French or German, and *parole* refers to speech. He proposed this distinction in *Cours de linguistique générale* (1916), which was not published until after his death. Today the major distinction between the three terms no longer hold up, partly because there is the idea that all languages are based on principles that govern all languages, such as Merge discussed in the introduction. However, there is the strong divide between language and speech ("parole"). The distinction between language as the core language system and speech as the input/output system is one of large theoretical importance as it implies that language, as a cognitive system in the human mind, is independent of its modality-specific realization (Berwick, Friederici, Chomsky, and Bolhuis, 2013).

Sign Language Studies

The empirical evidence for such a claim comes from studies on sign language, a language which provides an ideal testing ground because the input/output modalities are totally different from that of spoken language. Spoken language is expressed by the vocal system and perceived by the auditory system. Sign language, in contrast, is expressed by hand and body articulations and perceived by the visual system. Intensive analyses revealed that sign languages have the full range of complexity just like spoken languages. And as is in the case of spoken languages, there are several different sign languages. There is an American Sign Language, a German Sign Language, a French Quebec Sign Language, a Taiwan Sign Language, as well as other sign languages. The fact that users of different sign languages cannot necessarily understand each other is a first indication that sign languages should not be mistaken as consisting of iconic gestures. Sign languages have abstract phonologies, lexicons, and syntaxes of their own.

A first compelling proof that the relation between language and the brain is independent of modality was provided by case reports of deaf native signers who suffered from stroke-induced damage of the left hemisphere leading to language deficits (Poizner, Klima, and Bellugi 1987; Bellugi, Poizner, and Klima, 1989).[2] Depending on the location of the lesion, different language deficits were observed: lesions in the left inferior frontal cortex involving Broca's area caused non-fluent effortful production of sign language, with a simplified sentence structure sometimes being agrammatic (Poizner et al., 1987; Hickok, Kritchevsky, Bellugi, and Klima, 1996). Thus, the patterns in the production of sign language in these patients who had learned sign language early in their lives were very similar to the pattern observed in the spoken output of hearing patients with lesions in the inferior frontal cortex. A similar parallelism between signed and spoken language was reported for deaf and hearing patients after stroke-induced lesions in so-called Wernicke's area in the temporal lobe and partly also including the inferior parietal cortex (Chiarello, Knight, and Mandel, 1982; Poizner et al., 1987). These patients showed fluent signing, but with paraphasia and grammatically inappropriate sign sequences similar to paragrammatism in spoken language.

Nowadays, sign language studies cover a number of language aspects that have been already evaluated for spoken language, including language development and second language learning as well as the brain basis of semantic and syntactic processes.

Developmental sign language studies can tell us something about the biologically fixed developmental program for language independent of the input modality. It has been shown that infant babbling is not only linked to speech, but also occurs in sign language reflected in the infant's hand movement (Petitto and Marentette, 1991; Petitto, Holowka, Sergio, and Ostry, 2001). In a first study, two American deaf infants born to deaf parents were compared to three American hearing infants born to hearing parents in their manual activities between 10 and 14 months of age. Deaf infants used hand shapes when babbling in order to represent phonetic units of the American Sign Language. Although hearing infants and deaf infants produced a similar amount of hand movements, only deaf infants produced manual babbling movement (Petitto and Marentette, 1991). In a second study, two groups of hearing babies were compared: babies from one group were born to deaf parents, and babies from the other group were born to hearing parents. Recording and analyzing these babies' hand movements between 6 and 12 months of age revealed that only infants born to deaf parents showed linguistic hand movements (Petitto et al., 2000). These findings indicate that language develops in a biologically based universal manner. Crucially, the data indicate that language and language development are definitely not dependent on speech (*parole*).

At the behavioral level, categorical perception has been identified as a key perceptual principle in spoken language. It refers to the finding that speech stimuli are perceived categorically despite variations in acoustic realization, be it due to different speakers or different contexts in which a given phoneme appears. Experimentally, categorical perception is evidenced when continuously varying stimuli (from *ba* to *pa*) are perceived as discrete categories and when discrimination performance for adjacent stimuli in the continuum is better across a category boundary than within a category (Liberman, Harris, Hoffman, and Griffith, 1957). Now there is evidence for categorical perception of distinctive hand configurations in American Sign Language (Emmorey, McCullough, and Brentari, 2003). Only native deaf signers (and not hearing controls) showed evidence of categorical perception of distinctive hand configurations. From this it may be inferred that categorical perception of language-relevant information develops regardless of language modality.

At the neurophysiological level, an early ERP study investigated sign language processing in deaf signers who had learned American Sign Language from their deaf parents, in hearing native signers who were born to deaf parents learning sign language as their first language, and in hearing non-signers. Semantically normal and semantically anomalous sign language sentences were presented. The results suggested that the organization of the neural systems supporting language in the left hemisphere for sign language is similar to spoken language and thereby independent of the input modality (Neville et al., 1997). The well-known language-related ERP components—the N400 and the left anterior negativity (LAN) (compare sections 1.5 and 1.6) indicating semantic processes and syntactic processes, respectively—have also been reported for native signers. An N400 effect was reported for sentences with a semantically anomalous last word compared to those with a semantically appropriate sentence final word (Neville et al., 1997). Another ERP study reported a LAN effect for syntactic verb-agreement violations in American Sign Language similar to the effect observed in spoken language (Capek et al., 2009). These findings again indicate that sign language is processed similarly to spoken language once learned as a native language from birth.

Brain imaging studies on sign language tried to identify the neural network underlying the processing of sign language. Would the brain areas be the same or would they be different from those identified for spoken languages? Functional magnetic resonance imaging studies of deaf adults with sign language as their native language were tested viewing sentences in sign languages (Bavelier, Corina, and Neville, 1998; Neville et al., 1998). Activation was reported within the left-hemispheric language network including the left dorsolateral prefrontal cortex, the inferior precentral sulcus, and the left anterior superior temporal sulcus. However, activation was also found in the homologous areas in the right hemisphere. Another functional magnetic resonance imaging study used two different sign languages (American and French Quebec) in different perception tasks, an imitation and a verb generation task. The left inferior frontal gyrus was activated when signers generated meaningful signs and the planum temporale in the superior temporal gyrus was activated bilaterally when viewing signs or meaningful parts of signs (Petitto et al., 2000). This activation pattern was left hemispherically dominant similar to hearing participants in comparable experimental conditions. Both regions, the inferior frontal gyrus and the superior temporal gyrus with the planum temporale, are therefore seen as language areas that are independent of the input–output modality.

These and other brain imaging studies consistently report left hemispheric activations for sign language similar to those found for spoken language. These activations include those brain regions that are part of the neural language network identified on the basis of studies on auditory and written language input. These activations are found in the inferior frontal gyrus including Broca's area, the precentral sulcus, superior and middle temporal regions, and the angular gyrus (Newman, Bavelier, Corina, Jezzard, and Neville, 2002; MacSweeney, Capek, Campbell, and Woll, 2008).

One study contrasted brain activation and brain structural patterns in native signers and hearing non-signers naive to signing by using two different types of neuroscientific analyses (Meyer, Toepel, Keller, Nussbaumer, Zysset, and Friederici, 2007). In an analysis that takes into account the individual gross anatomical landmarks it was shown that only native signers activate the perisylvian language areas when viewing sign language. Hearing participants, in contrast, activate the visual cortex. The functional language network seen for sign language in native signers is generally less lateralized than that for spoken language in hearing individuals. Moreover, an analysis of the white matter structure shows a white matter reduction in the posterior portion of the left superior longitudinal fasciculus and the left uncinate fasciculus in signers compared to hearing non-signers. These findings were taken to imply that the neuroanatomy of deaf individuals displays a neural network similar to hearing individuals, but that this network undergoes some minor structural changes as a function of sign language input. This is an interesting conclusion since it suggests that a biologically determined universal language network is modulated by the particular language input during development.

The absence of oral speech in deaf signers who learnt sign language early offers the unique opportunity to disentangle speech from language circuits in the human brain. In newborns, the language network is not yet fully developed and we asked to what extent the connections within the language network might be modulated by the modality of the language used during childhood. To test this, a study measured diffusion-weighted MRI to reconstruct neural fiber tracts and cortical networks (Finkl et al., submitted). Quantitative comparison of the connectivity maps of 10 adult prelingual deaf signers with those of a matched hearing control group confirmed that all major language tracts within the network were present in both groups and did not differ qualitatively between signers and controls (see figure 5.3). However, pathways involved in the decoding and production of oral speech were weaker in the deaf group. These results provide structural evidence for the general concept of modality independence of the language network and separate this network from input and output pathways necessary for speaking and auditory decoding.

This conclusion is in line with the data of a more recent study on the structural language network in different languages, which found a similar structure network in all languages that

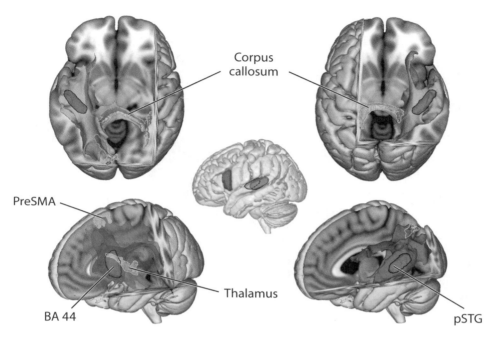

Figure 5.3
Speech and language network in deaf signers. The figure displays group differences between deaf signers and controls. Top: Connectivity profile of the primary auditory cortex (light blue, HG = Heschl's gyrus) with significant reduced interhemispheric connectivity (yellow) for left and right hemisphere, respectively. Bottom: Connectivity of left BA 44 showing the preserved syntax processing network (purple) and reduced connectivity to the thalamus and the preSMA (yellow, left) as well as to the posterior temporal cortex (pSTG) in the right hemisphere (red). Adapted from Finkl et al (submitted). Differential modulation of the brain networks for speech and language in deaf signers.

was modulated by the respective language used. In this study we investigated whether three languages with different processing demands—Chinese, English, and German—shape the structural connectivity within the language network differentially (Goucha, Anwander, and Friederici, submitted). It compared the connectivity of key cortical regions for language processing using tractography and found the general language network to be present in all languages, but differences across groups corresponding to the specific demands of these languages. German, with its predominant morphosyntax, showed higher fronto-temporal connectivity in the dorsal pathway, associated with the processing of sentence structure, whereas English, with more salient semantically guided processing, showed higher connectivity in the ventral pathway, associated with the processing of language content. In turn, Chinese, with its distinct phonology of tones and frequent homophones, showed enhanced connectivity in temporo-parietal and interhemispheric circuits, related to storage and discrimination of speech sounds. Crucially, these results highlight how the acquisition of a given language can shape brain structure during life long language use (see figure 5.4).

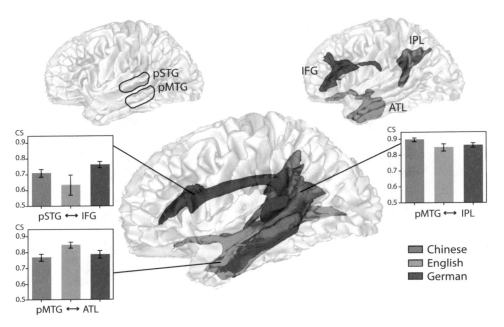

Figure 5.4
White matter fiber tracts in natives of different languages. Chinese (red), English (orange), and German (purple). Boxes display the strength of the fiber tracts (myelin) of the respected tracts between language-relevant region: pSTG to IFG, pMTG to anterior temporal lobe (ALT) and pMTG to inferior parietal lobe (IPL). Abbreviations: STG = superior temporal gyrus, MTG = middle temporal gyrus, ATL = anterior temporal lobe, and IPL = inferior parietal lobe. From Goucha, Anwander, and Friederici (submitted). Language shapes the brain: Cross-linguistic differences reflected in the structural connectivity of the language network.

These data provide impressive evidence for a general, possibly universal structural language network modulated by particular language inputs.

Summary

A comparison between brain activations of processing sign language in native signers to those reported for the processing of spoken language in hearing individuals reveals a large similarity in the ERP patterns and in the brain activation patterns. The brain regions that constitute the language network for sign language are also quite similar to those found for processing language in general, but reveal minor changes as well. Moreover, the general structural network known for hearing individuals was also found in prelinguistically deaf signers. This structural network was also observed for individuals coming from very different language backgrounds, although with a modulatory impact of the lifelong use of a specific language. These data point toward a universal neural language system largely independent of the input modality, even though it can be modulated slightly by the lifelong use of a given language.

6

Ontogeny of the Neural Language Network

Language seemingly starts with a newborn's first cry. But in fact, language actually starts before birth, since a fetus is able to perceive acoustic information in utero. The acoustic information perceivable for the fetus in the uterus is filtered by the surrounding water and tissue. The auditory information arrives at the fetus filtered by about 400 Hz, providing prosodic information but no phonetics details. Thus, the fetus already learns prosodic patterns of the mother tongue. Research findings, which indicate that a baby's first cry is language specific, support this view about language learning before birth. For example, French babies cry with a melody that is different from German babies, each mirroring the speech melody of their mother tongue (Mampe, Friederici, Christophe, and Wermke, 2009).

Language develops as a function of the language input, which differs considerably across the world. A normally developing child is even able to acquire those languages we as foreigners might find very difficult. This holds as long as adequate language input is provided. Language does not develop normally once the language input is missing, as in "abandoned children" (Curtis, 1977). In addition to language input, the child must be equipped with a system that allows language acquisition. It is assumed that this is a biological system (Chomsky, 1986) that follows a fixed series of developmental steps, as these are reported for each child independent of their mother tongue (van der Lely and Pinker, 2014). There are hundreds of behavioral studies on language acquisition that were published over the past 70 years. *The Language Instinct*, the book by Steven Pinker (1994), provides a sophisticated view of the available language acquisition literature at the time and discusses the findings in relation to Chomsky's view of a universal grammar (1986). *Language Acquisition* by Maria Teresa Guasti (2002) covers more recent studies on the acquisition of the different aspects of language from birth to the age of 6 years. I refer the reader to these books to learn about the linguistic and psycholinguistic aspects of language acquisition.

It is interesting to note that although there are some variations in the speed of language acquisition across individuals, the sequence of the different developmental phases is invariant. Even children with some developmental delay in language learning usually do not demonstrate language impairments after puberty.

In rare cases, however, the child does not develop language normally, despite adequate language input. This developmental disorder is called Specific Language Impairment (SLI), since it occurs in otherwise normal children. It is a genetic disorder, with many genetic variants contributing to affect different components of the language system. There is a specific subtype that selectively affects the grammar system, called grammatical SLI. Children with this disorder have difficulty interpreting and producing syntactic structures, in particular hierarchical structures (van der Lely and Pinker, 2014). This deficit is evidenced both in behavioral studies (van der Lely, Rosen, and McClelland, 1998) and in ERP studies (Fonteneau and van der Lely, 2008).

While the ERP responses to semantic anomalies are comparable to typically developing teenagers evidenced in the N400 component, ERP responses to syntactic violations are not. For syntactic violations typically developing teenagers show an adult-like ERP pattern, the ELAN–P600 pattern, whereas grammatical SLI teenagers show no ELAN, but an N400-like response co-occurring with a typical P600 instead (Fonteneau and van der Lely, 2008). In their review article van der Lely and Pinker (2014) cite a then-unpublished meta-analysis by Michael Ullman and colleagues that reports the neuroanatomical correlates of SLI; they also cite an older article (Ullman et al., 1997) showing that two brain regions were found to be affected: Broca's area and its right-hemisphere homolog, and the caudate nucleus as part of the basal ganglia—two regions involved in the processing of grammatical rules. Referring to Friederici's model (Friederici, 2009a, 175–181; 2012b), van der Lely and Pinker (2014) discuss the dorsal language system including the arcuate fasciculus as the possible system to be impaired in SLI. They admit, however, that the brain functional and neuroanatomical data on SLI are still too coarse to strongly support the view that grammatical SLI is caused by deficits in the anatomically defined dorsal system.

In the following pages I will focus on the neural underpinning of normal language acquisition and ask not only at which age certain milestones in language acquisition are achieved, but moreover to what extend is this achievement dependent on the maturation of particular brain structures. In our recent model, the neural basis of the developing language system is described to reflect two major phases: the first phase covers the first three years of life and a second phase extends beyond age 3, and beyond this until adolescence (Skeide and Friederici, 2016).

Since the adequate description of the developing language system toward the mature system in adults and its neural basis requires the respective knowledge about the adult language system. I will briefly summarize the adult model here. For more details, I refer the reader to chapter 1.

The neurocognitive model of adult language comprehension assumes different networks consisting of specific brain areas and the connections between them. These different networks support different aspects of language processing such as phonology, syntax, and semantics. The analysis of acoustic speech input is performed within 100 ms by the primary auditory cortex and the planum temporale. Phonetic processes as a part of phonology involve the left anterolateral and superior temporal gyrus and sulcus. The early syntactic process of initial phrase structure building between 120 and 200 ms and the later process of relating different phrases in a complex syntactic hierarchy between 300 and 500 ms are supported by two different syntactic networks, a ventral one and a dorsal one respectively. The left anterior superior temporal gyrus connected via the uncinate fasciculus to the left frontal operculum is involved in local combinatorics of two elements with ventral BA 44 being recruited for local syntactic phrase structure building. Complex hierarchical syntactic information is computed in the left BA 44 and left posterior superior temporal gyrus/superior temporal sulcus, connected by the long branch of the superior longitudinal fasciculus including the arcuate fasciculus. Semantic processes taking place between 300 and 500 ms are based in a temporo-frontal network consisting of the left superior temporal gyrus and middle temporal gyrus and left BA 45/47 connected by the inferior fronto-occipital fasciculus. Syntactic and semantic information is integrated after approximately 600 ms in the left posterior superior temporal gyrus/superior temporal sulcus. A shorter branch of the dorsal tract connecting the temporal cortex to the premotor cortex is a component of the sensorimotor interface. This direct connection between the auditory cortex in the temporal lobe and the premotor cortex in the frontal lobe is crucial when reproducing aloud what has been heard. Prosodic processes are mainly located in the right hemisphere (superior temporal gyrus and BA 44 in the inferior frontal gyrus), as are discourse processes (superior temporal gyrus and BA 45/47 in the inferior frontal gyrus).

Similar to the adult studies, neuroscientific methods such as electroencephalography and magnetoencephalography as well as magnetic resonance imaging have also been used in developmental studies, although at different ages. While electroencephalography and magnetoencephalography as well as near-infrared spectroscopy can be applied in very young children, the method of functional magnetic resonance imaging in awake children is mostly only used in children at and beyond age 3. These methods are described in section 1.1 in some detail when discussing the adult language network. Remember, that near-infrared spectroscopy (also called optical imaging) is a method that registers the hemodynamic response of the brain with a low spatial resolution, but which is much easier to use than functional magnetic resonance imaging in infants and young children as the registration system is mounted directly on the child's head. Functional and structural magnetic resonance imaging can be applied to awake children age 3 and older, but the success of the measurement and the quality of the data crucially depend on how well the young participants are familiarized with the experimental procedures beforehand.

6.1 Language in the First Three Years of Life

The model proposed by Skeide and Friederici (2016) is based on those ERP, near-infrared spectroscopy and magnetic resonance imaging studies that contribute to a fundamental question: Starting from phonological discrimination and continuing with the emergence of lexical, semantic, and syntactic knowledge, how does the language processing system develop over the first few years of life until young adulthood?[1]

Speech Perception in Infants

When entering the auditory world, a hearing newborn—as a first step toward language—has to distinguish between speech and other sounds and between different speech sounds.

A near-infrared spectroscopy study of speech recognition in 2- to 5-day-old infants indicated that newborns already show a significantly stronger hemodynamic response to forward speech compared to backward speech in the left temporal cortex (Peña et al., 2003). The authors concluded that the observed left-hemispheric activation asymmetry reflects a language-specific aspect and hence is not involved in non-linguistic sound processing. Another near-infrared spectroscopy study was able to investigate very young infants (28th week of gestation) (Mahmoudzadeh et al., 2013). The authors of this particular study found that at this age preterm infants are already able to discriminate a deviant speech sound in a sequence of otherwise identical speech sounds. A brain mismatch response was observed bilaterally in the posterior superior temporal and additional inferior frontal cortices, suggesting that premature babies use an early biological mechanism that enables them to discriminate speech-relevant acoustic parameters.

In a functional magnetic resonance imaging study with 2-day-old infants we found a bilateral activation in the superior temporal gyri in response to speech, with a maximum in the right auditory cortex (Perani et al., 2011) (see figure 6.1). This finding suggests that during speech processing newborns rely more on suprasegmental (prosodic) information, known to be processed in the right hemisphere, than on segmental (phonological) information, known to be processed in the left hemisphere.

Infants at age 3 months are reported to show a different speech-related hemodynamic activity pattern. According to two functional magnetic resonance imaging studies, the activation for these infants covered large portions of the bilateral superior temporal cortices, but it was most pronounced in the left hemisphere (Dehaene-Lambertz, Dehaene, and Hertz-Pannier, 2002; Dehaene-Lambertz et al., 2006). However, when directly contrasting natural forward speech with speech played backward, brain activity was only found in the left angular gyrus and the precuneus whereas the effect in the temporal cortex entirely disappeared. These data suggest that precursors of adult cortical language areas already support speech perception, although the left temporal cortex has not yet acquired its full language functionality at 3 months of age.[2]

Discriminating Phonological Information

Right from the start of language acquisition, infants can recognize acoustic and phonological cues in the speech input. In order to extract these cues and regularities from the auditory input, they must first be able to discriminate between different phonological parameters at the segmental and suprasegmental levels. Moreover, infants must possess an auditory

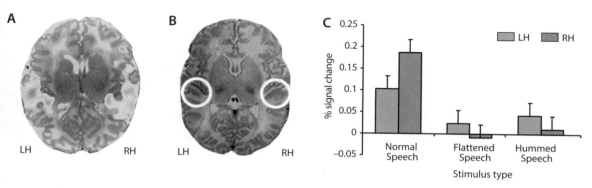

Figure 6.1
Neural disposition for language in the newborn brain. (A) Brain activation in newborns for normal speech (n = 15, random effects group analysis, significance threshold P < 0.05 at the voxel level, uncorrected) overlaid on a T2-weighted image from a single newborn. Regions significantly more active for the speech condition compared with silence are shown in orange/yellow. (B) Region-of-interest (ROI) analysis, including primary and secondary auditory cortex, was conducted on subjects' average for three different auditory conditions: normal speech; speech that was filtered such that only segmental phonetic information, but no suprasegmental prosodic information, was present (flattened speech); and speech that was filtered such that only prosodic information, but no segmental information, was presented (hummed speech). (C) Percent signal change in activation in the defined ROIs in left hemisphere (LH) and right hemisphere (RH) (circled area in B) are displayed for the three auditory conditions (normal speech, flattened speech, and hummed speech). The histograms show the percent signal change measured in the left and right ROI during each of the three stimulus types. Adapted from Perani et al. (2011). The neural language networks at birth. *Proceedings of the National Academy of Sciences of the United States of America*, 108 (38): 16056–16061.

sensory memory that allows for storage of auditory information in order to be mapped onto new incoming auditory information and thereby "recognized." Impressive evidence for such an early auditory memory comes from a near-infrared spectroscopy study showing that newborns are able to memorize the sound of a word after at least two minutes after hearing it and to differentiate it from a phonologically similar one (Benavides-Varela, Hochman, Macagno, Nespor, and Mehler, 2012).

Quite a number of ERP studies have investigated infants' phonological abilities during their first year of life. In these studies one ERP paradigm that has proven to be particularly useful in investigating young infants' abilities to distinguish between phonetic features is the so-called Mismatch Negativity paradigm. In this paradigm a rarely occurring (deviant) stimulus is presented within a sequence of standard stimuli. Deviant and standard stimuli usually differ in one crucial feature. In adults, the discrimination of these two stimulus types is reflected in a negative deflection with a peak latency of 100 to 200 ms following change onset. This negative deflection is labeled Mismatch Negativity (for a review, see Näätänen, Tervaniemi, Sussman, and Paavilainen, 2001, and section 1.2 of this book). The amplitude of the mismatch negativity is mainly modulated by the ease of discrimination and on physical difference between deviant and standard stimuli; the latency of the mismatch negativity primarily depends on the deviance onset and is related to the demands of sensory discrimination (for reviews, see Näätänen, Paavilainen, Rinne, and Alho, 2001; Garrido, Kilner, Stephan, and Friston, 2009; Näätänen et al., 2007; Picton et al., 2000).

Negative mismatch responses to phonologically varying stimuli have been reported in infants, and even in preterm newborns (e.g., Cheour et al., 1997; Morr, Shafer, Kreuzer, and Kurtzberg, 2002). The negative response in newborns, however, typically does not reveal the sharp negative deflection of the adult mismatch negativity, but rather a long-lasting negative wave or a late negative response occurs (Cheour-Luhtanen et al.,1995, 1996; Cheour et al., 2002; Martynova, Kirjavainen, and Cheour, 2003). Several studies even reported a broad positive response in the infants' ERPs that was more prominent for the deviant stimulus (Dehaene-Lambertz and Dehaene, 1994; Leppänen, Richardson, and Lyytinen, 1997; Dehaene-Lambertz, 2000; Dehaene-Lambertz and Peña, 2001; Friederici, Friedrich, and Weber, 2002). There are several reasons that may contribute to whether we observe a negative or a positive deflection as a mismatch response, including differences in an infant's state of alertness (Friederici et al., 2002), methodological differences such as filtering the data (Trainor et al., 2003; Weber, Hahne, Friedrich, and Friederici, 2004), and the coexistence or overlap of two types of mismatch responses (He, Hotson, and Trainor, 2007). In general, the available studies suggest a developmental transition from mismatch-triggered positive deflection during early developmental stages toward negative deflection and mismatch negativity in later developmental stages. Given the differences in ERP patterns of the mismatch response in young infants and adults, the discrimination process could be viewed as being qualitatively different, possibly more acoustically based early in development and phonemically based later on. Independent of these considerations, the mismatch response can be taken to functionally indicate discrimination in the auditory domain.

Discrimination between different language-relevant sounds that inform mismatch negativity responses have been observed for phonetic features in different languages such as Finnish, German, and English, for vowel contrasts (Cheour et al., 1997; Leppänen, Pihko, Eklund, and Lyytinen, 1999; Pihko et al., 1999; Friederici et al., 2002) and for consonant contrasts (Dehaene-Lambertz and Baillet, 1998; Rivera-Gaxiola, Silva-Pereyra, and Kuhl, 2005; Mahmoudzadeh et al., 2013). The different studies indicate that infants are able to discriminate different phonemes independent of their target language between ages 1 and 4 months or even preterm in the 28th week of gestation. Evidence for language-specific phonemic discrimination only seems to be established somewhat later between ages 6 and 12 months (Cheour et al., 1998, Rivera-Gaxiola et al., 2005). Younger infants, aged 6 and 7 months, discriminate between phonemic contrasts that are both relevant and not relevant to their target language; older infants, aged 11 and 12 months, only distinguish those phonemic contrasts that are relevant to their target language. These results are in agreement with behavioral data reporting language-specific reactions during the second half of the first year of life (Aslin, Pisoni, Hennessy, and Perey, 1981; Werker and Tees, 1984). A comparison between full-term and preterm infants suggests that the time point of establishing phonemic discrimination skills during this developmental period is determined by biologically driven brain maturation factors rather than by the onset of speech input (Peña, Werker, and Dehaene-Lambertz, 2012).

Language-specific brain responses for stress patterns of words were shown in an ERP study with infants as young as 4 months old (Friederici, Friedrich, and Christophe, 2007). When German and French infants were presented with bisyllabic pseudowords, they reacted according to the dominant stress pattern of their mother tongue. German infants reacted more strongly to items with stress on the second syllable, a stress pattern which is infrequent in their target language. French infants, in contrast, reacted more strongly to items with stress on the first syllable, a stress pattern infrequent in their target language (see figure 6.2). These data thus provide evidence for word-level language-specific brain reactions at the age of 4 months. The results clearly indicate the infants' sensitivity to phonological features early during development.

From Auditory Input to Semantic Meaning
During the next step infants acquire words and their meanings. To do this, infants have to segment words from their auditory stream. Before lexical knowledge is established, segmentation might be aided by knowledge of a given language's prosody and its potential word forms, such as dominant stress patterns of words or possible phonotactic structures (e.g., possible beginnings and endings) of words.

Prosody helps 6-month-old infants to segment auditory word forms from continuous speech as indicated by eye movement analysis (Shukla, White, and Aslin, 2011). The fast mapping of the auditory word form onto visual objects occurs only when the word is aligned with the prosodic phrase boundary. Behavioral studies suggest that stress information for speech segmentation is used at around the age of 7.5 (Jusczyk, Houston, and Newsome, 1999) to 9 months (Houston, Jusczyk, Kuijpers, Coolen, and Cutler, 2000). In languages with dominant stress on the first syllable, such as English or Dutch, the ability to segment bisyllabic words with stress on the first syllable from speech input was found at the age of 7.5 months, but word segmentation effects for bisyllabic words with stress on the second syllable are only reported at 10.5 months of age (Jusczyk et al., 1999; Kooijman, Hagoort, and Cutler, 2009; Polka and Sundara, 2012).

The neurophysiological data can be summarized in the following way. Infants are sensitive to stress information when words are presented in isolation as early as 4 months of age (Friederici et al., 2007). The ability to use word stress for word recognition during speech perception, however, is observed only in 10-month-old infants (Kooijman et al., 2005). A further ERP study evaluated when and to what extend phonological stress on words in speech (i.e., accentuation) plays the only role in the recognition of words or whether repetition of the word might, too (Männel, Schipke, and Friederici, 2013). This ERP study showed that infants' word recognition ability in sentential context changed over the course of development with accentuation and word repetition playing crucial roles at different

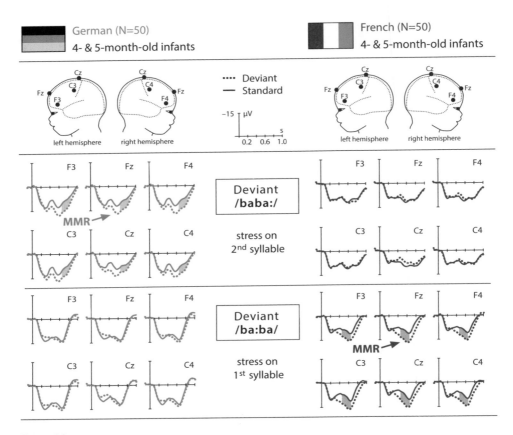

Figure 6.2
Language-specific perception of stress patterns in infants. Event-related brain potentials (ERPs) for 4-month-old German and French infants. The stimuli were two-syllable items with either intonational stress on the first syllable (typical for German two-syllable words) or on the second syllable (typical from French two-syllable words). Infants heard a sequence of frequent (standard) items interspersed with other (deviant) items. There were two conditions for each child: either the item with the stress on the second syllable served as the deviant, or the item with the stress on the first syllable served as the deviant. Averaged ERPs per condition (standard: solid line; deviant: dotted line) for each language group (German: left panel; French: right panel) and item type: for items with stress on the second syllable (top panel) and items with stress on the first syllable (bottom panel). The shaded area indicates the time window chosen for statistic analysis in which the effect was statistically significant MMR = mismatch response. The data indicate different responses for the two language groups, indicating a language-specific effect. Adapted from Friederici, Friedrich, and Christophe (2007). Brain responses in 4-month-old infants are already language specific. *Current Biology*, 17 (14): 1208–1211, with permission from Elsevier.

points during development. At 6 months, infants only recognized previously heard words when these were accentuated; at 9 months both accentuation and repetition played a role, while at 12 months only repetition was effective. These data indicate very narrow and specific input-sensitive periods in infant word learning and recognition, with accentuation being effective prior to repetition. This is an interesting experimental finding because parents often use both the accentuation and the repetition of a word as means to enhance infants' attention when speaking to them and when using new words during infant-directed speech.

From Lexical Form to Word Meaning Once word forms can be recognized, lexical knowledge has to be acquired. This means that the infant has to learn the relation between a word form and its semantic meaning. Before such a clear relation is established, and before a child is able to recognize the phonological word form as referring to a specific meaning, there may be a phase that can be described as "familiarity" with a phonological form.

Behavioral studies cannot ascertain the age at which infants know about the phonological rules of word formation in their native language, but neurophysiological studies can do so by using ERP measures. The applicable correlate in the ERP is the so-called N400 component, a negative waveform peaking at around 400 ms. In adults the N400 effect is observed for aspects of lexical form, and also for aspects of lexical meaning. The amplitude of the N400 is larger for pseudowords than for real words and larger for semantically incongruous words than for congruous words in a given context (for reviews, see Kutas and Van Petten, 1994; Lau, Phillips, and Poeppel, 2008). The N400 can thus be used to investigate lexically relevant knowledge both at phonotactic and semantic levels.

Some ERP studies suggest that children are able to map language sounds onto meaning at about 11 months of age. Such a mapping has been proposed based on a negative deflection observed around 200 ms post-stimulus onset at 11 months of age in response to listening to familiar versus unfamiliar words (Thierry, Vihman, and Roberts, 2003). Although there are methodological concerns about this study, thus challenging the authors' interpretation, it provides a first hypothesis. Investigating 12-, 14- and 19-month-olds with a picture-word-priming paradigm, a familiarity-related early fronto-central negativity between 100 and 400 ms in all age groups was found when the auditory word was congruous with a picture compared to the situation in which the word was incongruous (Friedrich and Friederici, 2005a). This effect was taken to be too early to represent a semantic N400 effect and was, therefore, interpreted as a phonological-lexical priming effect reflecting the fulfillment of a phonological (word form) expectation built up after seeing the picture of an object.

In order to evaluate when and to what degree young infants possess lexical knowledge a next study used phonologically legal (possible) and phonologically illegal (impossible) words in the child's native language. The ERPs recorded during word processing revealed that 12- and 14-month-olds showed a difference between known words and phonetically illegal words, but not between known words and phonetically legal words (Friedrich and

Friederici, 2005a). Thus at an early developmental phase, infants seem to have acquired some lexical knowledge and knowledge about possible word forms, but the relation between word form and the respective meaning might not yet be sharply defined; allowing phonetically similar words still to be considered as possible word candidates. At the age of 19 months, but not earlier, infants were able to discriminate phonotactically legal words from real words. This can be taken as an indication that there is a developmental change between 14 and 19 months of age. The data support the idea of a transition from a familiarity stage to a recognition stage during the development of lexical knowledge.

Word Level Semantics Early ERP studies on lexical-semantic processes investigated the processing of words whose meanings infants either did or did not know prior to the experiment, and found that infants between 13 and 17 months showed a bilateral negativity for unknown words, whereas 20-month-olds showed a negativity lateralized to the left hemisphere (Mills, Coffey-Corina, and Neville, 1997). This result was interpreted as a developmental change toward a left hemispheric specialization for language processing. In another study using a word-learning paradigm, 20-month-olds had to learn novel words either paired with a novel object or presented alone (Mills et al., 1997). The observed N200–500 for the condition with the object was linked to the processing of word meaning. Data showing that the semantic effect (N400) of adults are usually observed later in development challenge the interpretation of this early effect as a semantic effect. It is possible that the early onset of the effect in infants reported by Mills et al. (1997) is a phonological effect due to infants' relatively small vocabularies. A small vocabulary results in a low number of phonologically possible alternative word forms, allowing the brain to react early, that is to say, after hearing a word's first phonemes.

To investigate the acquisition of lexical semantics, a paradigm appropriate for both adults and young children is desirable, such as the one in which the participant is shown a picture of an object (e.g., duck) and at the same time is presented with an auditory stimulus. This auditory stimulus may be a word matching the object's name (e.g., *duck*) or a word not matching the object's name (e.g., *lamp*), or it may be a pseudoword that is phonotactically legal or illegal, that is, one that follows the phonological rules of the given language or not. Using this paradigm, developmental changes were observed between the ages of 12, 14, and 19 months compared to adults (Friedrich and Friederici, 2005a, 2005b) (figure 6.3). Only at 19 months an adult-like N400 effect signaling lexical processing was found for semantically non-matching words and for pseudowords, when these were phonotactically legal, but not when phonotactically illegal. At the age of 14 months, a late N400-like negativity effect was found for semantically non-matching words, suggesting the beginning of semantic representations.

Word learning and memory for words was evaluated in infants 9–16 months old (Friedrich, Wilhelm, Born, and Friederici, 2015). In a word-picture-mapping paradigm, infants learned the names of objects and were tested on these 1.5 hours later. Infants who had

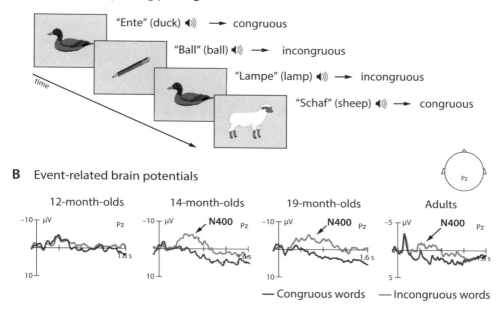

Figure 6.3
Processing word-level semantics in young children. Event-related brain potentials (ERPs) for word-level lexical-semantic information processing. (A) The picture-word-priming paradigm was used, presenting a word-picture combination that was either congruous or incongruous. (B) ERP average for different age groups. The N400 for incongruous (red) versus congruous (blue) words in different age groups. Note the microvolt scales differ between age groups. Adapted from Friedrich and Friederici (2005b). Phonotactic knowledge and lexical-semantic processing in one-year-olds: Brain responses to words and nonsense words in picture contexts. *Journal of Cognitive Neuroscience*, 17 (11): 1785–1802, © 2005 by the Massachusetts Institute of Technology. Also adapted with permission from Wolters Kluwer from Friedrich and Friederici (2005a). Lexical priming and semantic integration reflected in the ERP of 14-month-olds. *NeuroReport*, 16 (6): 653–656.

napped during this intermediate phase, compared to those who had stayed awake, were better able to remember specific word meanings and, moreover, successfully generalized word meanings to novel exemplars of the category. Sleep spindles measured during the nap indicate that these support the infant sleep-dependent brain plasticity. These findings demonstrate that the learning of words depends on and is enhanced by consolidation during sleep.

Sentence Level Semantics Semantic processes at the sentence level—as discussed in sections 1.6 and 1.7—require more than just the identification of the semantic meaning of each word. To understand the meaning of a sentence, we as listeners have to possess semantic knowledge about nouns and verbs as well as their respective relationship (for neural correlates of developmental differences between noun and verb processing, see Li, Shu, Liu, and

Li, 2006; Tan and Molfese, 2009; Mestres-Misse, Rodriguez-Fornells, and Münte, 2010). To investigate whether children already process word meaning and semantic relations in sentential context, the *semantic violation paradigm* can be applied with semantically correct and incorrect sentences such as *The pizza was eaten* and *The radio was eaten*, respectively (Friederici, Pfeifer, and Hahne, 1993; Hahne and Friederici, 2002). This paradigm uses the N400 as an index of semantic integration abilities, with larger N400 amplitudes reflecting higher integration efforts (i.e., for words that are semantically inappropriate for the given context). The semantic expectation of a possible sentence ending is violated in *The radio was eaten*, for example, because the verb at the end of the sentence (*eaten*) does not semantically meet the meaning that was set up by the noun in the beginning (*radio*). In adult ERP studies, an N400 has been found in response to such semantically unexpected sentence endings (Friederici et al., 1993; Hahne and Friederici, 2002).

Friedrich and Friederici (2005c) studied the ERP responses in sentences with semantically appropriate and inappropriate sentence final words in 19- and 24-month-old children. Semantically incorrect sentences contained objects that violated the selection restrictions of the preceding verb, as in *The child drinks the ball* in contrast to *The child rolls the ball*. For both age groups, the sentence endings of semantically incorrect sentences evoked N400-like effects in the ERP, with a maximum at central-parietal electrode sites (figure 6.4). In comparison to the adult data, the negativities in children started at about the same time (i.e., at around 400 ms post-word onset) but were longer lasting. This suggests that semantically unexpected, incorrect nouns that violate the selection restrictions of the preceding verb initiate semantic integration processes in children similar to adults, but that the integration efforts are maintained longer than in adults. The developmental ERP data indicate that even at the age of 19 and 24 months, children are able to process semantic relations between words in sentences in a similar manner to adults.

Silva-Pereyra and colleagues found that the sentence endings that semantically violated the preceding sentence phrases evoked several anteriorly distributed negative peaks in 3- and 4-year-olds (Silva-Pereyra, Klarman, Lin, and Kuhl, 2005; Silva-Pereyra, Rivera-Gaxiola, and Kuhl, 2005). These studies revealed differential responses to semantically incorrect and correct sentences in young children, but the distribution of these negativities did not match the usual central-parietal maximum of the N400 seen in adults. Other ERP studies on the processing of sentential lexical-semantic information have also reported N400-like responses to semantically incorrect sentences in older children, namely 5- to 15-year-olds (Holcomb, Coffey, and Neville, 1992; Hahne, Eckstein, and Friederici, 2004; Atchley et al., 2006).

In general, it appears that semantic processes at sentence level, as reflected by an N400-like response, are present, in principle, at the end of children's second year of life. The timing and the distribution of the effect, however, indicate that it takes a few more years before the neural network underlying these processes is established in an adult-like manner.

Ontogeny of the Neural Language Network

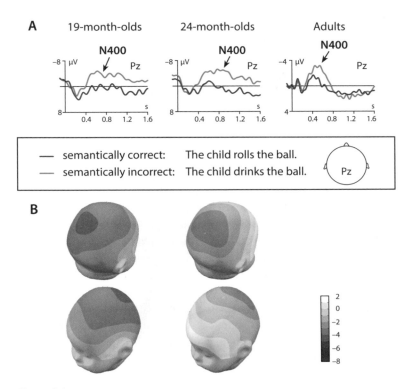

Figure 6.4
Processing sentence-level semantics in young children. Event-related brain potentials (ERPs) for sentence-level lexical-semantic information processing. (A) ERP data displayed for electrode PZ and (B) difference maps (incorrect/correct) of 19-month-olds, 24-month-olds, and adults in response to the sentence endings of semantically correct (blue) and incorrect (red) sentences in a semantic violation paradigm. Example sentences are displayed in the box middle of the figure. Adapted with permission from Wolters Kluwer from Friedrich and Friederici (2005c). Semantic sentence processing reflected in the event-related potentials of one- and two-year-old children. *NeuroReport*, 16 (6): 1801–1804.

From Auditory Input to Sentential Structure

In addition to lexical-semantic knowledge of words, a child must acquire syntactic rules according to which words are structurally related in sentences. In order to do so, children must structure the auditory input into relevant units (phrases, for example), and they must recognize regularities in the auditory input, as this allows them to determine the rule-based dependency between elements in a sentence.

From Prosodic Cues to Syntax One possible way of extracting structural information from auditory input lies in the fact that syntactic phrase boundaries and prosodic phrase boundaries largely overlap. Each prosodic phrase boundary is a syntactic boundary, although not every syntactic boundary is marked by prosody. Acoustically prosodic phrase boundaries are marked by three parameters: length of the preboundary syllable, pitch, and pause. It has been argued that prosodic information might aid in the acquisition of syntactic units and the relationships between them (Gleitman and Wanner, 1982). Behavioral studies have shown that 6-month-old English infants use converging cues to segment clauses, either pitch and pause together or pitch and preboundary length together (Seidl, 2007).

When investigating the processing of prosodic information during speech perception using electrophysiological measures in adults, a particular ERP component was found to correlate with the processing of prosodic boundaries, that is, intonational phrase boundaries. This ERP component is a positive shift occurring at the intonational phrase boundary called the Closure Positive Shift, located in the inferior parietal lobe (Steinhauer, Alter, and Friederici, 1999). This ERP component was observed in the adult not only when the intonational phrase boundary was marked by preboundary length, pitch, and pause, but it was also present when the acoustic cue, the pause cue was deleted (Steinhauer et al., 1999). The Closure Positive Shift component in the ERP is distributed over left and right parietal recording sites for spoken sentences in which segmental and suprasegmental information is present. For hummed sentences in which only suprasegmental information is present, it is lateralized to the right hemisphere (Pannekamp, Toepel, Alter, Hahne, and Friederici, 2005). This suggests that suprasegmental information is primarily processed in the right hemisphere, which is supported by functional magnetic resonance imaging studies in adults (Meyer, Alter, Friederici, Lohmann, and von Cramon, 2002; Meyer, Steinhauer, Alter, Friederici, and von Cramon 2004; Sammler, Kotz, Eckstein, Ott, and Friederici, 2010).

In infants, an adult-like right hemispheric dominant activation in the processing of sentential prosody was found for 3-month-olds in an imaging study using near-infrared spectroscopy for normal compared to flattened speech (Homae, Watanabe, Nakano, Askawa, and Taga, 2006) (see figure 6.5). This finding suggests that the neural mechanism processing prosodic information appears to be in place quite early during development.

When investigating infants' brain responses to prosodic boundaries, Pannekamp, Weber, and Friederici (2006) we found that 8-month-old infants (see figure 6.6) demonstrate a clear ERP effect at the intonational phrase boundary, which appears to be similar to a Closure Positive Shift. A similar effect was also found by Männel and Friederici (2009) with 5-month-old infants. This latter finding raised the question whether the observed effect in infants is an adult-like Closure Positive Shift effect, which in adults can be elicited independent of the presence of the most prominent acoustic cue, that is, the pause. In a follow-up study it was demonstrated that an adult-like Closure Positive Shift independent of the low-level cue of an acoustic interruption (pause) cannot be found in infants, but is only present in 3-year-old children—at an age when first syntactic knowledge is acquired.

This finding is of particular interest as it indicates a strong dependence of the full identification of the intonational phrase boundary on the syntactic development. The study by Männel and Friederici (2011) we set out to investigate prosodic boundary processing at different stages of syntax acquisition by measuring the associated ERPs in 21-month-old,

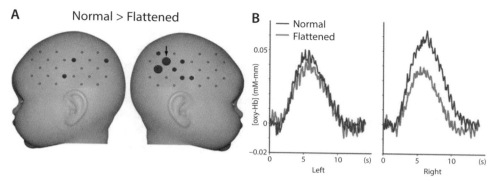

Figure 6.5
Processing prosody in infants. Prosodic processing in 3-month-old infants as measured by near infrared spectroscopy (NIRS). Comparisons between the normal speech and flattened prosody. (A) A direct comparison of the normal condition / flattened condition. Large- and small-filled circles indicate measurement channels (CH) that surpassed $p < 0.005$ and 0.05, respectively. The arrow indicates the CH16 in the right hemisphere. (B) The averaged time course of [activation] changes of the left CH16, and of the right CH16 for normal (blue) and flattened (green) conditions, respectively. Adapted from Homae et al. (2006). The right hemisphere of sleeping infant perceives sentential prosody. *Neuroscience Research*, 54 (4): 276–280, with permission from Elsevier.

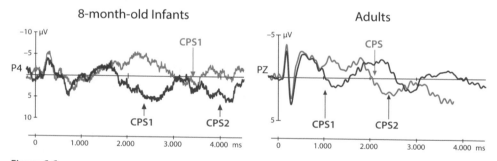

Figure 6.6
Processing intonational phrase boundaries in infants and adults. Event-related brain potentials (ERPs) during processing intonational phrase boundaries. The Closure Positive Shift (CPS) as an index of processing intonational phrase boundaries (IPh) for infants and adults. Left: Grand-average ERP for adults at electrode Pz. Right: Grand-average ERP for 8-month-old infants at electrode P4. Vertical line indicates sentence onset. Two-IPh sentences are displayed in blue and one-IPh sentences in green. Arrows indicate the CPS. Adapted from Friederici (2005). Neurophysiological markers of early language acquisition: From syllables to sentences. *Trends in Cognitive Sciences*, 9: 481–488, with permission from Elsevier. Also adapted from Steinhauer, Alter, and Friederici (1999). Brain potentials indicate immediate use of prosodic cues in natural speech processing. *Nature Neuroscience*, 2 (2): 191–196.

3-year-old, and 6-year-old children. The rationale of this approach was that—as children acquire a great deal of syntactic knowledge between age 2 and 3—children at age 3, and in particular children at age 6, possess crucial syntactic knowledge, while 21-month-olds do not. Children at the age of 21 months, 3, and 6 years of age were presented with sentences with and without an intonational phrase boundary. For 21-month-olds, ERP analyses revealed an obligatory acoustic component (due to the pause) in response to intonational phrase boundaries but no Closure Positive Shift, suggesting that at this age speech boundaries can only be detected based on low-level acoustic processes. In contrast, 3- and 6-year-old children showed both obligatory auditory components and an adult-like Closure Positive Shift in response to intonational phrase boundaries, and, moreover, also demonstrated a Closure Positive Shift in the absence of the pause. Taken together, the results of these studies demonstrate developmental changes in intonational phrase boundary processing to be dependent upon the level of syntax acquisition. The Closure Positive Shift in response to intonational phrase boundaries in the absent of the salient pause was only observed at an age when children are linguistically more advanced and have gained some syntactic (phrase structure) knowledge.

From Syntactic Regularities to Syntax In order to learn the structure of a language into which an infant is born, he or she must identify the positional regularities of elements in the speech input. Behavioral studies have shown that by the age of 8 months, infants calculate transitional probabilities within three-syllable strings in a miniature, artificial grammar (Saffran, Aslin, and Newport, 1996). A study with 7-month-olds, however, suggested that infants' learning at that age might go beyond statistical learning, possibly involving the extraction and representation of algebraic rules (Marcus, Vijayan, Rao, and Vishton, 1999). With a somewhat more complex artificial grammar, learning of transitional probabilities was demonstrated in 12-month-olds (Gómez and Gerken, 1999). In natural languages crucial grammatical information is not necessarily encoded in adjacent elements, for example, for subject-verb agreement (*he* look*s* vs. *we* look). The learning system has to recognize the relationship between the pronoun (*he/we*) and the inflection (*+s/-s*) by abstracting from the intervening verb stem.

In an artificial grammar–learning paradigm, we have shown that even 3-month-old infants can learn similar non-adjacent dependencies in an AxB pattern for three-syllable strings if their auditory processing system is already sensitive to pitch information (Mueller, Friederici, and Männel, 2012). In a language learning study that used a natural language (Friederici, Mueller, and Oberecker 2011), the ERP pattern indicated that 4-month-old German infants from monolingual German families were able to learn the non-adjacent dependency relation between an auxiliary and a verb inflection of a novel language, namely Italian (*sta* verb-*are* vs. *puo* verb-*ando*) in which these combinations are correct. This dependency between the auxiliary and the verb inflection is similar to the English dependency between *is* and *–ing* in *is* run*ing*. The learning of these non-adjacent dependencies was reflected in

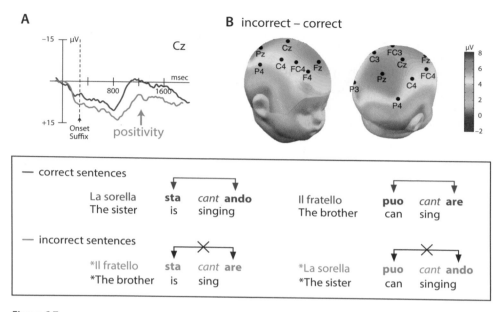

Figure 6.7
Learning phonology-based syntactic dependencies in infants. Event-related brain potentials (ERPs) of German infants for the learning of phonology-based syntactic dependencies in Italian sentences. Structure and examples of Italian stimulus sentences (see box at the bottom of the figure). The box displays the grammatical dependency between the auxiliaries (*sta/is* and *puo/can*) and the respective Italian verb inflections (*-ando* and *-are*). Correct structure of grammatical relation between *sta* and *-ando* as well as *puo* and *-are* are presented in the top row. Incorrect structure of grammatical relation between *sta* and *-are* as well as *puo* and *-ando* are presented in bottom row. Relation between crucial non-adjacent elements is indicated by arrows. An asterisk indicates an incorrect sentence. (A) Grand average event-related potentials of 4-month-old infants for the processing of the verb averaged across the four test phases. The processing of the incorrect condition (red line) is plotted against the processing of the correct condition (blue line). The solid vertical line indicates the onset of the verb, the broken vertical line at the scale plot indicates the onset of the suffix (verb inflection). Negativity is plotted upward. (B) Isovoltage map showing the scalp distribution of the effect. Positive difference is color-coded in red. Adapted from Friederici, Mueller, and Oberecker (2011). Precursors to natural grammar learning: preliminary evidence from 4-month-old infants. *PLoS ONE*, 6 (3): e17920.

the ERP when comparing incorrect to correct combinations of auxiliary and verb inflection in a sentence (see figure 6.7). Incorrect combination led to a centro-parietal positivity. These results clearly demonstrate that 4-month-old infants have learned this non-adjacent dependency in a novel language. They may, however, have learned this syntactically relevant relation as a phonological dependency, by matching the auxiliary with the related verb inflection.

For natural language acquisition, in contrast to the learning studies, a behavioral study reported that only 18-month-old children learning English can track the relationship between *is* verb-*ing* (e.g., correct: *is digging* vs. incorrect: *can digging*) (Santelmann and Jusczyk, 1998). However, work by Tincoff, Santelmann, and Jusczyk (2000) indicated that the relationship between the auxiliary and the progressive (-*ing*) is represented only between specific items (*is _ing*) and is not generalized to other combinations (*are _ing* or *were _ing*) in 18-month-olds. Moreover, it was demonstrated that the children's capacity to recognize non-adjacent dependencies relied on their ability to linguistically analyze the material positioned between the two dependent elements. German children at 19 months were able to recognize dependency relationships only when the intervening material was clearly marked (e.g., as in English, where adverbs are marked by the inflection -*ly*, as in is *energetically digging*), but not in the absence of a clear morphological marker (Höhle, Schmitz, Santelmann, and Weissenborn, 2006). Thus, non-adjacent dependencies in natural languages are acquired under particular circumstances around 18–19 months of age, as demonstrated by behavioral studies.

The observed difference concerning the ages of successfully processing non-adjacent dependencies in those studies looking at natural acquisition and those using learning paradigms may have multiple reasons: first, the stimulus material in the learning study was much more restricted than natural language input; second, the learning study reflects immediate rather than long-term memory of the dependency relation. It would be especially interesting to have ERP data regarding this issue, as these may help to identify the type of processing mechanism underlying the children's behavior given that there are specific ERP components related to particular syntactic processes and subprocesses.

When consulting ERP data, one finds that for adults, different ERP components identified to correlate with syntactic processes will reflect specific subprocesses. For syntactic violations in a sentence, an early left anterior negativity (ELAN, 100–200 ms) was observed for local phrase structure violations, and a somewhat later left anterior negativity (LAN, 300–400 ms) was found for the morphosyntactic violations (e.g., incorrect inflection in subject-verb agreement). Both of these negativities reflect the automatic detection of a structural violation usually followed by a late positivity (P600) taken to reflect sentence level inconsistency (for a review, see Friederici, 2002, 2011, and chapter 1).

For children, the available ERP studies suggest that these ERP effects can in principle be observed during natural language acquisition in 2-year-olds. French children, according to Bernal, Dehaene-Lambertz, Millote, and Christophe (2010), demonstrate an early left-lateralized ERP response for word category violations in 24-month-olds listening to spoken sentences. For German children we observed a late positivity (P600) in 24-month-olds for local phrase structure violations for which, at this stage, no early negativity was present (Oberecker and Friederici, 2006, see figure 6.8A). It is only at the age of 32 months that an early anterior negativity (child-specific ELAN) together with a late positivity (P600) (Oberecker, Friedrich, and Friederici, 2005; see figure 6.8B) emerges in response to local phrase structure violations in simple active voice sentences. These data suggest a developmental change toward more automatic phrase structure processes between the age of 24 and 32 months in German infants.

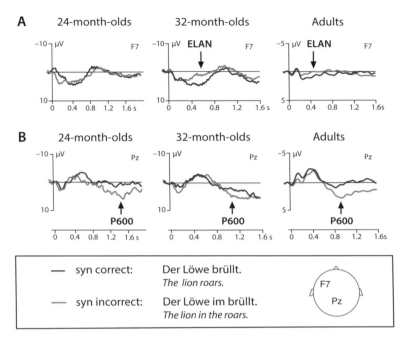

Figure 6.8
Processing of syntactic violations in young children. Event-related brain potentials (ERPs) in response to syntactic word category violation. Example sentences are displayed in the box at the bottom of the figure. The ELAN-P600 pattern is known to index syntactic processes. ELAN stands for early left anterior negativity whereas P600 stands for a late, centro-parietal positivity. Grand-average ERPs at selected electrodes (F7, Pz) across the different age groups are displayed in (A) for the ELAN effect and in (B) for the P600 effect. Note the different microvolt scales between children and adults. Adapted with permission from Wolters Kluwer from Oberecker and Friederici (2006). Syntactic event-related potential components in 24-month-olds' sentence comprehension. *NeuroReport*, 17 (10): 1017–1021. Also adapted from Oberecker, Friedrich, and Friederici (2005). Neural correlates of syntactic processing in two-year-olds. *Journal of Cognitive Neuroscience*, 17 (10): 1667–1678, © 2005 by the Massachusetts Institute of Technology.

Violations of non-adjacent dependencies in English (e.g., *will matching* instead of *is matching*) did not elicit a significant late positivity even in 30-month-old English children, but only in 36-month-olds (Silva-Pereyra, Rivera-Gaxiola, et al., 2005). The age difference, when compared to the German study, may be explained by the fact that the German study (Oberecker and Friederici, 2006) tested local dependencies (word category violation, e.g., *The lion in-the roars* instead of *The lion in-the zoo roars*), which may be easier to process than the non-local dependencies (modal verb-inflection agreement, e.g., *will matching*) tested in the English study. It is not surprising that the ELAN effect, which is taken to reflect highly automatic phrase structure building processes (Friederici, 2002), is present somewhat later during development than the P600, which is taken to reflect processes of syntactic integration. Thus, it appears that around the age of 2.5 to 3 years, first ERP evidence for syntactic processes can be identified. The data, moreover, suggest that local violations are detected earlier during development than non-adjacent dependency violations.

Functional Development of Language-Sensitive Brain Regions
So far I have discussed ERP studies on the processing of phonological, semantic, and syntactic information during early childhood. It would be interesting to know how the neural network underlying these processes develops during the first years of life. In recent years, some neuroimaging studies have contributed to our understanding of the developmental dynamics of regions within the language network. The findings indicate that although an adult-like pattern of functional activity underlying language processing is already established in general in young infants, the domain-specific cortical selectivities within the functional network emerge only gradually until they are fully established in late childhood.

The near-infrared spectroscopy study of speech recognition in 2- to 5-day-old infants by Peña et al. (2003) indicated that newborns already show a significantly stronger hemodynamic response to forward speech compared to backward speech in the left temporal cortex. The authors concluded that the observed left-hemispheric activation asymmetry is specific for language sounds. However, in our functional magnetic resonance imaging study with 2-day-old infants we found a bilateral brain response to normal speech in the superior temporal gyri with a maximum in the right auditory cortex (Perani et al., 2011). Because adults mainly involve the right hemisphere when it comes to processing suprasegmental information, this finding suggested that newborns seem to rely more on suprasegmental (prosodic) than on segmental (phonetic) information during speech processing.

This view is supported by a near-infrared spectroscopy study on prosodic processing in 3-month-olds by Homae et al. (2006) that compared normal speech with experimentally flattened speech quite similar to the manipulation used by us in an adult functional magnetic resonance imaging study (Meyer et al., 2004). Homae and colleagues (2006) found the prosodic effect of processing pitch in infants to be localized in the right temporoparietal region, quite similar to the effect reported for adults when processing suprasegmental information such as pitch.

As mentioned above two functional magnetic resonance imaging infant studies with 3-month-olds investigating passive listening to spoken sentences in general reported activity in large portions of the superior temporal cortices bilaterally, most pronounced in the left hemisphere (Dehaene-Lambertz et al., 2002, 2006). The effect in the temporal cortex, however, disappeared when directly contrasting forward speech with backward speech. These data suggest that the left temporal cortex has not yet reached its full language functionality at this age.

The strong involvement of the right hemisphere for speech processing in newborns (Perani et al., 2011) and the strong involvement of the left hemisphere in the adult brain beg a fundamental question: When and under what circumstances does lateralization of language to the left hemisphere occur during development? Minagawa-Kawai, Cristia, and Dupoux (2011) present a possible answer to this question. They propose a developmental scenario according to which lateralization of language emerges from the interaction between a biologically given hemispheric bias toward rapidly changing sounds such as phonemes (preferably being processed in the left hemisphere) and spectrally rich sounds such as sentential prosody (preferably being processed in the right hemisphere). As infants learn language with its different phonological categories being relevant for lexical learning, the left hemisphere comes into play more and more. Finally, with the acquisition of syntax the left hemisphere is almost exclusively recruited.

Summary
The available data from the neurocognitive studies reviewed in this section provide consistent evidence that very early on an infant is able to extract language-relevant information from the acoustic input. Infants appear to be equipped with the ability to identify those language sounds relevant for the phonetics of the target language, to perceive prosodic cues that allow them to chunk the input into phrases, and to recognize positional regularities and dependencies that are crucial for the syntax of the target language. Moreover, associative learning allows infants to rapidly acquire names of objects and actions, and the relation between them. All the processes mainly involve the temporal cortices of both hemispheres with a shift toward the left hemisphere with increasing age.

6.2 Language beyond Age 3

Up to the age of 3 years, major aspects of language are in place. A child has acquired the phonological aspects of his/her mother tongue, and has learned a lot of words. Around age 2–3 years, children acquire their first syntactic knowledge, reflected by a left anterior negativity in response to word category violations in simple sentences in French (Bernal et al., 2010) and in German (Oberecker et al., 2005). But before an adult-like status in language is reached, further developmental steps have to be made.

Apart from learning new words every day the child has to acquire knowledge about different types of syntactic structures. This includes the knowledge about constructions such as complement clauses and relative clauses, as well as object-first sentences, passive mode sentences, and questions. All the constructions mostly do not follow a simple subject-verb-object structure and thus require additional computational steps when producing and understanding sentences. Our knowledge about the brain basis underlying the development toward an adult-like status of the neural language network is still not complete. But based on ERP studies and functional magnetic resonance imaging studies available in the past decade, an initial picture has emerged.

The Time Course: ERP Studies

ERP studies provide a sketch of the developmental trajectory of time-sensitive syntactic processes toward the adult status. In adults, syntactic word category violation detection has been observed as a very fast process reflected in the ERP by the ELAN component followed by a P600. While the ELAN effect was taken to indicate early automatic phrase structure building, the P600 was taken to indicate processes of integration. In the adult model the early process was allocated to the first processing phase, whereas the later process was allocated to later processing phases.[3]

The processing of syntactic violations in children was investigated by us in a number of ERP studies. We examined children's reaction to phrase structure violation in simple declarative sentences (*The lion roars*). Phrase structure violations in simple sentences (*The lion in-the roars*) elicited a biphasic ELAN–P600 pattern by the age of 2.5 years (Oberecker et al., 2005). We also investigated the processing of phrase structure violations in syntactically more complex sentences (i.e., in passive sentences, *The pizza was in-the eaten*). In these passive sentences, the first noun was not the subject of the sentence, which made the interpretation more difficult than in active sentences (Hahne et al., 2004). When a syntactic violation occurred in passive sentences, the ELAN–P600 pattern was not evoked before the age of 7 to 13 years. Six-year-olds only displayed a late P600. The combined results from the two studies allow two conclusions: first, automatic syntactic processes, reflected by the ELAN, emerge later during language development than processes reflected by the P600. Second, there is a clear developmental difference in detecting a syntactic violation in simple active compared to more complex passive sentences. The developmental gap between syntax processing (ELAN–P600) in active sentences at age 2.5 years and at age 7 years is quite large. This raises the question of what could be the possible reason for this. Is it due to a genetically determined developmental program of acquisition, and if so can we find a neural correlate for this?

When looking at behavioral findings it is interesting to see that the ERP findings are in line with behavioral findings indicating that children are only able to process non-canonical sentences after the age of 5 years and, depending on the syntactic structure, this ability is found only around the age of 7 years (Dittmar, Abbot-Smith, Lieven, and Tomasello, 2008). The mastery of non-canonical object-first sentences in German is an example:

Den [accusative] *Hund zieht der* [nominative] *Vogel.*

The dog pulls the bird [literal translation].

The bird pulls the dog [sentence meaning].

Such object-first sentences, in which the object-noun phrase marked by accusative is in the initial position and the subject-noun phrase marked by the nominative [actor] is in the final position, were particularly challenging for these children. In such sentences the case-marked determiner (*der*, nominative; *den*, accusative) of the noun phrase is crucial for the identification of the noun phrase's grammatical role and thereby the thematic role assignment necessary for sentence interpretation. It is the nominative case that assigns the subject of a sentence and thereby the actor role, while the accusative case assigns the object of a sentence and thereby the patient or undergoer role.

Two ERP studies using German case-marked sentences were conducted. The first study used a violation paradigm with sentences that contained either two nominative-marked noun phrases or two accusative noun phrases (Schipke, Friederici, and Oberecker, 2011) thereby violating a grammatical principle called theta-criterion (Chomsky, 1981, 1986). This principle holds that to each of the noun phrases in a sentence a thematic role can only be assigned once. As the actor role is assigned to the nominative noun phrase and the patient role to the accusative noun phrase, two nominative noun phrases or two accusative noun phrases in one sentence are a clear violation of the theta-criterion. For both violation types, adults displayed a biphasic N400–P600 ERP response, reflecting thematic-semantic and syntactic processes. For double-nominative violations, 3-year-old children showed an adult-like processing pattern. For double-accusative violations, ERP results indicated that even 6-year-old children did not show an adult pattern. This suggests a late acquisition of the grammatical function of the accusative case. The second ERP study compared the processing of canonical subject-first to non-canonical object-first sentences (Schipke, Knoll, Friederici, and Oberecker, 2012). The ERP pattern in this study differed from adults at all age levels. Word order and case marking were manipulated in German main clauses. Adults' ERPs revealed a negativity for the processing of first accusative marked noun phrase in the object-initial structure. In contrast to adults, the ERPs of the children revealed a number of other ERP effects demonstrating that children in each age group used different strategies, which are indicative of their developmental stage. While 3-year-olds appear to detect differences in the two sentence structures, however, without being able to use this information for sentence comprehension, children aged 4.5 years proceed to use mainly a word-order strategy, processing the topicalized accusative marked noun phrase in both conditions in the same manner, which leads to processing difficulties upon detecting case marking cues at the second noun phrase in the sentence. At the age of 6, children are able to use case-marking cues for comprehension but still show enhanced effort for correct assignment of thematic roles. The data indicate a developmental progress from identification to the use of case marking information that stretches from 3 to 7 years.

Emergence of the Functional Representation: MRI Studies

The full syntactic ability necessary to process non-canonical sentences thus only emerges quite slowly between the age of 3 and 7 years. The ERP pattern observed in the different studies is not yet adult-like. Thus, this raises the hypothesis that the neural network underlying these processes has to reach an adult-like maturational status in order to allow adult-like behavior. Several neuroimaging studies seem to support this hypothesis.

The neural network underlying syntactic processes in the developing brain was investigated in a functional magnetic resonance imaging study with 5- to 6-year-olds using the syntactic violation paradigm (Brauer and Friederici, 2007). Sentences were canonical sentences containing a phrase structure violation similar to those used in earlier ERP studies. Processing of these sentences, as well as correct and semantically incorrect sentences, led to a bilateral activation of the superior temporal gyri and the inferior and middle frontal gyri, but syntactically incorrect sentences specifically activated left Broca's area. Compared with adults, this activation pattern was less lateralized, less specific, and more extended. A time course analysis of the perisylvian activation across correct and incorrect sentences also revealed developmental differences. In contrast to adults, children's inferior frontal cortex responded much later than their superior temporal cortex (figure 6.9). Additionally, in contrast to adults, children displayed a temporal primacy of right-hemispheric over left-hemispheric activation (Brauer, Neumann, and Friederici, 2008), suggesting that they rely heavily on right-hemisphere prosodic processes during auditory sentence comprehension. This is generally in line with the view that young children rely more on suprasegmental prosodic processes in the right hemisphere than older children.

An analysis focusing on the neural work underlying sentence-level semantic processes in 5- to 6-year-old children and adults also provides some evidence for the difference between the neural network recruited in children and adults (Brauer and Friederici, 2007). Activation in children was found bilaterally in the superior temporal gyri and in the inferior and middle frontal gyri when processing correct sentences and semantically incorrect sentences, whereas the language network of adults was more lateralized, more specialized with respect to different aspects of language processing (semantics vs. syntax), and engaged additional areas in the inferior frontal cortex bilaterally. Another functional magnetic resonance imaging study examined lexical-semantic decisions for semantically congruous and incongruous sentences in older children, aged 7 to 10 years, and adults (Moore-Parks et al., 2010). Overall, the results suggested that by the end of children's first decade, they employ a similar cortical network for lexical semantic processes as adults, including activation in left inferior frontal, left middle temporal, and bilateral superior temporal gyri. However, results also still revealed developmental differences, with adults showing greater activation in the left inferior frontal gyrus, left supramarginal gyrus, and left inferior parietal lobule as well as motor-related regions.

In a more recent cross-sectional functional magnetic resonance imaging study we investigated the brain activation during sentence comprehension in three groups of children covering the age of 3 to 10 years and in an adult group (Skeide, Brauer, and Friederici, 2014). The sentences in this study varied the factor semantics (semantic plausibility) and syntax (canonical subject-first vs. non-canonical object-first structure). Plausibility concerned real word knowledge (plausible sentences were those in which a tall animal (fox) carried a small animal (beetle), implausible sentences were those in which the small animal (beetle) carried a tall animal (fox). The functional magnetic resonance imaging data from

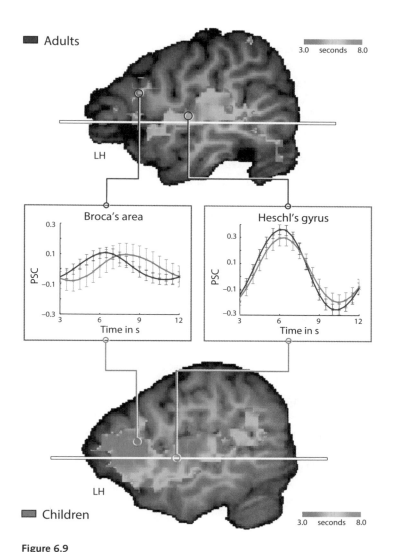

Figure 6.9
Time course of brain activation during auditory sentence processing. Temporal organization of cortical activation during auditory sentence comprehension. Brain activation of adults (top) and children (bottom) in sagittal section (x = −50) and horizontal section (z = 2). Data are masked by random-effects activation maps at z = 2.33 and display a color coding for time-to-peak values in active voxels between 3.0 and 8.0 seconds. The white horizontal lines indicate the cut for the corresponding section. Note the very late response in the inferior frontal cortex in children and their hemispheric differences in this region. Middle: diagrams demonstrate examples of activation to sentence comprehension in Broca's area and in Heschl's gyrus for adults (blue) and children (green). Adapted from Brauer, Neumann, and Friederici (2008). Temporal dynamics of perisylvian activation during language processing in children and adults. *NeuroImage*, 41 (4): 1484–1492, with permission from Elsevier.

this study provided evidence that 3- to 4- and 6- to 7-year-old children, in contrast to adults, do not process syntax independently from semantics as indicated by their activation pattern. Semantics and syntax interacted with each other in the superior temporal cortex. Interestingly, it is not until the end of the 10th year of life that children show a neural selectivity for syntax, segregated and independent from semantics, in the left inferior frontal gyrus similar to activation patterns seen in the adult brain. Thus, for processing the more complex object-first sentences compared to the less complex subject-first sentences, it is not until early adolescence that a domain-specific selectivity of syntax can be observed at the brain level.

In a functional magnetic resonance imaging study with 10- to 16-year-old children, Yeatman, Ben-Shachar, Glover, and Feldmann (2010) investigated sentence processing by systematically varying syntactic complexity and observed broad activation patterns in frontal, temporal, temporo-parietal, and cingulate regions. Independent of sentence length, syntactically more complex sentences evoked stronger activation in the left temporo-parietal junction and the right superior temporal gyrus, but not in frontal regions. Activation changes in frontal regions in their study correlated instead with syntax perception measures and vocabulary as tested in a standard language development test.

The adequate description of the neural basis of language processes requires additional information about the relation between brain structure and function. This holds true for both the gray matter of the relevant language regions and the white matter fiber bundles connecting these. Earlier studies on the relationship between structural brain maturation as indicated by changes in the gray matter and language development found that receptive and productive phonological skills of children between 5 and 11 years correlate with measurements of gray matter probability in the left inferior frontal gyrus (Lu et al., 2007). Another study found that the gray matter of the left supramarginal gyrus and left posterior temporal regions correlates with vocabulary knowledge in teenagers between 12 and 17 years (Richardson, Thomas, Filippi, Harth, and Price, 2010). In general, gray matter density decreases along development, in higher-order association regions (Gogtay et al., 2004; Brain Development Cooperative Group, 2012). Specifically, gray matter volume in those frontal and parietal brain regions that are involved in sentence processing in adults appear to decrease between the age of 7 and 12 years (Giedd et al., 1999; Sowell et al., 2003).

A more recent study focused on the gray matter of the language network and its relation to sentence processing and verbal working memory (Fengler, Meyer, and Friederici, 2016). In particular for the processing of syntactically complex sentences it has been argued that these may require a good verbal working memory capacity (Felser, Clahsen, and Münte, 2003) and that this capacity is most relevant in children's processing of complex sentences (Montgomery, Magimairaj, and O'Malley, 2008; Weighall and Altmann, 2011). Therefore in our study on the late functional development of syntactic processing we considered three aspects in different age groups (5–6 years, 7–8 years, and adults): first, functional brain activity during the processing of increasingly complex sentences; second, brain structure in the language-related areas (BA 44 and left temporo-parietal region); and third, the behavioral performance on comprehending complex sentences and the performance on an independent verbal working memory test (Fengler et al., 2016). All three factors contribute to the developmental difference between children and adults for the processing of complex center-embedded sentences. First, in children and adults, in the whole-brain analysis, brain-functional data revealed a qualitatively similar neural language network involving the left BA 44, the left inferior parietal lobe together with the posterior superior temporal gyrus (inferior parietal lobe/posterior superior temporal gyrus), the supplementary motor area, and the cerebellum. While the activation of BA 44 in the inferior frontal gyrus and the inferior parietal lobe/posterior superior temporal gyrus predicted sentence comprehension performance across age groups, only adults showed a functional selectivity in these brain regions, for more complex sentences. Second, the attunement of both BA 44 and the inferior parietal lobe/posterior superior temporal gyrus toward a functional selectivity for complex sentences is predicted by region-specific gray matter reduction. Third, the attunement of the latter region (inferior parietal lobe/posterior superior temporal gyrus) was additionally predicted by the verbal working memory capacity. These data are the first to indicate that both brain structural maturation and verbal working memory expansion lead to efficient processing of complex sentences.

A crucial study combined functional and structural MRT measures (Skeide, Brauer, and Friederici, 2016) focusing on the white matter of the language network in order to evaluate the contribution of the functional involvement of the network's gray matter and the white matter during language processing. Two subnetworks were defined: (1) a dorsal language network consisting of BA 44 (pars opercularis) in the inferior frontal gyrus, and the posterior superior temporal gyrus connected via the dorsal fiber tract arcuate fasciculus, and (2) the ventral language network consisting of BA 45 (pars triangularis) in the inferior frontal gyrus and the anterior superior temporal gyrus connected via the IFOF (compare figure 6.13). Children between the age of 3 and 10 years as well as adults underwent functional magnetic resonance imaging while performing a sentence comprehension task for canonical subject-first and non-canonical object first sentences. Functional activation within those two regions that constitute the dorsal language system, namely, the pars opercularis (BA 44) and the posterior superior temporal gyrus, increase their activation across age significantly, whereas the two regions that constitute the ventral language system, namely, the pars triangularis (BA 45) and the anterior superior temporal gyrus, do this to a much lesser degree (see figure 6.10).

For the same individuals the white matter maturation of the fiber tracts was determined using diffusion magnetic resonance imaging. The fractional anisotropy as a measure indicating the myelin density of the fiber tracts was shown to increase across age, in particular for the dorsal fiber tract targeting BA 44. Correlational analyses using a partial correlation approach including both information about the gray and the white matter indicate that both are relevant in language development. We showed activation increase in the gray matter

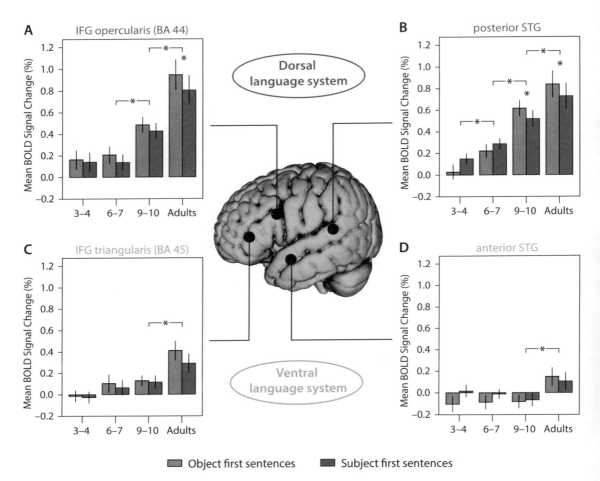

Figure 6.10
Functional activation in language-related brain areas across development. Functional activation in 4 syntax-relevant regions-of-interest (ROIs). (A) the pars opercularis (BA 44), (B) the posterior superior temporal gyrus (pSTG), (C) the pars triangularis (BA 45), and the anterior superior temporal gyrus (aSTG). Brain activation (mean BOLD signal change) is shown for simple-subject-first (blue) and complex-object-first (red) sentences in the four age groups. The two dorsally located ROIs that constitute the dorsal language system (A, B) revealed main effects of both sentence types against baseline with the pars opercularis (BA 44) showing an exclusive effect for the difference between both sentence types. In the ventrally located ROIs that constitute the ventral language system (C, D), only a main effect of object-first sentences was detected in pars triangularis (BA 45). Asterisks indicate significant differences between conditions. Error bars indicate standard errors of the means. From Skeide, Brauer, and Friederici (2016). Brain functional and structural predictors of language performance. *Cerebral Cortex*, 26 (5): 2127–2139, by permission of Oxford University Press.

can predict accuracy of comprehension performance. But, the density of the white matter of the dorsal fiber tract predicts accuracy and speed of sentence comprehension (Skeide et al., 2016). These data provide an answer to the question of what determines the late emergence of the ability to process syntactically complex sentences. This is the first study demonstrating that the achievement of certain milestones in language development, such as the processing of syntactically complex sentences, depends on the maturational state of the brain's structure (see section 6.3 below for details).

The direction of the relation between the increase of myelin and behavioral performance still has to be determined. Is it the increased use of complex syntax that leads to increased myelination of the dorsal fiber tract targeting BA 44, or is it that the biologically determined increase of the myelin of this fiber tract enhances syntax performance? If the former view would hold, syntax training in children below the age of 6 years should have a positive effect on the dorsal white matter fiber tract if the latter holds, this should not be the case. So far this cannot be reported, although there were only few attempts to do so. In my view biological maturation and language training interact, and it may be that the neurobiological program includes time windows during which the brain is particularly plastic and reactive to a certain input and thus training. This has been shown for the sensory systems as well as for language acquisition in general (Werker and Hensch, 2015), and it may also hold for specific aspects of language such as complex syntax and the maturation of the white matter fiber tract targeting BA 44 in particular.

Summary
The electrophysiological findings reviewed here reveal clear developmental changes for syntactic processes between the age of 3 years and later developmental stages. The magnetic resonance imaging data show that the development toward full language performance is dependent on maturational changes in the gray and white matter. While the gray matter in the language-relevant regions decreases with age, the white matter connectivity between these regions increases with age. The findings provide suggestive evidence that the full language capacity can only be reached once the brain has fully matured.

6.3 Structural and Functional Connectivity during Development

In the previous section I alluded to the development of the structural language network and its relation to language function in childhood. Here I will put this into context with data from infancy and data from adulthood and discuss it in the context of the few existing language-relevant developmental functional connectivity studies.

The general frame for the functionality of the neural language network is set by its adult-like structural maturity of the fronto-temporal white matter fiber tracts (refer back to part II). Given that functional connectivity between brain regions within a functional network cannot be totally independent from the structural connectivity between the respective brain regions, we assume that the maturation of the structural connectivity has an impact on the functional connectivity of the language network.

With respect to the structural connectivity, diffusion-weighted magnetic resonance imaging data from newborn infants indicate that a dorsal connection between the inferior frontal gyrus and the superior temporal gyrus via the superior longitudinal fasciculus/ arcuate fasciculus, which has been associated with complex syntax processing in the adult literature (Friederici, 2011; Wilson et al., 2011), is not yet myelinated at birth (Perani et al., 2011), nor at 1 to 4 months of age (Leroy et al., 2011). But the fiber tract of the dorsal pathway linking the premotor cortex and the superior temporal gyrus, which supports sensory-motor mapping of sound to articulation as seen in adults (Saur et al., 2010), is already present at birth (Perani et al., 2011),[4] and was also demonstrated in infants at 1 to 4 months of age (Dubois et al., 2009). As indicated in figure 6.11, diffusion weighted data in newborns and adults differ in their dorsal pathways connecting the posterior superior temporal gyrus to the frontal cortex, with the pathway reaching the premotor cortex being present in both groups, but with the pathway reaching into Broca's area not yet being matured.

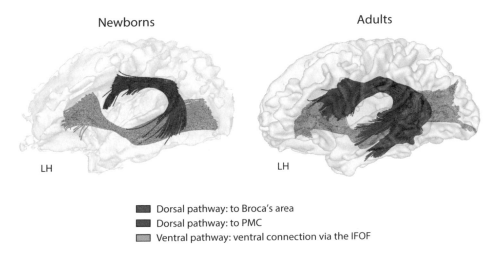

Figure 6.11
Fiber tract pathways in newborns and adults. Structural connectivity results. Fiber tracking of diffusion tensor imaging data for newborns and adults for speech-relevant regions with seed in Broca's area and seed in the precentral gyrus/premotor cortex. Two dorsal pathways are present in adults: one connecting the temporal cortex via the arcuate fasciculus (AF) and the superior longitudinal fasciculus (SLF) to the inferior frontal gyrus, i.e., Broca's area (purple); and one connecting the temporal cortex via the AF and SLF to the precentral gyrus, i.e., premotor cortex (PMC) (blue). In newborns, only the pathway to the precentral gyrus can be detected. The ventral pathway connecting the ventral inferior frontal gyrus via the inferior fronto-occipital fasciculus (IFOF) to the temporal cortex (orange) is present in adults and newborns. Adapted from Perani et al. (2011). The neural language networks at birth. *Proceedings of the National Academy of Sciences of the United States of America*, 108 (38): 16056–16061.

Thus, the superior longitudinal fasciculus/arcuate fasciculus, which has been related to support hierarchical syntactic processes in adults, is not yet myelinated in infants (Perani et al., 2011) and has to develop gradually during childhood. The structural data indicate that it is slowly maturing during childhood (Brauer and Friederici, 2007) and that it predicts the increase in syntax performance toward adulthood (Skeide et al., 2016).

Given these changes in the "strength" of structural connectivity between language-relevant areas, we expect developmental changes in the functional connectivity as well. Functional connectivity provides information about whether and how a particular region functionally cooperates with other regions in the brain. There are, however, only a few developmental studies that address this issue. The immature structural connection between BA 44 and the posterior superior temporal gyrus/superior temporal sulcus within the left hemisphere accompanies the absence of a functional connectivity between these regions. In newborns there is only a functional connectivity between left BA 44 and right BA 44, and between left posterior superior temporal gyrus/superior temporal sulcus and its right hemispheric homolog (Perani et al., 2011). When comparing the functional connectivity of BA 44 and the posterior superior temporal gyrus/superior temporal sulcus between infants and adults it was only found to be significant in adulthood—see figure 6.12 and Lohmann et al., 2010. That is at an age when the structural connectivity is fully matured. These

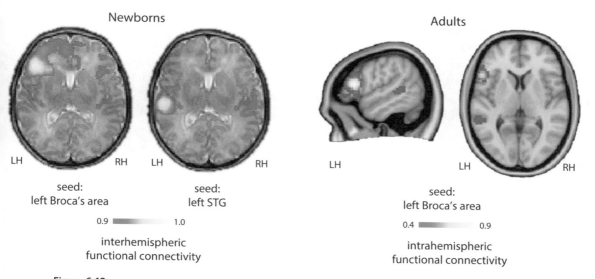

Figure 6.12
Functional connectivity in newborns and adults. Functional connectivity as revealed by analyzing the correlation value of low-pass-filtered residuals of language experiments in newborns and adults with seeds in Broca's area and in the left posterior superior temporal gyrus (STG). Newborns revealed a functional connectivity between the two hemispheres whereas adults showed a functional connectivity within the left hemisphere. Adapted from Perani et al. (2011). The neural language networks at birth. *Proceedings of the National Academy of Sciences of the United States of America*, 108 (38): 16056–16061.

functional connectivity analyses were based on data from language processing studies analyzing the low frequency fluctuations independently from the different language conditions. However, since analyses of low frequency fluctuation of the language studies differ from those on data from non-language studies, it was concluded that the functional connectivity observed on the data from language studies represents a basic language network (Lohmann et al., 2010).

Structurally, the dorsal fiber tract connecting the posterior superior temporal gyrus and pars opercularis in the inferior frontal gyrus (BA 44) as part of the dorsal language system slowly matures between the age of 3 and 10 years as indicated by the increase of fractional anisotropy (see figure 6.13A and B) (Skeide et al., 2016). This maturational increase is correlated with the performance of processing object-first sentences, both with respect to accuracy and speed of response (see figure 6.13D). No such correlation was observed for the ventral pathway, that is, the IFOF (see figure 6.13C and E).

The first task-related functional connectivity study (Vissiennon, Friederici, Brauer, and Wu, 2017) in 3- and 6-year-old children combined functional activation and functional connectivity analyses. Functional magnetic resonance imaging and behavioral measures were utilized to investigate functional activation and functional connectivity in 3-year-old and 6-year-old children during sentence comprehension. Transitive German sentences varying the word order (subject-initial and object-initial) with case marking were presented auditorily. Both age groups showed a main effect of word order in the left posterior superior temporal gyrus, with greater activation for object-initial compared to subject-initial sentences. However, age differences were observed in the functional connectivity between left posterior superior temporal gyrus and BA 44 in the left inferior frontal gyrus. The 6-year-old group showed stronger functional connectivity between the left posterior superior temporal gyrus and BA 44 in the left inferior frontal gyrus than the 3-year-old group. For the 3-year-old group, in turn, the functional connectivity between left posterior superior temporal gyrus and left BA 45 was stronger than its connectivity with left BA 44. This study demonstrates that while task-related brain activation was comparable in the two age groups, the behavioral differences between age groups were reflected in the underlying functional organization of the neural language network revealing the ongoing development.

Another way of gaining knowledge about the relation between functional connectivity and language function is a method that uses resting-state functional magnetic resonance imaging data and evaluates their correlation with language tasks. Resting-state functional magnetic resonance imaging measures brain activity while a participant is "doing nothing," that is to say, not involved in a task (Biswal et al., 1995) (see introduction, and section 4.2). This approach was applied in adults to describe changes in the intrinsic functional connectivity in participants suffering from various cognitive diseases (Alaerts et al., 2015; Tadayonnejad, Yang, Kumar, and Ajilore, 2015) and in participants showing different levels of cognitive performance (Zou et al., 2013).

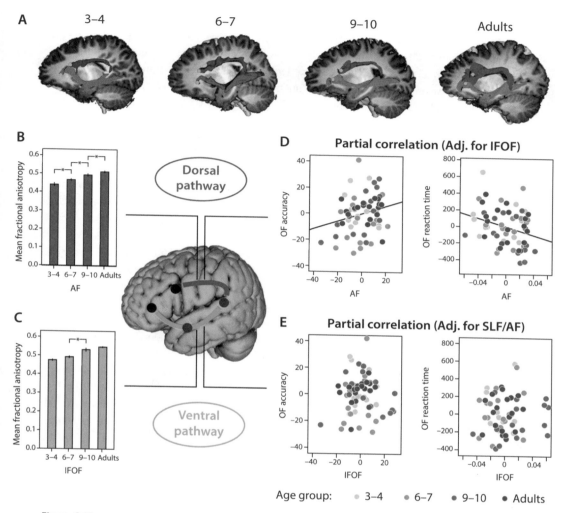

Figure 6.13
Maturation of dorsal and ventral fiber tract pathway. The maturation of the dorsal and ventral fiber pathways and their involvement in sentence processing. (A) Seeding in the left IFG inferior frontal gyrus (IFG) in the pars opercularis (BA 44) we reconstructed the left arcuate fasciculus (AF) (purple), and seeding in the left pars triangularis (BA 45) we reconstructed the left inferior-fronto-occipital fasciculus (IFOF) (orange). Depicted are 3-D renderings of all tracts for each age group. (B, C) Mean fractional anisotropy profiles of both tracts differing significantly between 3–4 and 6–7, 6–7 and 9–10, and 9–10 and adults in the AF and between 6–7 and 9–10 in the IFOF. (D, E) The individual FA values were significantly more strongly related to the individual performance on object first (OF) sentences in the dorsal, but not in the ventral tract when adjusting for the effect of the respective other tract by parietal correlation analyses. Asterisks indicate significant differences between age groups. Error bars indicate standard errors of means. Adapted from Skeide, Brauer, and Friederici (2016). Brain functional and structural predictors of language performance. *Cerebral Cortex*, 26 (5): 2127–2139, by permission of Oxford University Press.

Resting-state functional magnetic resonance imaging has also been used to describe developmental changes. Resting-state functional connectivity data from German 5-year-old children and adults were analyzed. For children, data were correlated with the performance on sentence processing (Xiao, Friederici, Margulies, and Brauer, 2016a). While adults show a strong intrahemispheric functional correlation between the left inferior frontal gyrus and the left posterior superior temporal gyrus, a strong interhemispheric functional correlation between the left inferior frontal gyrus and its right homolog was predominant in children. The correlational analysis between resting-state functional magnetic resonance imaging and sentence-processing performance in children revealed that a long-range functional connectivity between the left posterior superior temporal sulcus and the left inferior frontal gyrus is associated with the success in processing syntactically complex sentences. Thus, it appears that there is already an intrinsic functional connectivity within the left hemispheric language network by the age of 5, although there are still significant differences between the child and the adult system.

Another resting-state functional magnetic resonance imaging study (Xiao, Friederici, Margulies, and Brauer, 2016b) took a longitudinal approach and showed developmental effects in German preschool children at age 5 and age 6. A voxel-based network measure was used to assess age-related differences in connectivity throughout the brain. To explore the connection changes, a seed-based functional connectivity analysis was performed based on the left posterior superior temporal gyrus/superior temporal sulcus. The correlations between resting-state functional connectivity and language performance as tested by a standard German comprehension test revealed that the development of language ability was positively correlated with the resting-state functional connectivity between the left posterior superior temporal gyrus/superior temporal sulcus and the right inferior frontal sulcus as well as the posterior cingulate cortex in the group as a whole. High performance on processing syntactically complex sentences, in particular, was positively correlated with functional connectivity between the left posterior superior temporal gyrus/superior temporal sulcus and the left inferior frontal gyrus.

Summary

This section reviewed findings indicating that increased language ability is correlated with an increase in structural and functional connectivity between language-related brain regions in the left hemisphere, the inferior frontal gyrus, and the posterior superior temporal gyrus/superior temporal sulcus. Specifically, we have seen that the dorsal fiber tract connecting BA 44 in the inferior frontal gyrus to the posterior temporal cortex is crucial for the full achievement of syntactic abilities.

6.4 The Ontogeny of the Language Network: A Model

Humans are born with a brain that already contains connections between neurons, and between brain regions. Learning a language involves modulating and partly restructuring these connections as a function of the speech input. The auditory sensory systems in the temporal cortex required to deal with the speech input functionally are in place at birth. Higher order language processing systems only functionally develop later, and so do those of the long-range fiber tracts that connect relevant processing regions.

The brain-based language acquisition model by Skeide and Friederici (2016) reviews the available data and postulates two main developmental stages: a first stage covering the first three years of life during which bottom-up processes mainly based on the temporal cortex and the ventral language network are at work, and a second stage beyond the age of 3 until late childhood during which top-down processes emerge gradually under the involvement of a specialization and increasing connectivity of the left inferior frontal gyrus as part of the dorsal language network (see figure 6.14).

During the first stage, infants mainly activate the temporal cortices bilaterally during speech perception. Newborns show hemodynamic activity in the mid and posterior superior temporal gyrus stronger in the right than in the left hemisphere (Perani et al., 2011). By the age of 3 months, normal speech compared to backward speech activates the left temporal cortex (Dehaene-Lambertz et al., 2006). Early speech segmentation and the detection of phonological word forms can be observed at the age of 6 months (Bergelson and Swingley, 2012), and by 9 months infants are able to generalize word meaning into lexical categories (Friedrich et al., 2015). Between 12 and 18 months of age, the phase during which children learn a number of words every day, a clear N400 ERP effect for lexical-semantic processes emerges (Friedrich and Friederici, 2010). This N400 effect has been localized in the temporal lobe, that is, in central and posterior temporal regions extending from the left superior temporal gyrus to the inferior temporal gyrus (Travis et al., 2011).[5]

In order to extract lexical elements and possible phrases from the continuous speech stream, the speech input has to be chunked into subparts by the age of 5 months. The speech input can be chunked into phonological phrases on the basis of a dominant acoustic cue, namely the pause, (Männel and Friederici, 2009), but it takes until age 3 before children are able to recognize a phonological phrase boundary without this acoustic cue of a pause (Männel and Friederici, 2011). This ability relates to first syntactic knowledge, which is acquired at that time. While the acoustic cues are processed in the temporal cortex, the impact on these processes from syntactic knowledge must be seen as top-down processes from the prefrontal cortex that develop during this period. At the age of 2.5 years, children's first neurophysiological indications of syntactic processes can be found. Children at this age show an ERP response pattern similar to adults when they encounter a word category error in a prepositional phrase, reflected in a biphasic pattern of an ELAN and a P600 (Oberecker and Friederici, 2006). The ELAN has been found to originate from left anterior temporal sources (Herrmann, Maess, and Friederici, 2011) and possibly also from additional inferior frontal sources (Friederici, Wang, Herrmann, Maess, and Ortel, 2000). In children these processes, as well as the build-up of syntactic phrases, are supported by

A

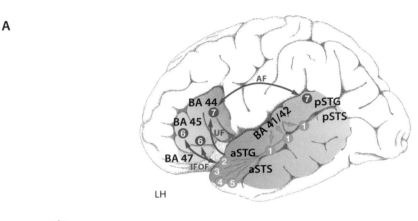

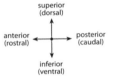

▭ inferior frontal gyrus (IFG)	① Phonological word form detection
▬ superior temporal gyrus (STG)	② Morphosyntactic categorization
▭ middle temporal gyrus (MTG)	③ Lexical semantic categorization
	④ Lexical access & retrieval
AF .... arcuate fasciculus	⑤ Phrase structure reconstruction
UF .... uncinate fasciculus	⑥ Analysis of semantic relations
IFOF .. inferior fronto-occipital fasciculus	⑦ Analysis of syntactic relations

bottom-up / top-down

B Development in years

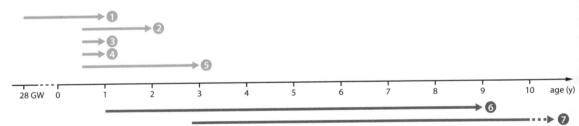

Figure 6.14
Development of the language network. The evolving cortical circuit underlying language comprehension during development. (A) Neural implementation of the main processing steps (1–7) in both hemispheres schematically projected onto the adult brain, left hemisphere only. (B) Language acquisition timeline (age in years). Earliest manifestation of a precursor (starting point of the lines) and first adult-like appearance (end point of the lines) of a certain neural processing milestone at the group level. Adapted from Skeide and Friederici (2016). The ontogeny of the cortical language network. *Nature Reviews Neuroscience*, 17: 323–332.

the ventral language system involving the temporal cortex and the ventral pathways to the ventral portions of the inferior frontal gyrus.

During the second developmental stage, top-down processes come into play more and more. Processing sentences involves identifying the semantic and syntactic relations between words and phrases and using this information effectively. In 3-to-7-year-old children, semantic- and syntax-related hemodynamic activity largely overlaps (Brauer and Friederici, 2007; Nunez et al., 2011), and moreover, the syntax and semantic factors statistically interact in the middle and posterior positions of the left superior temporal gyrus. It is not until the age of 9 to 10 years that semantic and syntactic domains are separated into distinct neural networks. Evidence of a functional selectivity of a given brain region for syntax in the left inferior frontal gyrus is first seen in 9- to 10-year-old children (Skeide et al., 2014). But at this age the selectivity pattern is still different from adults. While in adults BA 44 is found to be selective for complex syntactic processes, functional selectivity occurs in BA 45 in children by the age of 9 to 10. This indicates that at this age BA 44 has still not reached full efficiency for processing syntax.

The ordering of phrases at the sentence level to achieve comprehension involves the left inferior frontal gyrus and the left posterior superior temporal gyrus/superior temporal sulcus connected by a dorsally located fiber bundle (Wilson et al., 2011). The maturation of this dorsal language system and, in particular, the dorsal fiber bundle connecting the posterior superior temporal gyrus/superior temporal sulcus and BA 44 is crucial for processing syntactically complex non-canonical sentences (Skeide et al., 2016).

Although this model is based on a sparse empirical database, it appears that the first phase of language is mainly based on bottom-up processes recruiting the temporal cortices; the second phase employs top-down processes, thereby involving the frontal cortex and its backward projections to the posterior temporal cortex. Compelling evidence for a view that infants, in contrast to adults, initially learn from the speech input without the involvement of top-down processes located in the prefrontal cortex comes from a comparison on language learning in infancy and adulthood. In this study healthy adults had to learn a novel language under two conditions: a normal listening condition and under a condition where the prefrontal cortex known to house controlled processes was downregulated by neurostimulation. The study applied a language-learning paradigm previously used with infants (Friederici et al., 2011) (see figure 6.7). The comparison between infants and adults revealed a clear difference between 4-month-old infants and adults under normal listening: infants showed an ERP effect indicating acoustic-phonological structure-learning, whereas adults did not (Friederici, Mueller, Sehm, and Ragert, 2013). Adults, only demonstrated a structure-learning ERP effect similar to infants when the left prefrontal cortex was downregulated in its activity by neurostimulation. Thus, phonological learning, as observed in infants, appears to be more successful when controlled processes from the prefrontal cortex do not affect acoustic-phonological processes in a top-down manner.

Summary

The present model of development of the neural language network holds that during the first months of life, language processing is largely input-driven and supported by the temporal cortex and the ventral part of the language network. Beyond the age of 3 years, when top-down processes come into play, the left inferior frontal cortex and the dorsal part of the language network are recruited to a larger extent.

IV

In the last part of the book I will consider evolutionary aspects of the language network by providing comparisons between species based on behavioral as well as neuroanatomical aspects. These data make a strong suggestion concerning the crucial differences between the human who easily learns and uses language and the non-human primates who only are able to learn simple sequences but not complex syntactic sequences. The crucial neuroanatomical differences appear to be the neuroarchitectonic structure of the posterior part of Broca's area, namely BA 44, and its connection to the posterior temporal cortex—structures that we have seen to be involved in syntactic processes.

7

Evolution of Language

Historically, the debate on the evolution of language goes back to Plato. In one of his dialogues he questions whether language is a system of arbitrary signs or whether words have an intrinsic relation to those things to which they refer. This question has persisted over the centuries and is still discussed in the context of language evolution. If words are arbitrary, where do they come from?

Darwin's *The Descent of Man* (1871), in which cognitive abilities were compared between human and non-human primates, turned the philosophical question into a biological one. A brief review of major theories of language evolution will provide some insight into the discussion.

7.1 Theories of Language Evolution

For centuries, philosophers and other scientists have thought about the origins of language, formulating a variety of different theories. Only a few of these theories have survived the time. Historically, one of the first essays on the topic stems from Johann Gottfried Herder (1772), who submitted his piece, entitled *Treatise on the Origin of Language*, to the Berlin Academy of Science, which had offered a prize for the best essay on this subject. In his essay he claims that language is a human creation. Humans' first words were the origin of consciousness. The phonological forms of these words were primarily close to the sounds of objects they denominated, they were onomatopoetic. As humans developed, words and language developed and became more and more complex. In his view grammar is not at the origin of language—as "grammar is no more than a philosophy about language"—but rather it is a "method of its use" that evolved only as the need to talk about abstract entities increased. Herder's theory stands in direct contrast to modern views such as those of Chomsky, who considers syntax to be the core of language.

In *The Descent of Man*, Darwin (1871) sets forth his theory of human language. He clearly states that language is what differentiates humans from other primates, and that articulation is not what makes the crucial difference, but rather the brain. He argues that language is not a true instinct, as it must be learned, but that humans have an instinctive tendency to do so. Although he accepts the role of gesture in this context, he rejects the idea of a gestural origin of language; instead he considers musical aspects to be possibly relevant. This is an interesting thought given that prosody, known as the sentence melody of language, has been shown to support language learning.

A number of theories of language evolution do not consider any aspects related to the brain. In *The Evolution of Language*, W. Tecumseh Fitch (2010) describes them as "lexical protolanguages." For an excellent discussion of these theories—which appear in works such as *Language and Species* (Bickerton, 1990); *The Cultural Origins of Human Cognition* (Tomasello, 1999); *Grooming, Gossip and the Evolution* (Dunbar, 1996); and *Foundations of Language* (Jackendoff, 2002)—I refer the reader to Fitch (2010).

In the context of brain-related theory of the evolution of language, Rizzolatti and Arbib (1998) based their theory on experimental work with non-human and human primates. Their claim is that language evolved from motoric gestures. It is based on the finding that in macaque monkeys viewing a grasping movement activated neurons not only in the visual cortex but also neurons in the hand-motor cortex, which fired in response to the visual stimulus (Rizzolatti et al., 1996). These neurons were subsequently called *mirror neurons*. Similar effects were then observed in humans at the level of brain systems (rather than single neurons) and were called the *mirror system* (Iacoboni et al., 1999, 2005). In humans the mirror system involves Broca's area, which is found to activate together with the visual input system and the parietal cortex. These three brain regions are taken to constitute the cortical mechanisms of imitation (Iacoboni et al., 1999). Since Broca's area is part of this mechanism, it has been proposed that the mirror system in non-human and human primates might serve as the crucial evolutionary link between non-human and human primates, suggesting a direct step from gesture to language. This conclusion, however, has triggered a debate that is still ongoing, which includes the following two counterarguments: (1) gestures are not arbitrary, whereas words in a language are (Hurford, 2004); and (2) languages contain function words and syntax that allow the displacement of an element out of a language sequence, but not in a gesture sequence (Moro, 2014).

With respect to the neural basis of motor action and language the following can be stated: Although action imitation and language both involve BA 44 as part of Broca's area, a meta-analysis shows that the two domains involve different subclusters within BA 44 (Clos, Amunts, Laird, Fox, and Eickhoff, 2013). Moreover, for the language domain I have argued (Friederici, 2002) and reviewed ample evidence (Friederici, 2011) that Broca's area is part of a neural network including the posterior superior temporal gyrus/superior temporal sulcus, which does not hold for gestures. I acknowledged the involvement of Broca's area in processing sequences in other domains such as music and action, but argued that the specificity of Broca's area for a particular domain comes from the respective network. The neural network differs as a function of the domain, and it clearly differs between action and language.

The evolution of language has been discussed since Darwin (1871), mainly focusing on non-human primates and their cognitive abilities in comparison to human primates. Due to the difficulty of finding evidence for syntax-like abilities in non-human primates, a description of the phylogenetic evolution of language remains challenging. Today, two ways of approaching this issue are pursued. One focuses on a functional model independent of the consideration of close phylogenetic continuity. This approach uses songbirds as an animal model for language learning because these animals exhibit vocal learning and are able to learn auditory sequences. The second avenue is to also look at our closest relatives, the non-human primates. In a recent review, Petkov and Jarvis (2012) propose that both vocal learners (songbirds) and non-vocal learners (monkeys) should be considered when tracing the origins of spoken language (i.e., speech). While songbirds could serve as a model of vocal sequence learning, monkeys—as our closest relative—can provide information about the evolution of more general cognitive abilities of learning, including memory and attention.

Summary

The past and current views of language evolution all focus on a crucial question: What led to the human faculty of language, and can it be explained by continuity of phylogenesis from non-human to human primates? The view presented in this book holds that the difference between human and non-human primates lies in the structure of their brains, particularly in the way the relevant brain areas are connected by white matter fiber tracts.

7.2 Processing Structured Sequences in Songbirds

Since the processing of structured sequences is considered to be a precursor of processing syntax, a number of studies have investigated to what extent other animals can learn and process rule-based structured sequences. Two types of animals have been examined as non-human models for language learning: songbirds and non-human primates. Here I will briefly review studies on auditory sequence processing in songbirds and discuss the respective studies on non-human primates section 7.3.

Songbirds are clearly not our direct ancestors and are further away from us in the evolutionary tree, but they are equipped with a vocal learning system that has been said to bear a certain similarity to the human system. Song learning in birds and speech learning in humans has two phases: an auditory learning phase and a sensory-motor vocal learning phase. Behaviorally, it has been shown that both human infants and songbirds learn through a combination of a certain predisposition and auditory input and that there is a sensitive period for auditory-vocal learning. Moreover, both human infants and young songbirds go through a phase during which the young individual's vocalization gradually approaches the adult production. This phase is called *babbling* in humans and *subsong* in songbirds (Bolhuis and Moorman, 2015).

The auditory-vocal learning system allows birds to learn sequences of elements in a certain temporal order (Berwick, Okanoya, Beckers, and Bolhuis, 2011; Berwick, Friederici, Chomsky, and Bolhuis, 2013; Bolhuis, Tattersall, Chomsky, and Berwick, 2014). As these sequences may be governed by some kind of syntactic rule (Okanoya, 2004), it has been called "phonological syntax" (Marler, 1970). This phonological syntactic ability does not imply the ability to acquire context-free syntactic rules that generate hierarchical structures characteristic for natural languages. However, researchers have claimed that songbirds can even acquire such rules (Gentner, Fenn, Margoliash, and Nusbaum, 2006; Abe and Watanabe, 2011). This claim is highly controversial and has been challenged based on the stimulus material used in these studies (Beckers, Bolhuis, Okanoya, and Berwick, 2012; ten Cate and Okanoya, 2012).

Gentner and colleagues (2006) and Abe and Watanabe (2011) reported that songbirds are able to process A^nB^n grammars. However, although the grammar used by Abe and Watanabe (2011) can be described as being asymmetric hierarchical structure, similar to natural languages, the detection of incorrect sequences in the experiment could, in principle, be performed based on the following computation that does not require hierarchy building: the incoming sequence of (adjacent) elements and, upon detection of the center element, reverse-and-match the following sequence to the initial sequence (figure 7.1). The underlying mechanisms used to process A^nB^n grammars thus remain speculative, for both songbird studies (Gentner et al., 2006; Abe and Watanabe, 2011).

However, it is clear from these data that songbirds can learn simple rule-based sequences. In birds the underlying neuroanatomy for this ability involves an auditory system (secondary auditory regions), an anterior forebrain pathway essential for sensorimotor learning mediated by subcortical regions, and a song motor pathway involved in song production. It has been demonstrated that lesions of one of these systems, in particular the mediating pathway, lead to learning deficits in songbirds (for a review, see Bolhuis and Moorman, 2015). This suggests that the path from sensory-auditory system to motor-output system is crucial for song learning. A similar pathway from the auditory system to the motor system is already present at birth in the human infant, providing a good basis for auditory learning and babbling in humans. This pathway is the dorsal pathway connecting the posterior superior temporal gyrus and the premotor cortex (Perani et al., 2011), which must be distinguished from the dorsal pathway relevant for syntax targeting BA 44.

Summary

During the evolution of language two crucial abilities had to evolve: these are first, sensory-motor learning, and second, the ability to process hierarchical structures. Sensory-motor learning of simple rule-based sequences is an ability that is present songbirds. However, the ability to process hierarchical structures is not present in songbirds. Therefore, it is conceivable that the ability to process structural hierarchies is what should be considered as a crucial step toward the language faculty.

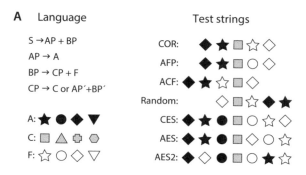

Figure 7.1
Artificial grammar used in songbird study. (A) Left panel: description of the grammar as a center-embedded structure and members of categories A, C, and F; Right panel: examples of test strings; the test strings can also be described differently: (B) Symmetrical description of center-embedded structure (CES) and violating test strings (AES and AES2). In (A) and (B) the center element is represented as a gray cube. Adapted from Friederici (2012). Language development and the ontogeny of the dorsal pathway. *Frontiers in Evolutionary Neuroscience*, 4: 3. doi: 10.3389/fnevo.2012.00003.

7.3 Comparing Monkeys and Humans

In recent years a number of studies have compared human and non-human primates on the behavioral level as well as on various neuroscientific levels going from the cytoarchitectonic to the network level. In the following subsections, we will look at the behavioral differences reported and also at the brain structural differences.

I will review the success of sequence learning in different species and discuss the possible underlying neural basis. The data from these studies suggest that the grammar learning mechanisms described above can be related to two different neural circuits. One circuit connects the auditory system in the temporal cortex to the motor system in the frontal cortex, and subserves the sensory-to-motor mapping. This circuit can support the learning of $(AB)^n$ grammars as well as artificial grammars that do not require the build-up of structural hierarchy. A second circuit, which connects the posterior temporal cortex to the posterior portion of Broca's area (BA 44) in the frontal cortex, may be needed for learning natural grammars with their hierarchical structures. The latter may only be active in humans as suggested by findings indicating that natural grammars (for a review, see

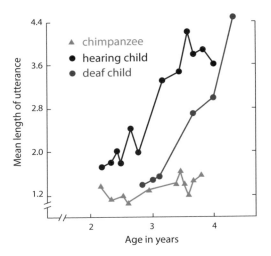

Figure 7.2
Utterance length across species. Mean length of utterance between 2 and 4 years in a hearing child (black), a deaf child (blue), and a chimpanzee (red). The term *utterance* refers to responses generated by speech in the hearing child, gestures in the deaf child, and visual symbols in the case of the chimpanzee. Adapted from Pearce (1987). Communication and Language (Chapter 8). In *An Introduction to Animal Cognition*, 251–283. Lawrence Erlbaum Associates.

Friederici, 2011) and artificial grammars requiring the buildup of syntactic hierarchies (Musso et al., 2003; Opitz and Friederici, 2004) activate BA 44.

Behavioral Differences
Several studies have focused on the behavior of primates—human and non-human. These indicate that the main difference between human and non-human primates lies in the ability to combine words or meaningful elements into larger sequences. In one of the first behavioral studies it was demonstrated that humans differ from the chimpanzee in their ability to generate word sequences (Pearce, 1987). Both a hearing and a deaf child compared to the chimpanzee showed a clear increase of the length of utterances (see figure 7.2).

Central to the discussion on sequence processing in primates is not only whether sequences can be learned, but also more crucially the question pertaining to what type of syntactic sequence can be learned.

In this context a fundamental distinction has been made between two grammar types, namely finite state grammars following an $(AB)^n$ rule and phrase structure grammar following an A^nB^n rule (Hauser, Chomsky, and Fitch, 2002; Fitch and Hauser, 2004). The important difference between these two types of grammars is that sequences based on the $(AB)^n$ rule contain adjacent dependencies between an A-element and a B-element, whereas sequences based on the A^nB^n lead to non-adjacent dependencies (compare figure 1.6).

There are at least three possible mechanisms by which grammatical sequence learning for these grammar types can take place: (1) Adjacent dependencies, as in $(AB)^n$ grammars, could be learned by extracting phonological regularities from the auditory input and memorizing these for further use. (2) Non-adjacent dependencies between A and B in A^nB^n artificial grammars that do not involve higher-order hierarchies could be learned through the same mechanism described in (1), as long as no buildup of hierarchy is required. (3) A^nB^n dependencies in a natural grammar, however, require the buildup of hierarchies guaranteed by the computation "Merge," which binds two elements into a minimal hierarchical structure (Chomsky, 1995b). This is the basic mechanism through which a natural grammar with its higher-order hierarchical structure can be processed.

Fitch and Hauser (2004) were the first to investigate grammar learning in human and non-human primates using a finite state grammar and phrase structure grammar. Testing cotton-top tamarins and human adults in a behavioral grammar learning study, they found that humans could learn both grammar types easily, whereas monkeys were only able to learn the finite state grammar with its adjacent dependencies. The neural basis for this ability in cotton-top tamarins is unknown, since there are no functional or structural brain-imaging studies on this type of monkey.

Artificial grammar learning was also investigated in two types of monkeys that differ in their evolutionary distance to humans: marmosets and macaques (Wilson et al., 2013). Evolutionarily, macaques are closer to humans than marmosets, and the comparison between these therefore allows one to follow possible phylogenetic traces of grammar learning. In the experiment both species had to learn an artificial grammar with non-deterministic word transitions. Marmosets only showed sensitivity to simple violations in the sequence, whereas macaques showed sensitivity to violations of a higher complexity. This suggests an evolutionarily interesting result with monkeys, namely, that those that are closer relatives to us demonstrate a more advanced artificial grammar processing ability than those that are evolutionarily more distant.

However, when it comes to processing pitch information, marmosets—which are evolutionarily more distant from humans than macaques—demonstrate abilities in pitch processing that are comparable to humans (Song, Osmanski, Guo, and Wang, 2016). Pitch processing is relevant for marmosets, as pitch information is part of their vocalization and thus their communication. The data from the two studies suggest that pitch processing as an acoustic processing mechanism may have emerged early during evolution, whereas mechanisms allowing the processing of complex auditory sequences only evolved later.

Human and non-human primates clearly differ in their abilities to process complex rule-based sequences. And so far there is no evidence that any other species except humans can process and learn hierarchically structured sequences as they appear in syntactic structures of natural languages. The genetic difference between human and non-human primates is less than 2 percent, and only minor differences exist in the basic neuroanatomy. These minor differences, however, may be crucial and deserve a closer look, both with respect to brain structure and brain function.

Brain Structural Differences
With respect to brain structure we have to consider the brain's gray matter (with its region-specific cytoarchitectonics) and the brain's white matter connectivity (which guarantees the communication between different brain regions).

Discussing the evolution of the neural association networks with respect to its overall volume in the human and non-human primate, Buckner and Krienen (2013) state that "the human cerebral cortex is over three times the size of the chimpanzee's," with a "disproportionate growth of the association cortex including both anterior and posterior regions." This concerns, in particular, those regions that are involved in human language.

A first across-species comparative look focusing on the white matter within the prefrontal cortex is already very instructive. The development of the white matter in the prefrontal cortex between human and non-human primates is the key to understanding the evolution of complex cognitive abilities in humans. The developmental pattern of white matter representing the connections within the prefrontal cortex and the reciprocal connections with posterior regions differs markedly between macaques and humans (Fuster, 2002; Knickmeyer et al., 2010). In contrast to humans, macaques already show an adult-like mature pattern during pre-puberty. This is not the case for our closest relative the chimpanzee, in whom the white matter volume develops during pre-puberty, similarly, but not identically, to humans. In humans the rate of development during infancy is faster than in chimpanzees (Sakai et al., 2011). This suggests that humans together with their last common ancestors demonstrate a late white matter maturation of the prefrontal cortex, meaning that it remains "open" to postnatal experience. The faster maturation of the prefrontal white matter in humans compared to chimpanzees may support the development of complex cognitive abilities in humans early in life (see also Sakai et al., 2013).

A more focused across-species look at the language-related structures as defined for humans may even allow more specific conclusions. These language-related brain regions in humans are the inferior frontal gyrus and the temporal cortex, in particular, the superior temporal gyrus. The latter structures were analyzed with respect to their volume and white matter across eleven primate species including apes, monkeys, and humans (Rilling and Seligman, 2002). For the entire temporal lobe, the overall volume, surface area, and white matter were found to be significantly larger in humans than in non-human primates. These data were interpreted by the authors to reflect the neurobiological adaptions supporting "species-specific communication" across different primates.

In humans, language is lateralized to the left hemisphere. Neuroanatomically it has long been reported that in the human brain the posterior temporal cortex is larger in the left than in the right hemisphere (Witelson, 1982). For the planum temporale, a region that lies posterior to Heschl's gyrus and encompasses Wernicke's area, which has long been identified to support speech and language processing, a hemispheric asymmetry is consistently reported for humans (Geschwind and Levitsky, 1968; Steinmetz et al., 1989; Watkins et al., 2001). It has been shown in a recent meta-analysis that the anatomical asymmetry of the posterior temporal cortex is necessary for optimal verbal performances (Tzourio-Mazoyer and Mazoyer, 2017). A cross-species comparison involving chimpanzees and three other non-human primate species, including macaques, focused on the gray matter asymmetry of the planum temporale. Analyses revealed that only chimpanzees demonstrate an asymmetry of the planum temporale similar to humans (Lyn et al., 2011). This suggests that during evolution brain the asymmetry of the planum temporale was established before the division between our closest relative, the chimpanzee. The planum temporale in the temporal cortex is a brain region that—as we discussed in section 1.1—supports auditory language processing. The other important language-relevant region is Broca's area in the inferior frontal cortex.

Broca's area, which is crucially involved in the human ability to process syntax, also deserves a detailed neuroanatomical evaluation. It has been demonstrated that a leftward asymmetry of Broca's area evidenced by a cytoarchitectonic analysis exists in the adult brain (Amunts, Schleicher, Ditterich, and Zilles, 2003). No such asymmetry can be found in Broca's area 44 or 45 in adult chimpanzees (Schenker et al., 2010; figure 7.3). The observed neurobiological difference between the human and the non-human primate brain as well as the development trajectory of its subparts BA 45 and BA 44 may be viewed as a crucial parameter for the evolution of language.

In humans this asymmetry has a different developmental trajectory for the anterior portion of Broca's area (BA 45) and its posterior portion (BA 44). The left-larger-right asymmetry for BA 45 is present by the age of 5 years, whereas the left-larger-right asymmetry for BA 44 only emerges by the age of 11 years. This is an interesting observation since BA 45 and BA 44 serve different language functions in the adult brain, with BA 45 supporting semantic processes and BA 44 subserving syntactic processes. These processes also have different behavioral and neurophysiological trajectories in child language development. Semantic processes are established much earlier than syntactic processes, which only reach adult-like behavioral performance much later (Friederici, 1983; Dittmar, Abbot-Smith, Lieven, and Tomasello, 2008), and neurophysiological patterns for processing complex syntax only appear to be adult-like after the age of 10 (see Hahne, Eckstein, and Friederici, 2004, for electrophysiological data; see Skeide, Brauer, and Friederici, 2014, for functional magnetic resonance imaging data).

Another functionally relevant neuroanatomical parameter is the dendritic structure that neurons display. The dendritic structure refers to the branching projections of a neuron through which electrical signals are received from other neurons. These constitute the building blocks of information transfer at the micro level. A very early histological study investigated the Broca's area in the human with respect to its dendritic structure. The study analyzed the dendritic structure of the neurons in cortical layer III of Broca's area and the

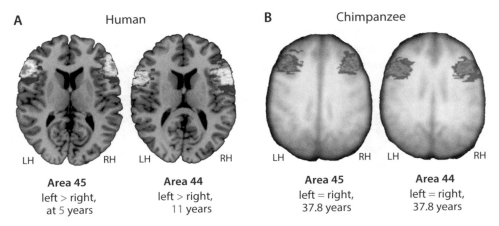

Figure 7.3
Cytoarchitectonic maps of Broca's area in humans and monkeys. Cytoarchitectonic probability maps of the location of areas 44 and 45 (A) in the adult human brain (Amunts et al., 1999) and (B) on a template chimpanzee brain (Schenker et al., 2010). Axial series through area 44 and area 45 are shown. Colors indicate the number of individuals in which the area is occupied by the region of interest. Warmer colors (more red) indicate greater numbers of individuals overlapping, and cooler colors indicate fewer numbers of individuals overlapping. No difference between the left and the right hemisphere is indicated by =, leftward asymmetry is indicated by >. The information about the age during which asymmetry emerges is taken from Amunts et al. (2003). (A) Figure by courtesy of Michiru Makuuchi based on data by Roland and Zilles (1998), the indication "larger than," "smaller than" are taken from Amunts et al. (2003). (B) Adapted from Schenker et al. (2010). Broca's area homologue in chimpanzees (pan troglodytes): Probabilistic mapping, asymmetry, and comparison to humans. *Cerebral Cortex*, 20 (3): 730–742, by permission of Oxford University Press.

adjacent motor cortex in the precentral gyrus of both the left and the right hemisphere in post mortem brains from individuals aged 47 to 72 years (Scheibel, 1984). The author's working hypothesis was that more sophisticated levels of processing demanded by language would lead to more elaborate dendritic trees in layer III of the cortex. And indeed, the dendritic structure was most elaborate in the posterior portion of Broca's area, the left pars opercularis (BA 44) outperforming its right hemispheric homolog as well as left and right orofacial regions in the precentral gyrus (BA 6), thereby supporting the hypothesis (figure 7.4). Whether the more complex dendritic structure in the left pars opercularis (BA 44) is a result of lifelong language processing or whether it is already preprogrammed and present early during development could not be determined on the basis of these data.

Additional analyses on the dendritic growth during development provided an answer to this issue. These revealed that during the second and third years of life the dendritic systems in Broca's area—pars opercularis (BA 44) and triangularis (BA 45)—begin to exceed those of the adjacent motor cortex (BA 6) and extend their primacy in the following years (Simonds and Scheibel, 1989). By the age of 3 to 6 years, "the neurons of the left hemisphere seem generally to have gained primacy both in total dendrite length and in distal segment length" (54). This observation can be mapped onto behavioral data as well

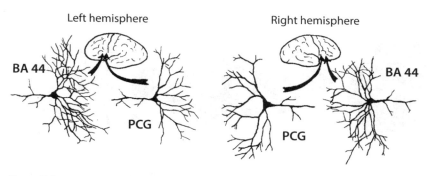

Figure 7.4
Dendritic structure of Broca's area (BA 44) in humans. Schematized drawing of typical dendritic ensembles from cells for the left and the right hemisphere. Arrows indicate the locations of the area studied. Note the increased number of higher-order segments in the left pars opercularis (BA 44) compared to all other areas, and the relatively greater length of second- and third-order branches in the right pars opercularis (BA 44) and precentral (PCG) areas. Reprinted from Scheibel (1984). A dendritic correlate of human speech. In *Cerebral Dominance: The Biological Foundations*, ed. Norman Geschwind and Albert M. Galaburda, 43–52. Cambridge, Mass.: Harvard University Press, Copyright © 1984 by the President and Fellows of Harvard College.

as functional neurophysiology data that reveal two critical periods during language acquisition. A first period up to the age of 3, during which language can be learned in a native manner, and a second period between the age of 3 and 6, during which language learning is still easy, although not necessarily in the same native manner (see chapters 5 and 6). These data indicate a certain specificity of BA 44 at the dendrite level that may have consequences for language acquisition. Unfortunately, no data concerning the dendritic structure of a respective area in the non-human primate are available.

There are, however, a number of structural imaging studies on long-range white matter connections in macaques, chimpanzees, and humans that suggest interesting differences between human and non-human primates (Catani, Howard, Pajevic, and Jones, 2002; Anwander, Tittgemeyer, von Cramon, Friederici, and Knösche, 2007; Rilling et al., 2008; Saur et al., 2008; Makris and Pandya, 2009; Petrides and Pandya, 2009). These studies indicate differences in the strength of the fiber bundles connecting the frontal and temporal regions known to be involved in language processing in humans. In the human brain, Broca's area is connected to the posterior superior temporal gyrus/superior temporal sulcus via a dorsal white matter pathway (Catani, Jones, and Ffytche, 2005; Rilling et al., 2008). In non-human primates this dorsal pathway is much weaker than in humans (Rilling et al., 2008; see figure 7.5). A direct comparison, moreover, revealed differences between humans and non-human primates: macaques and chimpanzees display a strong ventral and a weak dorsal pathway, whereas humans display a strong dorsal pathway and a well-developed ventral pathway. The dorsal pathway was therefore discussed as the crucial pathway for the language ability in human adults (Rilling et al., 2008; see also Rilling, Glasser, Jbabdi, Andersson, and Preuss, 2012).

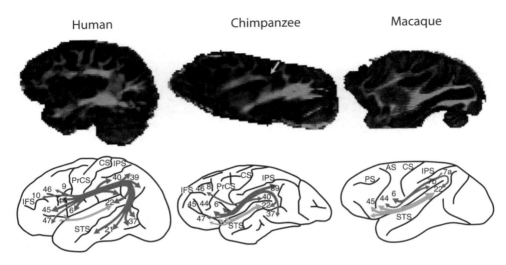

Figure 7.5
White matter fiber tracts across species. Top: Color maps of principal diffusion direction in one in vivo human, one postmortem chimpanzee, and one postmortem rhesus macaque brain. Yellow arrow points to mediolaterally oriented fibers in chimpanzee brain. Bottom: Schematic summary of results shown in Figure 2a in Rilling et al. (2008). Center of gravity of human MTG projections at × ¼ ± 48 are at Montreal Neurological Institute coordinates, × ¼ −48, y ¼ −42, z ¼ −3 and × ¼ −48, y ¼ −36, z ¼ −7. Key to abbreviations: abSF, ascending branch of the Sylvian fissure; AS, arcuate sulcus; CS, central sulcus; IFS, inferior frontal sulcus; IPS, intraparietal sulcus; PrCS, precentral sulcus; PS, principal sulcus; SF, Sylvian fissure; SFS, superior frontal sulcus; STS, superior temporal sulcus. Numbers indicate Brodmann areas. Adapted by permission from Nature Publishing Group: Rilling et al. (2008). The evolution of the arcuate fasciculus revealed with comparative DTI. *Nature Neuroscience*, 11: (4): 426–428.

This difference in the strength of these fiber tracts is of particular interest in light of a functional imaging study in humans (Friederici, Bahlmann, Heim, Schubotz, and Anwander, 2006), which applied an artificial grammar type similar to those used in the behavioral study by Fitch and Hauser (2004) and which also analyzed structural connectivities. In humans, processing the $(AB)^n$ grammar, with its adjacent dependencies, activated the frontal operculum, whereas processing the more complex A^nB^n grammar, additionally recruited the phylogenetically younger Broca's area (Friederici, Bahlmann, et al., 2006). The structural imaging analyses took the functional activation peaks as seeds for fiber tracking and found that the frontal operculum for the $(AB)^n$ grammar was connected to the temporal cortex via a ventral pathway. In contrast, the posterior Broca's area computing the A^nB^n grammar was connected to the posterior temporal cortex via a dorsal pathway (Friederici, Bahlmann, et al., 2006). These data were taken to suggest that posterior Broca's area (BA 44) and its dorsal connection to the temporal cortex, in particular, support the processing of higher-order hierarchically structured sequences relevant to language.

Brain Functional Studies

There are a number of recent studies comparing brain functional data between human and non-human primates. These include electrophysiological measures and functional magnetic imaging measures.

In electrophysiological studies we directly compared sequence processing in prelinguistic human infants and in non-human primates (Milne et al., 2016). These focused on the processing of rule-based dependency of non-adjacent elements in an auditory sequence (Mueller, Friederici, and Männel, 2012). Such non-adjacent dependencies in their simplest form are AxB structures in which A and B are constant and the x varies (e.g., as in syllable sequences *le* no *bu*, *le* gu *bu*). The ability to learn such non-adjacent dependencies is believed to be a precursor to learning syntactic dependencies in a natural language. The first study conducted with 3-month-old infants and adults revealed that violations in AxB sequences were reflected in a mismatch negativity (Mueller et al., 2012). Violations were either acoustic (pitch violation) or rule-based (violation of syllable sequence, AxC instead of AxB). In infants the ERP response in the acoustic condition was predictive for the rule-based syllable condition (pitch), suggesting a correlation between the ability to detect acoustic violations and the ability to detect violation in rule-based syllable sequences.

Infants, as we have learned in chapter 6, clearly differ from adults in their strength of the dorsal fiber tract targeting BA 44. We therefore concluded that their ability to process AxB structures cannot be based on this fiber tract. The respective dorsal fiber tract is also not well developed in the monkey (Rilling et al., 2008), therefore, monkeys may not be in the position to process complex syntax, but similar to infants, may be able to process those simple AxB structures.

In order to examine to what extent human prelinguistic infants and non-human primates show processing similarities, a second electroencephalography study with macaques applied the very same stimulus material used with human infants and adults, as presented above (Mueller et al., 2012). In this study (Milne et al., 2016) we tested the processing of non-adjacent dependencies (AxB structures) for two different stimulus types either in the auditory domain (pitch) or in the rule domain (rule) in macaques. In this paradigm correct and incorrect sequences with pitch deviants or rule deviants were presented auditorily. The ERP results of the monkeys to the incorrect elements reveal an ERP pattern similar to those of human infants, but different to those of human adults (figure 7.6). These data provide support for the view that the monkeys' processing abilities for rule-based sequences represent a stage comparable to those of human infants.

Another across-primate comparison study focused on functional activation pattern during tone sequence processing in humans and macaques (Wang, Uhrig, Jarraya, and Dehaene, 2015). Sequences focused either on the number of tones or on the tone-repetition pattern. In addition to a number of interspecies similarities, the data revealed that in humans, in contrast to monkeys, the inferior frontal area (mainly BA 44) and the posterior

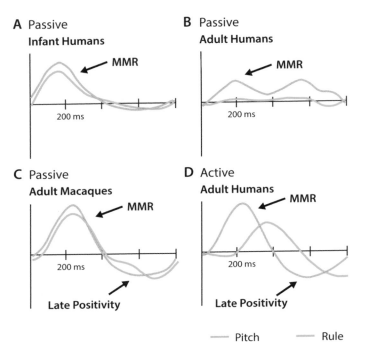

Figure 7.6
Schematized ERP components between species. Human event-related potential (ERP) results schematized from Mueller, Friederici, and Männel (2012), depicting the mature ERP components of (A) infants who acquired the non-adjacent rule with passive habituation, (B) adult ERPs after passive exposure, and (D) ERPs from adults that acquired the non-adjacent rule through an active task. (C) The macaque ERP components illustrated are those found in the late sessions of the study. In macaques, mismatch response (MMR) was produced for the pitch deviant at ~ 200ms. Only infants and macaques showed similar MMRs for the rule deviant as for the pitch deviant. In human adults no rule MMR is observed under passive conditions (B), and a negativity is observed in adults under the active rule conditions, evidenced as an N2 component that occurs later than the MMR (D). In human adult and macaque ERPs, later positivities are also observed for the rule and pitch violation conditions (C and D), which are not evident in the infant results (A). See text for discussion. From Milne et al. (2016). Evolutionary origins of non-adjacent sequence processing in primate brain potentials. *Scientific Reports*, 6: 36259. doi:10.1038/srep36259.

temporal sulcus show correlated effects of number and sequence change in the sequences. The authors consider this result to be "most compatible with the hypothesis that evolution endowed the human brain with novel inferior frontal cortical and superior temporal circuits capable of regular sequence patterns ... and that this enhanced learning ability relates to inferior frontal gyrus–pSTS circuitry" (Wang et al., 2015, p. 1972). Anatomically, these areas are connected by the dorsal fiber tract, which is strong in humans but weak in non-human primates (Rilling et al., 2008) (compare figure 7.5).

The processing of simple rule-based sequences in monkeys and humans was evaluated in a functional magnetic-resonance imaging study (Wilson et al., 2015). Activation differences were recorded for Broca's area (BA 44 and BA 45) as well as the ventral frontal opercular cortex in the left and the right hemisphere. Monkeys revealed an involvement of the opercular cortex and of Broca's area, whereas human adults only activated the opercular cortex but not Broca's area. The activation pattern for humans can be related to that reported by Friederici, Bahlmann, and colleagues (2006) also revealing activation of the frontal operculum, but not Broca's area, for the processing of simple rule-based sequences.

A comparison of resting-state brain activity in humans and chimpanzees reveals some similarities, but also interesting differences (Rilling et al., 2007). Humans, compared to chimpanzees, show a higher level of activation in left-hemispheric cortical areas known to be involved in language and conceptual processes. The combined functional data suggest that during evolution a novel circuit connecting the language-relevant areas emerged phylogenetically. It appears that Broca's area in the inferior frontal gyrus and the posterior temporal cortex crucially evolved from monkey to human brains. These two regions are structurally connected by a dorsal fiber tract, which also developed in strength during phylogeny.

There is also an across-primate comparison study reporting functional connectivity data (Neubert, Mars, Thomas, Sallet, and Bushworth, 2014). Besides a number of similarities, the study revealed interesting differences between the species with respect to functional connectivities. The human connectivity of the posterior temporal cortex to the ventral inferior frontal cortex (including Broca's area) is systematically stronger than in the macaque for almost all regions in the ventral inferior frontal cortex, in particular the posterior Broca's region (BA 44), which in humans is involved in syntactic hierarchy building.

Taken together, the data suggest a clear evolutionary trajectory across primates with respect to the strength of the dorsal pathway that connects two syntax-relevant brain areas, namely, the posterior portion of Broca's area and the posterior superior temporal gyrus/superior temporal sulcus. This phylogenetic change parallels the ontogenetic trajectory of the emerging strength of the dorsal fiber tract connecting the posterior portion of Broca's area (BA 44) and the posterior superior temporal gyrus/superior temporal sulcus from the newborn (Perani et al., 2011) through childhood up to adulthood (Skeide, Brauer, and Friederici, 2016) (figure 7.7). This fiber tract has been demonstrated to be crucially involved in processing syntactically complex sentences (Skeide et al., 2016).

Summary

Across-species comparisons between the human and non-human primate brain reveal differences in the cytoarchitectonic analyses of Broca's area and connectivity of Broca's area to the temporal cortex. First, cytoarchitectonic analyses demonstrate a leftward asymmetry of Broca's area in the inferior frontal gyrus in humans, but not in non-human primates. Second, the dorsal connection between BA 44 in Broca's area and the superior temporal cortex is stronger in the human brain than in the non-human primate brain. These structures

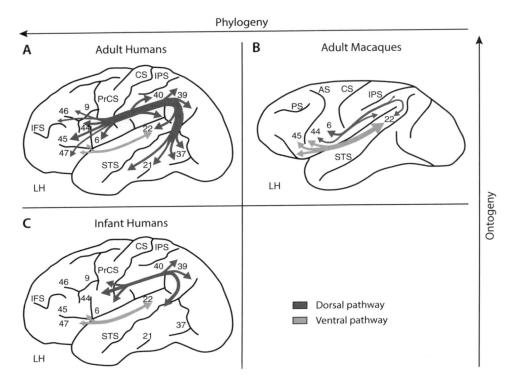

Figure 7.7
Schematized white matter fiber tracts in phylogeny and ontogeny. Structural connectivity. Schematic view for phylogeny (A, B) and for ontogeny (A, C). Dorsal fiber tract (purple), ventral fiber tract (orange). (A) Human adult. (B) Macaque. Center of gravity of human MTG projections at $x = \pm 48$ are at Montreal Neurological Institute coordinates, $x = -48$, $y = -42$, $z = -3$ and $x = 48$, $y = -36$, $z = -7$. abSF, ascending branch of the Sylvian fissure; AS, arcuate sulcus; CS, central sulcus; IFS, inferior frontal sulcus; IPS, intraparietal sulcus; PrCS, precentral sulcus; PS, principal sulcus; SF, Sylvian fissure; SFS, superior frontal sulcus; STS, superior temporal sulcus. Adapted by permission from Nature Publishing Group: Rilling et al. (2008). The evolution of the arcuate fasciculus revealed with comparative DTI. *Nature Neuroscience*, 11: (4): 426–428. (C) Schematized view based on the fiber tracking of diffusion tensor imaging data for human infants for speech-relevant regions compare figure 6.11 in this book. In newborns, the dorsal pathway (purple) only target the precentral gyrus but not Broca's area (BA 45/44). The ventral pathway connecting the ventral inferior frontal gyrus via the extreme capsule to the temporal cortex (orange) is present in newborns. Adapted from Perani et al. (2011). The neural language networks at birth. *Proceedings of the National Academy of Sciences of the United States of America*, 108 (38): 16056–16061.

may have evolved to subserve the human capacity to process syntax, which is at the core of the human language faculty.

7.4 Paleoanthropological Considerations of Brain Development

In the absence of any direct evidence on the brain tissue of our ancestors, information about hominin brain development during evolution can only be taken from fossil bones—from the casts of the internal bony braincase (Holloway, 1978; Falk, 1980, 2014; Holloway, Broadfield, and Yuan, 2004). These casts provide evidence about brain size and shape; the reconstructed imprints, moreover, can provide information about the brain's convolutions (i.e., the prominent folds on the surface of the brain). This approach allows the majority of the earlier studies in this field to only focus on the development of brain size relative to body size. Considering these data, it can be concluded that until at least 600,000 years ago, the increase in brain size was correlated with an increase in body size, and only thereafter did brain size increase independent of body size (Hublin, Neubauer, and Gunz, 2015).

During ontogeny, human brain size increases during early development and reaches adult size later than other primates (Leigh, 2004). The main brain development takes place after birth in high interaction with the environment. The development concerns the gray matter (Gogtay et al., 2004) and the white matter (Miller et al., 2012) and continues until late adolescence (Lebel, Walker, Leemans, Phillips, and Beaulieu, 2008; Brauer, Anwander, and Friederici, 2011). It is, in particular, the white matter in its long-range and local networks that differentiates macaques, as has been shown for the prefrontal cortex across species (Sakai et al., 2011).

Information from fossil endocasts—which are often fragmentary—is growing, but with new methods it may increase over the coming years. Recently, details have been discussed with respect to specific sulci in the human and non-human primate brain as these are imprinted in the endocasts. One of these is the so-called lunate sulcus, marking the anterior limit of the occipital lobe, which corresponds to the primary visual cortex (Smith, 1904; Falk, 2014). The evolutionary expansion of the parietal and temporal cortex is taken by some researchers to cause a shift of this sulcus during phylogeny. This may have led to enlarged parietotemporal-occipital association cortices in humans (Holloway, Clarke, and Tobias, 2004). Other researchers, however, debate these results and conclusions. Cross-species differences in the sulcus structure have also been discussed with respect to the prefrontal cortex, in particular the middle frontal sulcus (Teffer and Semendeferi, 2012; Falk, 2014). The observed differences have been taken to indicate an expansion of the prefrontal cortex in humans.

Summary
Although still under discussion, the available paleoanthropological findings suggest a reorganization of the brain during phylogeny, and a possible rewiring that, due to the prolonged ontogeny in humans, is shaped by environmental input.

8

The Neural Basis of Language

Language is an integral part of our daily life. We use it effortlessly, and when doing so, we usually think about what we want to say rather than how to say it in a grammatically correct way. Thus we use language quite automatically. The intriguing question is what puts us humans in the position to learn and use language the way we do. I started out in the beginning of the book by saying that this is due to an innate language faculty. Over the course of the book it became clear that the language system is a complex system.

In the natural sciences complex phenomena and systems are usually broken down to their basic elements or operations in order to make them approachable. These basic elements—once identified—can be put together and thereby explain a complex system. In chemistry, for example, we are dealing with different basic elements that can be reduced further to a number of atoms, but which can also bind together with other elements to build more complex structures. In this book I have taken a similar approach by breaking the complex language system into its elements and basic operations: these are the lexical elements and the syntactic rules. For the domain of syntax that approach works well due to a clearly formulated linguistic theory that defines a most basic computation, namely Merge (Chomsky, 1995b). This computation was localized in the brain with a high consistency across individuals. For the domain of lexical and sentential semantics this is more difficult, partly due to the indeterminacy of a word's meaning caused by the fact that a word has many mental associations that differ from individual to individual. Linguistic theory can describe the semantic relation between a noun and a verb by means of selectional restrictions in a given language that holds for all individuals using this language, but is not in the position to describe the associations at a cross-individual level, as these may differ substantially from person to person. Given this it is not surprising that the localization of these processes in the brain shows more inter-individual variance.

I also stated that language is a uniquely human trait characterized by the ability to generate and process hierarchically structured syntactic sequences that differentiate us from non-human primates. A comparison with our phylogenetically close relatives, therefore, could provide some insight into what this difference is based upon. Such a comparison can be conducted at different levels: the behavioral level, the brain level, or the genetic level.

With respect to the genetic level, we know that the genetic difference between human and non-human primates is less than 2 percent, but so far we do not know which genetic aspect is responsible for the language faculty. It has been proposed that *FOXP2* is a gene that plays a major role in speech and language because the mutation of this gene was identified in a family with speech and language problems (Fisher, Lai, and Monaco, 2003), although they were more speech-related rather than language problems as such. The view, however, has also been challenged for several reasons. One reason is that *FOXP2* can also be found in non-human primates, mice, birds, and fish, thus in animals that do not speak (Scharff and Petri, 2011). A second reason is the finding that *FOXP2* as such regulates a large number of other genes and thereby the coding of proteins—a relevant aspect in detailed genetic analyses. Novel methods of genetic analyses can inform us about the time trajectory of human evolution and thereby possibly the emergence of language. Using such analyses it could be shown that the *FOXP2* codes a particular protein in the Neanderthals that is the same protein as in the modern human but different from that in the chimpanzee (Krause et al., 2007, Prüfer et al., 2014). These findings suggest that the genetic basis for language was laid down early during evolution—after the separation from the chimpanzee, but before the emergence of the Neanderthals—about 300,000 to 400,000 years ago.[1]

Today geneticists agree that the language faculty can not be traced down to a single gene, but rather to a number of related genes that jointly contribute to neural pathways that are important for normal brain maturation and possibly the development of language. With respect to brain development it is known that genetic programs help to guide neuronal differentiation, migration, and connectivity during development, together with experience-dependent contribution of the environment (Fisher and Vernes, 2015). These results provide evidence for a relation between genes and brain development in general, but they do not speak directly to the genetics of language. Most recently it has been shown that a specific gene controls the number of gyri in a particular brain region that includes Broca's area (Bae et al., 2014). Using a combined genetics and brain imaging approach, these researchers examined 1000 individuals with respect to their cortical folding. They found five individuals of three different families that suffered intellectual and language difficulties but had no motor disability. The genetic analysis revealed abnormalities to mutations in the regulatory region of the GPR6 gene; GPR6 encodes a protein that functions in neuronal guidance and thereby cortical maturation. This study is the first to show a relation between genetics, brain structure of Broca's area, and language behavior. It is a promising case showing how a tripartite relation between genetics-brain-behavior can help to determine the biological basis of language.

In this book I claimed that the structural language network is genetically predetermined. Under this hypothesis one would expect to find genetic-brain-language relations most likely with respect to the brain-structural parameters. The study by Bae et al. (2014) provides support for the assumption that language emerged during evolution in a genetically yet to be defined step that left its traces in the human brain.

The remaining discussion and the view proposed here will, therefore, be restricted to the brain-behavior level. In the next section I will bring together evidence from the different research areas I discussed in the previous chapters. I will try to provide an integrative view of the language network by taking into account empirical data from human and non-human primates across the various neuroscientific levels.

8.1 An Integrative View of the Language Network

The integrative view of the neural basis of the language network that I propose here will not primarily concern the integration of syntax and semantics within the human language system, but will rather try to integrate language-relevant data across two biologically relevant dimensions: a neurophysiological dimension and an evolutionary dimension. The neurophysiological dimension will focus on the human brain and try to identify within the human brain to what extend particular brain structures for language are characterized by specific parameters at different neurophysiological levels from the neuron to large-scale neural networks. The evolutionary dimension will compare the human and the non-human primates and consider the changes of brain structure and function between primates with the goal of determining possible differences crucial for language. The formulation of such a view, however, must remain incomplete at this point in time, since the required data base is still sparse, but the available data do provide interesting information in particular for the domain of syntax.

With respect to the neurophysiological dimension, I have focused the discussion in this book on the core language system and its representation in the human brain, directing my efforts toward a consideration of the neural basis of language at the functional and macrostructural level, and moreover, across the different levels of neuroscientific analyses—from the neuron up to the large-scale network.

Here I will aggregate the topics discussed in prior chapters. Starting at the system level, I have described the different language functions such as phonology, syntax, and semantics and localized these in the brain. More importantly I have described the dynamics of functional interplay between the different subsystems within the neural network in time and identified the white matter brain structure that could guarantee this interplay. We saw that the general language network could, moreover, be characterized as a specific network at the molecular level based on the participating regions' neurotransmitter similarity. This provides a strong signal that language function and brain structure stand in a certain dependency relation.

In this book I have argued that syntax is at the core of human language, as it provides the rules and computational mechanisms that determine the relation between words in a sentence. In chapter 1, I identified a left-hemispheric fronto-temporal network for language and a particular neural network for syntax. This specific network guarantees (a) the syntactic binding of two words (or elements) into phrasal units, based on an operation called

Merge and, moreover, (b) the syntactic binding of phrasal units into sentences based on the multiple, recursive application of Merge. Within this network the posterior portion of Broca's area together with the adjacent frontal operculum represent a local circuit subserving the basic syntactic computation of Merge. The posterior portion of Broca's area (BA 44) as part of the core language network is connected to the posterior superior temporal cortex via a particular white matter fiber tract, the arcuate fasciculus (see figure 0.1). The two brain regions together with the connection between them represent a long-range circuit which supports the assignment of syntactic and thematic relations of phrasal units within a sentence. The two brain regions, and in particular the fiber tract connecting them, have been shown to be crucial for the development of syntactic abilities and possibly for the human faculty of language. This means that parts of the gray matter together with a part of the white matter appear to be the neurobiological basis of the unique human faculty of language.

However, the description of the neural network with its functional gray and white matter parts remains incomplete unless we add information about the temporal dynamics within the network. After all, language processing takes place in time. Electrophysiological measures, which allow us to register brain activation millisecond by millisecond, have provided information about the temporal course of phonological, syntactic, and semantic processes—but only with a coarse brain localization of these dynamics, except for the few instances where intracranial recordings were used. These intracranial recordings provided interesting results on word processing but so far only concerned single neurons in one brain region and neglected the rest of the network. Thus there is room for speculations of how different parts of the language network code, share, and integrate pieces of information. I briefly touched upon this issue and speculated the information exchange between different brain regions may be possible by so-called *mirror neural ensembles* that encode and decode information in the sending and receiving brain regions, respectively.

In the remainder of this section I will recapitulate what we know about the relevant parts of the language network functionally and structurally at the different neuroscientific levels within the human brain. I will start with the larger fronto-temporal language network in the adult human brain and then zoom into particular parts that can be backtracked through different neuroscientific levels down to their neuroanatomical detail and, moreover, can be followed back in phylogeny.

The Language Network in Our Brain

The human language network constitutes areas in the inferior frontal gyrus and in the temporal cortex that are connected by white matter fiber tracts. This larger network holds responsibility for the processing of language. Within this larger network, distinct brain areas and specific sub-networks support the processing of the sound of words, their meanings, and their relations in phrases and sentences as well as the melodic parameters of these entities.

The sound of language concerns the word forms and the melodic parameters, words, phrases, and sentences. Words are composed of different phonemes that are initially processed in the primary and secondary auditory cortex located in the middle portion of the superior temporal cortex. Regions in the posterior portion, as well as in the anterior portion of the superior temporal cortex, are responsible for the processing of the phonological language-specific information. These latter regions are connected to the primary auditory cortex by short-range fiber bundles guaranteeing the transmission of the information from the lower level–processing region of the auditory system to the next level of phonological processing.

Once the phonological word form is identified the next processing levels and their respective neural networks become active. It is assumed that the system initially uses syntactic word category information to build up phrase structure online. This process appears to involve two steps: a first step during which adjacent elements are identified and a second step during which a local syntactic is hierarchically built. The first step involves the anterior superior temporal gyrus and the frontal operculum connected via a ventrally located fiber tract called uncinate fasciculus, color-coded in red in figure 8.1. The second step during which the syntactic hierarchy is built involves the posterior portion of Broca's area (BA 44). Processing of the syntactic relation in a sentence involves a network consisting of Broca's area (BA 44) and the posterior superior temporal gyrus/sulcus. These two brain regions are connected via a dorsally located fiber tract called the arcuate fasciculus, color-coded in purple in figure 8.1. Lexical-semantic information of words stored in the middle temporal gyrus is accessed through the anterior portion of Broca's area (BA 45/BA 47). These two brain regions are connected via a ventrally located fiber pathway with the abbreviation inferior fronto-occipital fasciculus (IFOF), color-coded in orange in figure 8.1. The posterior superior temporal gyrus/sulcus is assumed to serve as a region where syntactic and thematic information are integrated. The right hemisphere comes into play when prosodic information is processed and the corpus callosum (as the fiber bundle connecting the two hemispheres) supports the interaction between syntactic information processed in the left hemisphere and prosodic information processed in the right hemisphere.

The white matter fiber tract connections are the basis of the information transfer from one region to the next and the cooperation between them. This cooperation has been demonstrated by functional connectivity results and oscillatory measures, both of which reflect the joint activation of these regions.

During language perception semantic and syntactic information processed in different brain regions and subsystems have to be integrated to achieve comprehension. At this point the question arises about which brain area is responsible for the integration of syntactic and semantic information. I discussed the posterior temporal cortex as the major region of integration, based on the results that we observe activation in this region when both semantic and syntactic information in natural sentences has to be processed, but not when processing syntactic sequences without any semantic information. A similar activation

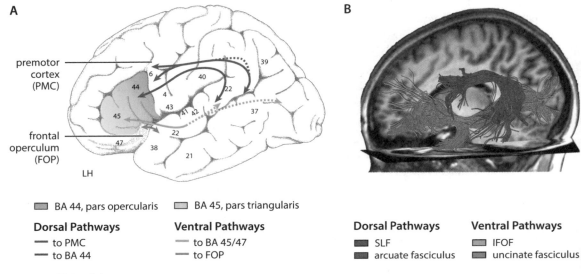

Figure 8.1
Core language fiber tracts in the human brain. (A) Systematic view; (B) anatomical view. There are two dorsally located pathways and two ventrally located pathways. The dorsal pathway connecting the dorsal premotor cortex (PMC) with the posterior temporal cortex [the posterior superior temporal gyrus and sulcus (pSTG/STS)] involves the superior longitudinal fasciculus (SLF) (depicted in light blue); the dorsal pathway connecting Brodmann area (BA) 44 with the posterior STG involves the arcuate fasciculus (depicted in purple). The ventral pathway connecting the inferior frontal cortex—that is, BA 45/47 and others—with the temporal cortex involves the inferior fronto-occipital fasciculus (IFOF) (depicted in orange); the ventral pathway connecting the anterior inferior frontal cortex—that is, frontal operculum (FOP) with the anterior superior temporal gyrus (aSTG) involves the uncinate fasciculus (depicted in red). Adapted from Friederici and Gierhan (2013). The language network. *Current Opinion in Neurobiology*, 23 (2): 250–254, with permission from Elsevier.

pattern depending on the presence of semantic information, however, was also found for BA 45 in the inferior frontal gyrus (Goucha and Friederici, 2015). This leaves the option open that BA 45 may also play a role in syntactic-semantic integration. And indeed the larger region of the inferior frontal cortex has been proposed as the integration region by Hagoort (2005). Here I suggest that if the inferior frontal cortex is involved in integration processes, it may rather be BA 45 in particular, given the data by Goucha and Friederici (2015), which revealed an involvement of BA 45 together with BA 44 for the processing of syntactically structured sequences as long as minimal semantic information was present therein. BA 45 can qualify as an integration area not only on the basis of the functional data, but, moreover, on the basis of receptorarchitectonic data (Amunts et al., 2010). These data indicate a subdivision of BA 45 into two subregions: a posterior part (45p) and an anterior part (45a) (see the introduction, figure 0.2). The posterior part of BA 45 (45p) is adjacent to BA 44, known to be a syntax-relevant area, and is sometimes seen to be active together with BA 44 (at least when minimal semantic information is still present in the stimulus material). The anterior part of BA 45 (45a) is adjacent to BA 47 and often seen active together with 47 when semantic information is processed. The two subparts of BA 45, the anterior part (45a) and the posterior part (45p), may thus provide a possible basis for integration processes in addition to the posterior temporal cortex. It remains to be determined whether the posterior temporal cortex and BA 45 serve different subfunctions of integration or whether they perform integration jointly.

The Syntactic Network in Our Brain

In this subsection I concentrate on the syntactic network for two reasons. First, I argued that syntax is the core of the human language capacity. Second, there is ample evidence showing that non-human animals such as non-human primates and dogs can learn the meaning of words and frozen phrases but cannot learn syntax.

When decomposing the human language network at the different neuroscientific levels, it seems that those brain structures involved in the human specific ability to process syntax have specific characteristics at the neurostructural and neurofunctional level. There are two structures, relevant for syntactic processes: first is BA 44, the posterior portion of Broca's area in the inferior frontal gyrus; and second is the fiber tract connecting this area to the posterior temporal cortex, namely the arcuate fasciculus. This network has been identified to support the processing of complex syntax in humans. So what are the specifics of these brain structures?

Broca's area. Within the syntactic network involving BA 44 in Broca's area and the posterior superior temporal gyrus, BA 44 plays a special role. BA 44 is responsible for the generation of hierarchical syntactic structures, and accordingly has particular characteristics at the neurophysiological level. This area is different from other brain regions in many respects, both at the functional and various neurostructural levels.

At the cytoarchitectonic level, as one of the lowest structural levels, BA 44 differs from other regions in the frontal cortex in the granularity of the six layers of the cortex (Amunts et al., 1999), with BA 44 showing more pyramidal neurons in certain layers (dysgranular structure) than the more posteriorly located premotor cortex (BA 6) (agranular structure), but fewer of these neurons than the fully granular cortex of BA 45 (granular structure) (Brodmann, 1909). Moreover, in human adults BA 44 has been demonstrated to be larger in the language-dominant left hemisphere compared to the right hemisphere. This asymmetry has been related to the human language faculty, in particular as a similar difference in the cytoarchitectonics of the left and right hemisphere is not present in chimpanzees (Schenker et al., 2010). Thus BA 44 in the left hemisphere, compared to its right hemispheric homolog, appears to be specifically large in adult humans.

At the microstructural histological level it has been shown that BA 44's dendritic structure is more elaborate than that of the adjacent motor cortex and its right hemispheric homolog (Scheibel, 1984). Dendrites guarantee the information transfer between neurons and are thus relevant for those processes represented in the particular area. BA 44's intricate dendritic structure has been discussed as being related to the particularly complex functional processes this region supports—that is, language or more specifically, syntax. In the adult human brain we see a clear functional difference between BA 44 and the adjacent motor cortex, as indicated by local field potentials as measured by electrocorticography: while BA 44 is involved in speech planning, the motor cortex supports the motor act of speaking (Flinker et al., 2015). Thus BA 44 in the human left hemisphere is specific, since it has a more dense dendritic structure than adjacent brain regions or its homolog in the right hemisphere.

As one of the lowest functionally relevant levels we can consider the level of the neurotransmitters that were analyzed for the larger language network (Zilles, Bacha-Trams, Palomero-Gallagher, Amunts, and Friederici, 2014), and for Broca's area in the inferior frontal cortex in particular (Amunts et al., 2010). These analyses revealed (a) that BA 44 as the posterior portion of Broca's area can be differentiated from BA 45 as the anterior portion of Broca's area, (b) that area 44 can be subdivided receptorarchitectonically into a dorsal and a ventral part, and (c) that BA 44 can be separated from the frontal operculum located more ventrally. These neuroreceptorarchitectonic differentiations can be nicely mapped onto the functional language data at the systems level. Let's consider the data point by point.

(a_1) BA 44 is cytoarchitectonically (Amunts et al., 1999) and receptorarchitectonically (Amunts et al., 2010) different from BA 45.

(a_2) BA 44 is responsible for syntactic processes and functionally separate from BA 45 responsible for semantic, propositional processes (Friederici, 2002; Goucha and Friederici, 2015).

(b_1) Receptorarchitectonic analyses suggest a subdivision of BA 44 into a dorsal and a ventral part (Amunts et al., 2010).

(b_2) Neuroimaging data found the dorsal part to be involved in phonological processes and the ventral part to be involved in syntactic processes (Hagoort, 2005).

(c_1) The ventral part of BA 44 is receptorarchitectonically different from the adjacent frontal operculum (Amunts et al., 2010).

(c_2) The ventral part of BA 44 can functionally be separated from the frontal operculum (Zaccarella and Friederici, 2015a).

These data indicate that at the receptorarchitectonic level BA 44 can clearly be differentiated from its adjacent regions, both from the more anterior located BA 45 and from the more ventro-medially located frontal operculum.

Thus in sum, the structural and molecular analyses reveal a number of parameters that differentiate BA 44 in Broca's area from adjacent areas in the left inferior frontal gyrus.

These differences map onto specific functions in the language network and may therefore be crucial for the language faculty.

The dorsal language pathway. Within the larger left hemispheric language network the subnetwork that specifically supports the ability to process human syntax involves BA 44 in the inferior frontal gyrus and the posterior temporal gyrus connected via a dorsally located fiber tract, in particular the arcuate fasciculus (figure 8.1).

At the macrostructural level we can relate the degree of white matter fiber density functionally to syntactic behavior, and we see that the degree of maturity of the arcuate fasciculus during development, in particular its myelination, predicts syntactic performance (Skeide, Brauer, and Friederici, 2016). This structure-function relation is furthermore supported by the report that the decrease in integrity of this dorsal pathway during progressive aphasia correlates with a decrease in syntactic performance (Wilson et al., 2011). This classical language network involving Broca's area and the posterior temporal cortex with its dorsal and ventral pathways is universally present in all human brains, but the strength of some of the fiber tracts in the language pathways is modulated by the particular language learned (Goucha, Anwander, and Friederici, 2015).

At the level of functional connectivity, BA 44 in Broca's area and the posterior superior temporal gyrus show a high functional cooperation during sentence processing (den Ouden et al., 2012; Makuuchi and Friederici, 2013). This connectivity between these two areas increases with age during childhood; moreover, correlational analyses with sentence comprehension tests revealed that the connectivity between these areas becomes increasingly relevant for sentence processing between ages 5 and 6 years (Xiao, Friederici, Margulies, and Brauer, 2016b).

Studies using oscillatory measures found a high synchronicity between left frontal and left parietal and temporal brain regions when processing the syntactic structure of phrases and sentences (Ding, Melloni, Zhang, Tian, and Poeppel, 2015; Meyer, Grigutsch, Schmuck, Gaston, and Friederici, 2015). These data suggest that neural ensembles in the left frontal region and neuronal ensembles in the left posterior regions are active coherently during sentence processing. It appears that the neuronal ensembles in the respective brain regions work together online.

In sum these data point toward a distinct structural connectivity between BA 44 in Broca's area and the posterior temporal cortex via the dorsal pathway and a strong functional connectivity between these regions. The functional relevance of the dorsal pathway for the processing of syntactic hierarchies has been demonstrated clearly in several studies.

The Language Network across Phylogeny and Ontogeny

Finally, when considering the language network across ontogeny and phylogeny we may learn more about the specifics of the neurobiology underlying the human faculty of syntax. The simple view is that non-human primates and young infants, in contrast to human adults, do not have the capacity of language nor syntax in particular. What is the difference in their respective structural networks? The available data suggest that the networks differ with respect to area BA 44 and, moreover, with respect to the dorsal fiber tract connecting BA 44 and the posterior superior temporal cortex.

Broca's area. We have seen that the cytoarchitectonic structure of BA 44 and its relative expression in the left and the right hemisphere is different in human adults and non-human primates. The cytoarchitectonic analysis for Broca's area including BA 44 shows a clear leftward asymmetry in human adults (Amunts, Schleicher, Ditterich, and Zilles, 2003), but not in chimpanzees (Schenker et al., 2010). In humans, the leftward asymmetry is not present early in life, but then follows a function-related developmental trajectory.

During human development we can observe an interesting differentiation between BA 44 known to support syntactic processes and BA 45 known to support semantic processes. While BA 45 shows this asymmetry by the age of 5 years, BA 44 only demonstrates this asymmetry by the age of 11 years (Amunts et al., 2003). These data mirror the functional development trajectory of both semantics and syntax at the behavioral (Friederici, 1983), the electrophysiological level (Strotseva-Feinschmidt, Schipke, Gunter, Friederici, and Brauer, submitted). At the brain imaging level the specificity of BA 44 for syntactic processes only develops in late childhood, and that prior to this BA 45 is also recruited (Skeide, Brauer, and Friederici, 2014), indicating a late emergence of the full syntactic language ability. The mature BA 44 then shows a clear specificity for syntax and appears to be indicative for the structural and functional difference observed between human adults compared to infants and monkeys and may thus be crucial for the faculty of language.

The dorsal language pathway. We have learned that BA 44 and the posterior superior temporal gyrus are connected by a dorsal fiber tract (the arcuate fasciculus), which together represent the syntactic network in the adult human. At this point we may ask whether there is a parallel between its emergence during phylogeny and ontogeny. Comparisons reveal that this fiber tract in human adults differs from both human infants (Perani et al., 2011) and non-human primates (Rilling et al., 2008). This dorsal pathway is very weak in macaques and chimpanzees (who do not have language) but strong in adult humans (who master language) (Rilling et al., 2008). Moreover, the arcuate fasciculus is very weak and poorly myelinated in newborns (who do not yet master language) (Perani et al., 2011). These data indicate that when myelin of the arcuate fasciculus is low, the syntax function is poor. This conclusion is supported by a result demonstrating that its increasing strength correlates directly with the increasing ability to process complex syntactic structures (Skeide et al., 2016). This fiber tract may thus be one of the reasons for the difference in the language ability in human adults compared to the prelinguistic infant and the monkey.

In conclusion these phylogenetic and ontogenetic data provide compelling evidence for the specific role of BA 44 and the arcuate fasciculus connecting BA 44 and the posterior superior temporal cortex for the language faculty. These brain structures may have emerged to subserve the human capacity to process syntax. This fiber tract could be seen as the missing link that has to evolve in order to make the full human language capacity possible.

8.2 Epilogue: Homo Loquens—More than Just Words

Humans differ from non-human primates in their ability to build syntactic structures. This statement has set the scene for our assessment of language in this book. Many sections of *Language in Our Brain* have focused on syntax, and yet our feeling—when considering language—may rather tend to favor semantics and meaning as the most important aspects of language. We do, after all, want to communicate meaning. In order to do so, however, we need syntax to combine words into phrases and sentences.

We use words to refer to objects and mental entities and to refer to actions and mental states, but we need syntax to determine the thematic relation between the words in a sentence. Syntax allows to express logical relations (e.g., the if-then relation: if he does X then Y will happen) and it permits us to formulate believes of ourselves and of others (e.g., I think he believes that …).

Language is what enables us to plan the future together with others, and to learn from the past through narratives. Language forms the basis of our social and cultural interaction not only in the "here and now," but most importantly also in the "there and then."

This evolutionary step is realized in a brain that—as far as we know—underwent only minor, although important changes compared to our closest relatives. I have argued that the syntactic specificity of Broca's area, in particular its posterior part, namely BA 44 and its connection to the temporal cortex granting the interplay between these two regions, is what accounts for one of these changes. From an evolutionary perspective it remains an open question whether humans develop language to cover those needs and whether the emergence of language determined the homo loquens as a cultural being.

Glossary

Language

Agrammatism Form of aphasia identifying patients who after brain lesion produce sentences in in telegraphic style and leave out function words from the message. It can also affect the comprehension of grammatically complex constructions whose comprehension depends on function words or inflectional morphology, for example in passive sentences, such as *The boy was pushed by the girl*, which are misinterpreted when content words are processed (*boy pushed girl*)

Animacy Semantic feature of a noun, which qualifies it as being sentient/alive. It is strongly linked to the role of the actor of an action in a sentence, as most actors are animate (except for robots doing an action)

Angular gyrus (AG) Posterior to the supramarginal gyrus, located in the parietal lobe, along the superior edge of the temporal lobe and bordered ventrally by the superior temporal sulcus (BA 39)

Anterior Insula (aINS) Anterior section of the insular cortex, which is a portion folded deep within the lateral sulcus.

Anterior superior temporal gyrus (aSTG) Located in the anterior section of the superior temporal gyrus. Medially, it is bound by the circular sulcus of the insula and extends laterally from the anterior limit of the superior temporal sulcus to the anterior limit of Heschl's sulcus.

Aphasia Any language impairment due to a dysfunction in a specific language-related area, generally caused by strokes, head traumas, or cortical atrophy.

Arcuate fasciculus (AF) Neural pathway connecting the prefrontal cortex to the posterior superior temporal gyrus dorsally. It partly runs closely in parallel with the superior longitudinal fasciculus from prefrontal to parietal regions, but not in its posterior portion curving into the posterior temporal cortex.

Artificial grammar learning (AGL) Class of experimental paradigms used in cognitive psychology and neuroscience, which consist of made-up rules generating meaningless artificial languages. In neurolinguistics, such grammars are used to investigate those areas selectively involved in syntactic processing and implicit/explicit learning, when semantic information is stripped away from the sequence.

Associative learning A process of learning based on an association between two separate stimuli that occur together.

Attention A process of selecting and concentrating on specific aspects of a stimulus, while ignoring other aspects.

Axon Connection of the nerve cells via which it forward signals to other neurons. The axons come into contact with other neurons via synapses at which the transmission of the signals is realized by neurotransmitters.

Basal ganglia Set of subcortical nuclei situated at the base of the forebrain. These include: dorsal striatum, ventral striatum, globus pallidus, ventral pallidum, substantia nigra, and subthalamic nucleus.

Brain lesion Damage to neural tissues through injury or disease.

Broca's area Region of the cerebral cortex in the inferior part of the frontal lobe of the brain. It consists of two cytoarchitectonically separable parts, an anterior part called pars triangularis (Brodmann area 45) and a posterior part called pars opercularis (Brodmann area 44).

Brodmann area (BA)   A region of the cerebral cortex that is defined based on its cytoarchitectonics, or organization of cells, according to the classification proposed by Brodmann (1909).

Case marking   Morphological property of language expressing the grammatical function of nouns, pronouns, determiners, and modifiers in a phrase or sentence. In many languages, case information of a specific word is overtly marked by morphological inflections, which are usually attached to the end of the word. Typical grammatical cases are: nominative, accusative, genitive, and dative. They are closely related to the thematic roles in a sentence.

Center-embedded sentences   A sentence (*the boy is my son*) having another sentence embedded within it (*the boy* (*who is running*) *is my son*).

Central sulcus   Fold in the cerebral cortex of brains in vertebrates. Prominent landmark of the brain, separating the parietal lobe from the frontal lobe and the primary motor cortex from the primary somatosensory cortex.

Cerebellum   Region of the brain located underneath the cerebrum in the metencephalon, behind the fourth ventricle, the pons, and the medulla.

Cloze probability   Probability of a word to complete a sentences frame. The cloze probability of a word depends on the congruency that the word has with respect to the prior sentential context, the higher the more congruent.

Conduction aphasia   Rare form of fluent aphasia showing intact comprehension and spared production, but poor capacity for repeating words and sentences. Frequent errors during spontaneous speech include phoneme transposing and substitutions.

Content words   Words that primarily carry lexical-semantic information, have descriptive content and referential weight, and are context-independent. They refer to events (either states or actions) and entities participating in them. Content words typically include nouns, adjectives, verbs, and adverbs.

Corpus callosum   Largest white matter structure of the brain connecting left and right hemispheres along the bottom of the medial longitudinal fissure.

Dendrite   Connection of the nerve cells receiving signals from other neurons.

Derivational morphology   Elements by which words can be changed into others, by adding morphological affixes carrying semantic meaning (*drink* vs. *drinkable*). They can change both the category of a word and the meaning of its stem.

Dysarthria   Neurological deficit in the motor-speech system with poor phoneme articulation.

Embedding   Possibility for a linguistic unit to have another linguistic unit included within it.

Emotional prosody   Emotional information that is conveyed in addition to what is uttered, by prosodic parameters of speech, such as intensity and pitch, but also lengthening of speech elements and pausing between these. Emotional prosody carries meaning (e.g., happy and sad, or fear and anger). In speech, this information is encoded in the auditory signal.

Fiber bundles   Nerve cell projections of the white matter connecting various gray matter areas across the brain.

Frontal lobe   One of the four cortical lobes of humans and other mammals, located at the front of each cerebral hemisphere, and positioned anterior to (in front of) the parietal lobe and superior and anterior to the temporal lobes.

Frontal operculum (FOP)   Brain region in the inferior frontal cortex covering the insula and positioned ventromedial next to the classical language-related brain region called Broca's area.

Function words   Elements that primarily carry syntactic information that allows structural assignment by anchoring, linking, or sequencing other items. Function words typically include prepositions, pronouns, determiners, conjunctions, auxiliary verbs, participles, and inflectional morphology.

Fusiform gyrus (FG)   Part of the temporal lobe bordered medially by the parahippocampal gyrus, laterally by the inferior temporal gyrus, and caudally by the lingual gyrus. (BA 37).

Grammar   The set of rules expressing how words can be combined together in a given language.

Grammatical encoding   The process of producing well-formed syntactic structures in which all grammatical relations are consistently expressed.

Gray matter The 2–3mm outer covering over gyri and sulci of the cerebral cortex, containing numerous cell bodies organized along different cortical layers.

Heschl's gyrus The transverse temporal gyrus in the temporal cortex, buried within the lateral sulcus and extending medially till the circular sulcus of the insular cortex.

Hierarchical structure In syntax, the fundamental structural organization of linguistic constituents, expressing the dominance relation between elements within a phrase and within a sentence.

Hippocampus Major component of the brains of humans and other mammals. In humans and other primates, the hippocampus is located inside the medial temporal lobe, beneath the cortical surface. Like the cerebral cortex, with which it is closely associated, it is a paired structure, with mirror-image halves in the left and right sides of the brain.

Inflectional morphology Elements determining the grammatical relations between constituents, by adding to the words morphological affixes carrying morphosyntactic information (e.g., *run* vs. *runs*). Prototypical inflectional categories include number, tense, person, case, and gender.

Inferior frontal gyrus (IFG) Gyrus of the frontal lobe of the human brain. Its superior border is the inferior frontal sulcus, its inferior border is the lateral fissure, and its posterior border is the inferior precentral sulcus. Above it is the middle frontal gyrus, behind it the precentral gyrus.

Inferior frontal sulcus (IFS) Sulcus located between the middle frontal gyrus and the inferior frontal gyrus.

Inferior-fronto-occipital fasciculus (IFOF) Ventral pathway connecting BA 45 and BA 47 with the posterior STG/MTG, and the occipital cortex, also called Extreme Capsule Fiber System (ECFS)

Intonation The aspect of prosodic information relative to the accentuation of the thematic focus of a sentence. Intonation serves to indicate the emotions and intentions of the speaker by differentiating, for example, between questions and statements.

Language faculty The mental predisposition to learn, produce, and understand language.

Layers Structural levels of cell organization of the gray matter. The cerebral cortex mostly consist of six layers, which can be distinguished according to the type of density of nerve cells.

Lexical selection In production, the process of retrieving and selecting a specific word from the mental lexicon.

Merge In human language, the computational mechanism that constructs new syntactic objects Z (e.g., ate the apples) from already-constructed syntactic objects X (ate) and Y (the apples).

Middle temporal gyrus (MTG) Gyrus in the temporal lobe located between the superior and the inferior temporal sulcus (BA 21).

Mirror neuron A neuron that fires both when observing an action and when performing it.

Morphemes The smallest linguistic unit carrying a meaning (in the word *books*, e.g., two morphemes, *book* and *-s*). Morphemes can be free (any individual word) or bound (affixes). They can also be distinguished into inflectional (*-s* in *he runs*) and derivational (*-hood* in *childhood*) morphology.

Morphology The study of the internal structures of words from more fundamental linguistic units (morphemes).

Motor cortex Area of the frontal lobe located along the dorsal precentral gyrus anteriorly to the central sulcus. It comprises the primary motor cortex (BA 4), the premotor cortex (BA 6) and the supplementary motor area (SMA).

Movement In some theories of syntax, the operation that generates non-canonical sentences from more basic ones. It displaces an element from its original position (where it leaves a silent, but active, copy for interpretation, called the gap) to a new position along the sentence (the filler).

Myelin Fatty white substance that surrounds the axon of nerve cells and works as insulating layer for electrically transmission.

Myelination The growth of myelin during development.

Neuron Electrically excitable cell of nervous system.

Neurotransmitters Chemical messengers enabling neurotransmission from one neuron to another.

Non-canonical A sentence in a human language in which the order of the elements diverges from the basic word order of that specific language.

Occipital lobe One of the four cortical lobes of humans and other mammals, located in the rearmost portion of the brain, behind the parietal lobe and the temporal lobe.

Ontogeny The course of development of an individual organism.

Ordering The process of linking the arguments of sentence (subject, direct object, and indirect object) to the argument structure imposed by the verb. Ordering is especially demanding for the processing of non-canonical structures.

Paragrammatism Speech disturbance leading to the production of utterances in which constituents are incorrectly crossed or unfinished.

Parietal lobe One of the four cortical lobes of humans and other mammals, located above the occipital lobe and behind the frontal lobe and the central sulcus.

Pars opercularis Opercular part of the inferior frontal gyrus in the posterior section of Broca's area (BA 44).

Pars orbitalis Orbital part of the inferior frontal gyrus located anterior-ventral to Broca's area (BA 47).

Pars triangularis Triangular part of the inferior frontal gyrus in the anterior section of Broca's area (BA 45).

Phoneme The abstract mental representations of a sound unit, which distinguish words from another in a given language (e.g., *hat* vs. *cat*).

Phonology The study of sounds as discrete, abstract elements in the speaker's mind that distinguish meaning.

Phrase A group of words, or a single word, forming a constituent in the syntactic structure of a sentence. Phrases are named according to the syntactic category of the main element within it (noun phrase, determiner phrase, adjectival phrase, etc.).

Phylogeny The evolutionary course of a kind of organism.

Pitch The perceptual correlate of the fundamental frequency (the frequency at which the vocal folds vibrate to produce a sound).

Pragmatics The study of the contextual knowledge during communication.

Precentral gyrus Gyrus laying between the central sulcus and the postcentral gyrus in the frontal lobe.

Precentral sulcus Sulcus located parallel to, and in front of, the central sulcus. It divides the inferior, middle, and superior frontal gyri from the precentral gyrus. In the majority of brains, the precentral sulcus is divided into two parts: the inferior precentral sulcus and the superior precentral sulcus. However, the precentral sulcus may also be divided into three parts or form one continuous sulcus.

Primary auditory cortex (PAC) Located on the superior surface of the temporal lobe bilaterally in the so-called Heschl's gyrus, also called primary auditory cortex (BA 41). In addition three regions can be identified adjacent to Heschl's gyrus: a region located posteriorly (the planum temporale), a region anterolateral to Heschl's gyrus (the planum polare), and a region at the lateral convexity of the cortex in the superior temporal gyrus extending inferiorly to the superior temporal sulcus.

Progressive non-fluent aphasia Type of aphasia following cortical atrophy in prefrontal regions bilaterally and peri-sylvian regions in the left hemisphere, and causing both production difficulties and optimal understanding of grammatically complex sentences.

Prosody The set of speech properties describing language at suprasegmental level (syllable, prosodic words, intonational phrases). These properties include intonation, tone, stress, and rhythm.

Receptorarchitecture Description of the cortex of the regional and laminar distribution of neurotransmitter receptors in different cortical regions.

Receptors Protein molecule receiving chemical signals (neurotransmitters) from outside the cell.

Scrambling Common term for word orders that differ from the basic order. In the Chomskyan tradition, every language is assumed to have a basic word order fundamental to its sentence structure, so languages that exhibit a wide variety of different orders are said to have "scrambled" them from their "normal" word order.

Segmental Any segment of speech that can be identified in the speech stream.

Semantics The study of the meaning of individual words and the meaning of words clustered together into phrases and sentences.

Speech errors In language production, any misspeaking deviating from the target form of an utterance. Speech errors can be classified according to the linguistic unit being modified (phoneme, morpheme, word) and the level of representation (phonological, semantic).

Subject-verb agreement Grammatical matching in gender, number, and case via morphological inflection between the subject and the verb of a sentence.

Superior longitudinal fasciculus (SLF) Neural pathway connecting the inferior parietal lobe to the premotor and frontal cortex.

Superior temporal gyrus (STG) One of three gyri of the temporal lobe of the human brain, bounded by the lateral sulcus above and the superior temporal sulcus below. It includes BA 41 and 42 (primary auditory cortex) and Wernicke's area posterior (BA 22).

Superior temporal sulcus (STS) Sulcus separating the superior temporal gyrus from the middle temporal gyrus in the temporal cortex.

Suprasegmental The features of a sound or sequence of sounds beyond a single speech sound, such as stress or pitch.

Syllable Group of sounds consisting of an optional onset (typically, consonants), a nucleus (generally, vowel) and an optional coda.

Syntax The rules for arranging items (words, word parts, and phrases) into their possible permissible combinations in a language.

Temporal lobe One of the four cortical lobes of humans and other mammals, located beneath the Sylvian fissure on both cerebral hemispheres.

Temporal pole The most anterior tip of the temporal lobe, located anterior to the anterior limit of the superior temporal sulcus (BA 38).

Temporo-parietal junction (TPJ) Region located between the temporal and the parietal lobes, toward the posterior end of the Lateral Sulcus (BA 39).

Thalamus Subcortical midline symmetrical structure of two halves located between the cerebral cortex and the midbrain.

Thematic relations Semantic descriptions expressing the relation between the function denoted by a noun and the meaning of the action expressed by the verb.

Thematic roles The type of function that a noun plays with respect to the verb of a sentence. The two major thematic roles are agent (i.e., actor) and patient. Actor: The person or thing performing the act described by the verb of the sentence. Patient: The person or thing undergoing the act described by the verb of the sentence.

Thematic role assignment The assignment of a function to a noun by the verb of a sentence.

Uncinate fasciculus (UF) Ventral pathway connecting the frontal operculum to the anterior temporal cortex.

Wernicke's aphasia Form of aphasia identifying patients with spared linguistic production, but with severe comprehension impairment, generally due to the incapacity to grasp the meaning of spoken words and sentences.

Wernicke's area Region of the cerebral cortex in the temporal lobe of the brain, located in the posterior section of the superior temporal gyrus (BA 22p).

White matter Brain structure that is mainly composed of fiber bundles that in mature state are surrounded by myelin, enabling the rapid propagation of an electrical signal.

Working memory (WM) Ability to actively hold information in the mind needed to do complex tasks such as reasoning, comprehension, and learning. Working memory is a theoretical concept central both to cognitive psychology and neuroscience.

Word category Grammatical class to which a word belongs (noun, verb, preposition, adjective, etc.).

Word order The way words are linearly organized in the sentences of a specific language.

Word substitutions Speech errors in which two words in a sentence are substituted with each other.

World knowledge Extra-linguistic knowledge driving conceptual-semantic interpretation.

Methods

Blood-oxygen-level dependent (BOLD) A measure of neural activity changes, based on the effect of neurovascular coupling, in response to external stimulation or to intrinsic fluctuations at rest.

Closure Positive Shift (CPS) ERP component reflected in a centro-parietally distributed positive shift to be taken as a marker for processing intonational phrase boundary. CPS appears to correlate with prosodic boundary both when realized openly in the speech stream and when realized covertly in written sentences (sometimes indicated by a comma).

Connectivity-based parcellation A technique that permits the division of brain regions into distinct subregions, based on the similarity of the axonal connectivity profiles.

Coordinate system Three-dimensional coordinate system of the brain, used to map the location of brain structures along the x, y, and z axes. The most common of such space systems are the Talairach space (created by the neurosurgeon Jean Talairach) and the MNI space (originated at the Montreal Neurological Institute).

Diffusion tensor imaging (DTI) A technique providing information on the internal fibrous structure based on the measure of water diffusion. Since water will diffuse more rapidly in the direction aligned with the internal structure, the principal direction of the diffusion tensor can be used to infer white matter connectivity in the brain. It can be used to identify the different fiber tracts in the human brain.

Dynamic Causal Modeling (DCM) A technique designed to investigate the influence between brain areas using time series from fMRI or EEG/MEG, by which various models are fit to the time series data and by using Bayesian model comparison to select the winning model.

Early left anterior negativity (ELAN) ERP component showing a negativity (N) deflection between 120 and 200 ms with a maximum over the left anterior scalp, which reflects initial structure building processes.

Effective connectivity A paradigm of analysis to determine the information flow between defined regions in the neural language network. The term *effective connectivity* refers to statistical-mathematical approaches to assess the direction of the data flow from one region to another in the neural network.

Electroencephalography (EEG) A technique that records at the scalp electrical activity as neural oscillations reflected in different frequency bands.

Event-related Potentials (ERP) Quantification of electrical activity in the cortex in response to a particular type of stimulus event with high temporal resolution in the order of milliseconds. Averages of electrocortical activity over events of similar types appear as waveforms in which so-called ERP components, which have either positive or negative polarity relative to baseline, a certain temporal latency in milliseconds after stimulus onset, and a characteristic but poorly resolved spatial distribution over the scalp. Both the polarity and the time point at which the maximum ERP component occurs, as well as partly its distribution, are the basis for the names of the different ERP components

Fractional anisotropy (FA) Scalar value describing the degree of anisotropy (the property of being directional dependent) of water diffusion in axonal bundles. FA is a measure often used in diffusion imaging to quantify the diffusion process. It is taken to reflect fiber density, axonal diameter, and myelination in white matter.

Frequency bands Delta wave: frequency range of brain activity between 1–4 Hz; theta wave: frequency range of brain activity between 4–8 Hz; alpha wave: frequency range of brain activity between 8–13 Hz; beta wave: frequency range of brain activity between 13–30 Hz; gamma wave: frequency range of brain activity between 30–100 Hz.

Functional connectivity A paradigm of analysis to assess brain dynamics. Methods that assess functional connectivity quantify the synchronicity of neuronal charge/discharge alternations in local and remote brain networks, using direct (EEG/MEG) and indirect (fMRI) data sources. The term *functional connectivity* refers to maps of synchronic brain oscillations of brain regions. These maps indicate which brain regions work together without providing information about the direction of the data flow between the regions.

Functional magnetic resonance imaging (fMRI) A technique to localize brain activity related to particular functions in the brain. It has replaced the partly invasive positron emission tomography by a non-invasive, state-of-the-art method for functional-anatomical reconstruction of the language network in the order of submillimeters. However, the temporal resolution of magnetic resonance imaging is limited, as it measures the hemodynamics of brain activity taking place in the order of seconds. Functional magnetic resonance imaging reveals precise information about the location and the magnitude of neural activity changes in response to

Glossary

external stimulation but also about intrinsic fluctuations at rest, that is, in the absence of external stimulation. These neural activity changes are reflected in blood-oxygen-level dependent (BOLD) signal changes based on the effect of neurovascular coupling.

Intracranial recordings An invasive technique that records brain activity by placing electrodes directly on the exposed surface of the brain. It is also known as electrocorticography (ECoG).

Left anterior negativity (LAN) ERP component showing negativity (N) deflection between 300 and 500 ms with a maximum over the left anterior scalp, which reflects morphosyntactic processes.

Magnetencephalography (MEG) A technique that records magnetic fields induced by electrocortical activity. Magnetoencephalography provides information about the amplitude, latency, and topography of language-related magnetic components with a temporal resolution comparable to ERPs but with an improved spatial resolution.

Mismatch Negativity paradigm Experimental paradigm, in which a rarely occurring (deviant) stimulus is presented within a sequence of standard stimuli. Deviant and standard stimuli usually differ in one crucial feature. In adults, the discrimination of these two stimulus types is reflected in a negative deflection with a peak latency of 100–200 ms following change onset. This negative deflection is labeled Mismatch Negativity (MMN).

Near infrared spectroscopy (NIRS) A technique that allows more flexible recording of the BOLD response, since the registration system is mounted directly on the participant's head, which means that the participant does not have to lie still, as is the case during fMRI. This advantage made it an important method for language acquisition research in infants and young children. However, the spatial resolution of NIRS is much lower than that of fMRI, whereas its temporal resolution is just as poor. For this reason this technique is mainly used with very young participants, who find it difficult to lie still for a longer period.

N100 ERP component showing a negativity (N) deflection at around 100 ms after stimulus onset. It has been associated with acoustic processes.

N400 ERP component showing a negativity (N) deflection around 400 milliseconds with a centro-parietal distribution, which reflects lexical-semantic processes.

Oscillation The rhythmic or repetitive firing of a population of neurons in the central nervous system. The number of firings per second represents the frequency of the oscillation. Depending on the number of firings, different frequency bands can be isolated.

P600 ERP component showing a positivity after 600 ms with a centro-parietal distribution, which is associated with late syntactic integration.

Resting-state functional magnetic resonance (rfMRI) A technique that measures fMRI data when participants are "at rest," that is to say, not involved in a task. This contrasts with task-related fMRI, for example from language studies, separating out the condition-related activation and only using the remaining data for analysis. Both approaches when combined with behavioral language data can provide valuable data concerning the functional connectivity between different brain regions in the language network.

Structural magnetic resonance imaging (sMRI) A technique that provides detailed morphometric and geometric features of the brain's gray and white matter such as its volume, density, thickness, and surface area.

Notes

Introduction

1. One has to be aware that the genetic basis will not be represented by a single gene as there are complex routes by which genomes contribute to brain development and behavior (Fisher and Vernes, 2015).

2. This theory does not go unchallenged. Jackendoff and Pinker (2005) provide an alternative conception in which combinatoriality is not restricted to narrow syntax, but expands to larger constructions.

3. Note that this is a hypothesis which momentarily awaits systematic empirical evaluation.

4. A thorough description of the history of psycholinguistics can be found in Levelt (2013).

5. The model in its initial version was published by Friederici (2002) and further developed in a further version published by Friederici (2011). These versions were based on reviewing the data available at that time. Here it serves as the backbone and reference for the data discussed in the book.

Chapter 1

1. Note, however, that there are other psycholinguistic models. Basically, two main classes of models have been proposed to account for the behavioral data on language comprehension: (1) serial, syntax-first and (2) interactive, constraint-satisfaction models. For alternative models see Mitchell (1994) and Marslen-Wilson and Tyler (1980).

2. Section 1.1 is partly based on a paper by Friederici (2011). This model will serve as the backbone for the empirical data discussed in sections 1.1 through 1.9.

3. Cytoarchitectonic studies have indicated that the primary auditory cortex usually covers the medial two-thirds of the anterior Heschl's gyrus (Morosan et al., 2001), and the identification of a subregion in the lateral convexity of the superior temporal gyrus has been confirmed by a receptorarchitectonic analysis (Morosan, Rademacher, Palomero-Gallagher, and Zilles, 2005).

4. Note other neuroimaging studies, however, that reported how the planum temporale or the supramaginal gyrus responded to speech compared with non-speech sounds (Dehaene-Lambertz et al., 2005; Jacquemot, Pallier, LeBihan, Dehaene, and Dupoux, 2003; Meyer et al., 2005). These studies, however, used attention-demanding tasks, which stand in contrast to passive listening paradigms used by others (Obleser, Zimmermann, et al., 2007). From these data, it appears that under specific task demands and under the influence of top-down processes, the differentiation between speech and non-speech sounds may occur at an earlier processing level, in this case to the planum temporale.

5. The following paragraphs are partly cited from Friederici and Singer (2015).

6. Examples of phrasal types. DP or NP: *the car*, *car*; VP: *drive*, *drive the car*; PP: *in the car*; AP: *big car*; AdvP: *drive slowly*.

7. There has been a discussion of the functional significance of the ELAN effect reported in response to word category violations in a critical review (Steinhauer and Drury, 2012). They suggest that context-driven top-down processing may play a larger role than assumed in syntax-first models, such as the one by Friederici (2002).

8. It should be noted at this point that the specific role of Broca's area in language processing is still under debate (Grodzinsky and Santi, 2008; Hagoort, 2005, 2008; Rogalsky and Hickok, 2011). This discussion takes place on multiple levels. At the most general level, the claim is made that Broca's region supports action observation and execution and that its part in language is related to motor-based speech production and comprehension processes (Rizzolatti and Arbib, 1998; Pulvermüller and Fadiga, 2010). At the next level, the claim is that Broca's region supports verbal working memory (Smith and Jonides, 1999) and that this is why this region shows activation when processing syntactically complex sentences (Caplan and Waters, 1999; Rogalsky, Matchin, and Hickok, 2008). At a linguistic level, subregions of Broca's area have been allocated to different aspects of language processing, either seeing BA 44 as supporting syntactic structure building, BA 44/45 as supporting thematic role assignment and BA 45/47 supporting semantic processes (Friederici, 2002); or specifying Broca's area (BA 44/45) as the region supporting the computation of syntactic movement (Grodzinsky, 2000); or defining Broca's region (BA 44/45/47) as the space for the unification of different aspects in language (Hagoort, 2005). Here I will not reiterate each of these studies, but will discuss recent studies that have contributed possible solutions to the open issues at the linguistic level and the related verbal working-memory processes.

9. The following paragraphs are based on Zaccarella and Friederici (2015b).

10. Note, however, that others suggest the inferior frontal gyrus as the candidate region of integration, since both semantic and syntactic information are processed in Broca's regions (Hagoort, 2005).

11. Here I only discuss the relation of speech and music with respect to prosodic processes, but not with respect to other aspects such as syntax and semantics. For a systematic comparison of language and music and their neural basis, see Patel (2008).

Chapter 2

1. This large and distributed activation pattern is possibly due to the fact that the difference between active sentences (*The woman strangles the man*) and a passive sentence (*The man is strangled by the woman*) implies more than just syntax. The different sentence structure (active/passive) also signals a different thematic focus on the first noun phrase as the actor or the undergoer. This may explain the involvement of temporo-parietal areas known to process aspects of the thematic aspects of argument structure.

Chapter 3

1. This section is partly based on Friederici (2015) and Friederici (2016).

2. This section is partly based on Friederici (2015).

3. A study that used intraoperative external stimulation to reveal the respective function of the two ventral fiber tracts suggests that the ECFS is more relevant for language than the uncinate fasciculus because verbal fluency (naming as many words as possible in a particular category) and performance on matching an article to the noun were found to be interrupted when externally stimulating the ECFS, but not when stimulating the uncinate fasciculus (Duffau, Gatignol, Moritz-Gasser, and Mandonnet, 2009). These tasks may have challenged the lexical-semantic system more than the combinatorial system and, thus, do not rule out the possibility that the uncinate fasciculus is involved in combinatorial processes.

4. This section is partly based on pp. 352–354 in Friederici (2016). The neuroanatomical pathway model of language: Syntactic and semantic networks. In *Neurobiology of Language*, ed. Steven L. Small and Gregory Hickok, 349–356. Elsevier Press.

Chapter 4

1. This section is largely based on Friederici (2012, pp. 262–266).

Part III Introduction

1. In the present context the term *native-like* means that the brain structures involved are similar to those observed under typical language learning conditions.

Chapter 5

1. This section is partly based on Friederici and Wartenburger (2010).
2. The term *native signer* refers to a person who has learned sign language as the first language early in life.

Chapter 6

1. This section is based on Friederici and Skeide (2015) and on Friederici and Männel (2014).
2. Note, however, that Dehaene-Lambertz and colleagues (2000, 2002) interpret their data to show that the system for processing speech is already in place shortly after birth.
3. This section is partly based on Friederici and Männel (2015).
4. Dubois et al. (2016), however, report a structural similarity of the white matter including the arcuate fasciculus between infants postnatally and young adults. The authors admit that there is variability in the white matter bundles with respect to their frontal terminating points, partly due to crossing fiber bundles. The infant data they present reveal arcuate fasciculus terminating in the precentral gyrus, similar to infant data presented by Perani et al. (2011).
5. Activation in the inferior frontal gyrus is also reported during speech perception (Dehaene-Lambertz et al., 2006; Perani et al., 2011), but its functional role remains elusive.

Chapter 8

1. Note that in this book I did not discuss reading and writing, which comprise a secondary system to the auditory language system. Reading and writing only developed late during evolution, with pictographic systems and ideographic systems emerging about 6000 BC, and with today's alphabetic systems in which one grapheme represents an elementary sound and which only developed around 3,000 years ago. This phylogenetically late emergence of reading and writing makes the evolution of a specific brain system unlikely—as such a development would require probably more than a couple of thousand years. It has, therefore, been argued that the brain might have "neuronally recycled" a brain region or circuit that originally evolved for another process (Dehaene and Cohen, 2007). In the case of reading, this would be an area that originally was involved in object recognition and became tuned late to recognize letters and words, such as the visual word form area in the left inferior temporal cortex (Dehaene, Cohen, Sigman, and Vinckier, 2005).

References

Abe, Kentaro, and Dai Watanabe. 2011. Songbirds possess the spontaneous ability to discriminate syntactic rules. *Nature Neuroscience* 14 (8): 1067–1074.

Abutalebi, Jubin. 2008. Neural aspects of second language representation and language control. *Acta Psychologica* 128 (3): 466–478.

Alaerts, Kaat, Kritika Nayar, Clare Kelly, Jessica Raithel, Michael P. Milham, and Adriana Di Martino. 2015. Age-related changes in intrinsic function of the superior temporal sulcus in autism spectrum disorders. *Social Cognitive and Affective Neuroscience* 10 (10): 1413–1423.

Amunts, Katrin, Marianne Lenzen, Angela D. Friederici, Axel Schleicher, Patricia Morosan, Nicola Palomero-Gallagher, and Karl Zilles. 2010. Broca's region: Novel organizational principles and multiple receptor mapping. *PLoS Biology* 8:e1000489.

Amunts, Katrin, Axel Schleicher, Uli Bürgel, Hartmut Mohlberg, Harry B. M. Uylings, and Karl Zilles. 1999. Broca's region revisited: Cytoarchitecture and intersubject variability. *Journal of Comparative Neurology* 412 (2): 319–341.

Amunts, Katrin, Axel Schleicher, Annette Ditterich, and Karl Zilles. 2003. Broca's region: Cytoarchitectonic asymmetry and developmental changes. *Journal of Comparative Neurology* 465 (1): 72–89.

Amunts, Katrin, and Karl Zilles. 2012. Architecture and organizational principles of Broca's region. *Trends in Cognitive Sciences* 16 (8): 418–426.

Andreewsky, Evelyne, and Xavier Seron. 1975. Implicit processing of grammatical rules in a classical case of agrammatism. *Cortex* 11 (4): 379–390.

Angrilli, Alessandro, Barbara Penolazzi, Francesco Vespignani, Marica De Vincenzi, Remo Job, Laura Ciccarelli, Daniela Palomba, and Luciano Stegagno. 2002. Cortical brain responses to semantic incongruity and syntactic violation in Italian language: An event-related potential study. *Neuroscience Letters* 322 (1): 5–8.

Anwander, Alfred, Marc Tittgemeyer, Detlef Yves von Cramon, Angela D. Friederici, and Thomas R. Knösche. 2007. Connectivity-based parcellation of Broca's area. *Cerebral Cortex* 17 (4): 816–825.

Ash, Sharon, Corey McMillan, Delani Gunawardena, Brian Avants, Brianna Morgan, Alea Khan, Peachie Moore, James Gee, and Murray Grossman. 2010. Speech errors in progressive non-fluent aphasia. *Brain and Language* 113 (1): 13–20.

Aslin, Richard N., David B. Pisoni, Beth L. Hennessy, and Alan J. Perey. 1981. Discrimination of voice onset time by human infants: New findings and implications for the effects of early experience. *Child Development* 52 (4): 1135–1145.

Atchley, Ruth Ann, Mabel L. Rice, Stacy K. Betz, Kristin M. Kwasney, Joan A. Sereno, and Allard Jongman. 2006. A comparison of semantic and syntactic event related potentials generated by children and adults. *Brain and Language* 99 (3): 236–246.

Bae, Byoung-Il, Ian Tietjen, Kutay D. Atabay, Gilad D. Evrony, Matthew B. Johnson, Ebenezer Asare, Peter P. Wang, et al. 2014. Evolutionarily dynamic alternative splicing of GPR56 regulates regional cerebral cortical patterning. *Science* 343 (6172): 764–768.

Baggio, Giosue, Travis Choma, Michiel van Lambalgen, and Peter Hagoort. 2009. Coercion and compositionality. *Journal of Cognitive Neuroscience* 22 (9): 2131–2140.

Bahlmann, Jörg, Ricarda I. Schubotz, and Angela D. Friederici. 2008. Hierarchical artificial grammar processing engages Broca's area. *NeuroImage* 42 (2): 525–534.

Bahlmann, Jörg, Ricarda I. Schubotz, Jutta L. Mueller, Dirk Köster, and Angela D. Friederici. 2009. Neural circuits of hierarchical visuo-spatial sequence processing. *Brain Research* 1298:161–170.

Basnakova, Jana, Kirsten Weber, Karl Magnus Petersson, Jos van Berkum, and Peter Hagoort. 2014. Beyond the language given: The neural correlates of inferring speaker meaning. *Cerebral Cortex* 24 (10): 2572–2578.

Bastiaansen, Marcel, and Peter Hagoort. 2006. Oscillatory neuronal dynamics during language comprehension. *Progress in Brain Research* 159:179–196.

Bastiaansen, Marcel, Lilla Magyari, and Peter Hagoort. 2010. Syntactic unification operations are reflected in oscillatory dynamics during on-line sentence comprehension. *Journal of Cognitive Neuroscience* 22 (7): 1333–1347.

Bastiaansen, Marcel C. M., Marieke van der Linden, Mariken ter Keurs, Ton Dijkstra, and Peter Hagoort. 2005. Theta responses are involved in lexical-semantic retrieval during language processing. *Journal of Cognitive Neuroscience* 17 (3): 530–541.

Bastos, Andre M., and Jan-Mathijs Schoffelen. 2016. A tutorial review of functional connectivity analysis methods and their interpretational pitfalls. *Frontiers in Systems Neuroscience* 9:175. doi:10.3389/fnsys.2015.00175.

Bates, Elizabeth, and Brian MacWhinney. 1987. Competition, variation, and language learning. In *Mechanisms of Language Acquisition*, ed. Brain MacWhinney, 157–193. Hillsdale, NJ: Lawrence Erlbaum Associates.

Bavelier, Daphne, David P. Corina, and Helen J. Neville. 1998. Brain and language: A perspective frown sign language. *Neuron* 21 (2): 275–278.

Beaucousin, Virginie, Anne Lacheret, Marie-Renee Turbelin, Michel More, Bernard Mazoyer, and Nathalie Tzourio-Mazoyer. 2007. FMRI study of emotional speech comprehension. *Cerebral Cortex* 17 (2): 339–352.

Beckers, Gabriel J. L., Johan J. Bolhuis, Kazuo Okanoya, and Robert C. Berwick. 2012. Birdsong neurolinguistics: Songbird context-free grammar claim is premature. *Neuroreport* 23 (3): 139–145.

Behrens, Tim E. J., Heidi Johansen-Berg, Mark W. Woolrich, Stephen M. Smith, Claudia A. M. Wheeler-Kingshott, Philip A. Boulby, Gareth J. Barker, E. L. Sillery, K. Sheehan, Olga Ciccarelli, Alan J. Thompson, J. Michael Brady, and Paul M. Matthews. 2003. Non-invasive mapping of connections between human thalamus and cortex using diffusion imaging. *Nature Neuroscience* 6 (7): 750–757.

Bellugi, Ursula, Howard Poizner, and Edward S. Klima. 1989. Language, modality and the brain. *Trends in Neurosciences* 12 (10): 380–388.

Bemis, Douglas K., and Liina Pylkkänen. 2011. Simple composition: A magnetoencephalography investigation into the comprehension of minimal linguistic phrases. *Journal of Neuroscience* 31 (8): 2801–2814.

Bemis, Douglas K., and Liiana Pylkkänen. 2013. Basic linguistic composition recruits the left anterior temporal lobe and left angular gyrus during both listening and reading. *Cerebral Cortex* 23 (8): 1859–1873.

Benavides-Varela, Silvia, Jean-Rémy Hochmann, Francesco Macagno, Marina Nespor, and Jacques Mehler. 2012. Newborn's brain activity signals the origin of word memories. *Proceedings of the National Academy of Sciences of the United States of America* 109 (44): 17908–17913.

Ben-Shachar, Michal, Talma Hendler, Itamar Kahn, Dafna Ben-Bashat, and Yosef Grodzinsky. 2003. The neural reality of syntactic transformations: Evidence from functional magnetic resonance imaging. *Psychological Science* 14 (5): 433–440.

Ben-Shachar, Michal, Dafna Palti, and Yosef Grodzinsky. 2004. Neural correlates of syntactic movement: Converging evidence from two fMRI experiments. *NeuroImage* 21 (4): 1320–1336.

Bergelson, Elika, and Daniel Swingley. 2012. At 6–9 months, human infants know the meanings of many common nouns. *Proceedings of the National Academy of Sciences of the United States of America* 109 (9): 3253–3258.

Bernal, Byron, Alfredo Ardila, and Monica Rosselli. 2015. Broca's area network in language function: A pooling-data connectivity study. *Frontiers in Psychology* 6:687. doi:10.3389/fpsyg.2015.00687.

Bernal, Savita, Ghislaine Dehaene-Lambertz, Severine Millotte, and Anne Christophe. 2010. Two-year-olds compute syntactic structure on-line. *Developmental Science* 13 (1): 69–76.

Berken, Jonathan A., Xiaoqian Chai, Jen-Kai Chen, Vincent L. Gracco, and Denise Klein. 2016. Effects of early and late bilingualism on resting-state functional connectivity. *Journal of Neuroscience* 36 (4): 1165–1172.

Berthier, Marcelo L., Matthew A. Lambon Ralph, Jesus Pujol, and Cristina Green. 2012. Arcuate fasciculus variability and repetition: The left sometimes can be right. *Cortex* 48 (2): 133–143.

Berwick, Robert C., and Noam Chomsky. 2015. *Why Only Us. Language and Evolution*. Cambridge, MA: MIT Press.

Berwick, Robert C., Angela D. Friederici, Noam Chomsky, and Johan J. Bolhuis. 2013. Evolution, brain and the nature of language. *Trends in Cognitive Sciences* 17 (2): 89–98.

Berwick, Robert C., Kazuo Okanoya, Gabriel J. L. Beckers, and Johan J. Bolhuis. 2011. Songs to syntax: The linguistics of birdsong. *Trends in Cognitive Sciences* 15 (3): 113–121.

Bickerton, Derek. 1990. *Language and Species*. Chicago: University of Chicago Press.

Bierwisch, Manfred. 1982. Formal and lexical semantics. *Linguistische Berichte* 80:3–17.

Binder, Jeffrey R., Julie Anne Frost, Thomas A. Hammeke, Patrick S. Bellgowan, Jane A. Springer, Jacqueline N. Kaufman, and Edward T. Possing. 2000. Human temporal lobe activation by speech and nonspeech sounds. *Cerebral Cortex* 10 (5): 512–528.

Binder, Jeffrey R., Rutvik H. Desai, William W. Graves, and Lisa L. Conant. 2009. Where is the semantic system? A critical review and meta-analysis of 120 functional neuroimaging studies. *Cerebral Cortex* 19 (12): 2767–2796.

Biswal, Bharat, F. Zerrin Yetkin, Victor M. Haughton, and James S. Hyde. 1995. Functional connectivity in the motor cortex of resting human brain using echo-planar MRI. *Magnetic Resonance in Medicine* 34 (4): 537–541.

Blesser, Barry. 1972. Speech perception under conditions of spectral transformation. I. Phonetic characteristics. *Journal of Speech, Language, and Hearing Research: JSLHR* 15:5–41.

Bloch, Constantine, Anelis Kaiser, Esther Kuenzli, Daniela Zappatore, Sven Haller, Rita Franceschini, Georges Luedi, Ernst-Wilhelm Radue, and Cordula Nitsch. 2009. The age of second language acquisition determines the variability in activation elicited by narration in three languages in Broca's and Wernicke's area. *Neuropsychologia* 47 (3): 625–633.

Blumenthal, P. J. 2003. *Kaspar Hausers Geschwister. Auf der Suche nach dem wilden Menschen*. Wien: Deuticke.

Blumstein, Sheila E., William E. Cooper, Harold Goodglass, Sheila Statlender, and Jonathan Gottlieb. 1980. Production deficits in aphasia—voice-onset time analysis. *Brain and Language* 9 (2): 153–170.

Bögels, Sara, Herbert Schriefers, Wietske Vonk, Dorothee J. Chwilla, and Roel Kerkhofs. 2010. The interplay between prosody and syntax in sentence processing: The case of subject- and object-control verbs. *Journal of Cognitive Neuroscience* 22 (5): 1036–1053.

Bolhuis, Johan J., and Sanne Moorman. 2015. Birdsong memory and the brain: In search of the template. *Neuroscience and Biobehavioral Reviews* 50:41–55.

Bolhuis, Johan J., Ian Tattersall, Noam Chomsky, and Robert C. Berwick. 2014. How could language have evolved? *PLoS Biology* 12 (8): e1001934. doi:10.1371/journal.pbio.1001934.

Bookheimer, Susan. 2002. Functional MRI of language: New approaches to understanding the cortical organization of semantic processing. *Annual Review of Neuroscience* 25:151–188.

Bornkessel, Ina, and Matthias Schlesewsky. 2006. The extended Argument Dependency Model: A neurocognitive approach to sentence comprehension across languages. *Psychological Review* 113 (4): 787–821.

Bornkessel, Ina, Stefan Zysset, Angela D. Friederici, D. Yves von Cramon, and Matthias Schlesewsky. 2005. Who did what to whom? The neural basis of argument hierarchies during language comprehension. *NeuroImage* 26 (1): 221–233.

Bornkessel-Schlesewsky, Ina, Franziska Kretzschmara, Sarah Tune, Luming Wang, Safiye Genç, Marcus Philipp, Dietmar Roehm, and Matthias Schlesewsky. 2011. Think globally: Cross-linguistic variation in electrophysiological activity during sentence comprehension. *Brain and Language* 117 (3): 133–152.

Bornkessel-Schlesewsky, Ina, and Matthias Schlesewsky. 2008. An alternative perspective on "semantic P600" effects in language comprehension. *Brain Research. Brain Research Reviews* 59 (1): 55–73.

Bozeat, Sasha, Matthew A. Lambon Ralph, Karalyn Patterson, Peter Garrard, and John R. Hodges. 2000. Nonverbal semantic impairment in semantic dementia. *Neuropsychologia* 38 (9): 1207–1215.

Brådvik, Björn, Christina Dravins, Stig Holtås, Ingmar Rosén, Erik Ryding, and David Ingvar. 1991. Disturbances of speech prosody following right hemisphere infarcts. *Acta Neurologica Scandinavica* 84 (2): 114–126.

Brain Development Cooperative Group. 2012. Total and regional brain volumes in a population-based normative sample from 4 to 18 years: The NIH MRI study of normal brain development. *Cerebral Cortex* 22:1–12.

Brauer, Jens, Alfred Anwander, and Angela D. Friederici. 2011. Neuroanatomical prerequisites for language functions in the maturing brain. *Cerebral Cortex* 21 (2): 459–466.

Brauer, Jens, and Angela D. Friederici. 2007. Functional neural networks of semantic and syntactic processes in the developing brain. *Journal of Cognitive Neuroscience* 19 (10): 1609–1623.

Brauer, Jens, Jane Neumann, and Angela D. Friederici. 2008. Temporal dynamics of perisylvian activation during language processing in children and adults. *NeuroImage* 41 (4): 1484–1492.

Breitenstein, Caterina, Andreas Jansen, Michael Deppe, Ann-Freya Foerster, Jens Sommer, Thomas Wolbers, and Stefan Knecht. 2005. Hippocampus activity differentiates good from poor learners of a novel lexicon. *NeuroImage* 25 (3): 958–968.

Brennan, Jonathan, Yuval Nir, Uri Hasson, Rafael Malach, David J. Heeger, and Liina Pylkkänen. 2012. Syntactic structure building in the anterior temporal lobe during natural story listening. *Brain and Language* 120 (2): 163–173.

Bresnan, Joan. 2001. *Lexical-Functional Syntax*. Oxford: Blackwell.

Broca, Paul. 1861. Remarques sur le siège de la faculté du langage articulé, suivies d'une observation d'aphémie (parte de la parole). *Bulletins de la Société Anatomique de Paris* 6:330–357.

Brodmann, Korbinian. 1909. Beiträge zur histologischen Lokalisation der Grosshirnrinde: VI. Die Cortexgliederung des Menschen. *Journal für Psychologie und Neurologie* 10:231–246.

Brown, Colin, and Peter Hagoort. 1993. The processing nature of the N400 - evidence from masked priming. *Journal of Cognitive Neuroscience* 5 (1): 34–44.

Bryan, Karen L. 1989. Language prosody and the right hemisphere. *Aphasiology* 3 (4): 285–299.

Buckner, Randy L., Jessica R. Andrews-Hanna, and Daniel L. Schacter. 2008. The brain's default network—Anatomy, function, and relevance to disease. *Annals of the New York Academy of Sciences* 1124:1–38.

Buckner, Randy L., and Fenna M. Krienen. 2013. The evolution of distributed association networks in the human brain. *Trends in Cognitive Sciences* 17 (12): 648–665.

Buzsáki, Gyoergy, Nikos Logothethis, and Wolf Singer. 2013. Scaling brain size, keeping timing: Evolutionary preservation of brain rhythms. *Neuron* 80 (3): 751–764.

Capek, Cheryl M., Giordana Grossi, Aaron J. Newman, Susan L. McBurney, David Corina, Brigitte Roeder, and Helen J. Neville. 2009. Brain systems mediating semantic and syntactic processing in deaf native signers: Biological invariance and modality specificity. *Proceedings of the National Academy of Sciences of the United States of America* 106 (21): 8784–8789.

Caplan, David, Evan Chen, and Gloria S. Waters. 2008. Task-dependent and task-independent neurovascular responses to syntactic processing. *Cortex* 44 (3): 257–275.

Caplan, David, and Gloria S. Waters. 1999. Verbal working memory and sentence comprehension. *Behavioral and Brain Sciences* 22 (1): 77–94.

Capone, Nina C., and Karla K. McGregor. 2004. Gesture development: A review for clinical and research practices. *Journal of Speech, Language, and Hearing Research: JSLHR* 47 (1): 173–186.

Caramazza, Alfonso, and Edgar B. Zurif. 1976. Dissociation of algorithmic and heuristic processes in language comprehension—evidence from aphasia. *Brain and Language* 3 (4): 572–582.

Catani, Marco, and Valentina Bambini. 2014. A model for Social Communication And Language Evolution and Development (SCALED). *Current Opinion in Neurobiology* 28:165–171.

References

Catani, Marco, Matthew P. G. Allin, Masud Husain, Luca Pugliese, Marsel M. Mesulam, Robin M. Murray, and Derek K. Jones. 2007. Symmetries in human brain language pathways correlate with verbal recall. *Proceedings of the National Academy of Sciences of the United States of America* 104 (43): 17163–17168.

Catani, Marco, Robert J. Howard, Sinisa Pajevic, and Derek K. Jones. 2002. Virtual in vivo interactive dissection of white matter fasciculi in the human brain. *NeuroImage* 17 (1): 77–94.

Catani, Marco, Derek K. Jones, and Dominic H. Ffytche. 2005. Perisylvian language networks of the human brain. *Annals of Neurology* 57 (1): 8–16.

Catani, Marco, and Marsel M. Mesulam. 2008. What is a disconnection syndrome? *Cortex* 44 (8): 911–913.

Catani, Marco, and Michel Thiebaut de Schotten. 2008. A diffusion tensor imaging tractography atlas for virtual in vivo dissections. *Cortex* 44 (8): 1105–1132.

Chait, Maria, Steven Greenberg, Takayuki Arai, Jonathan Z. Simon, and David Poeppel. 2015. Multi-time resolution analysis of speech: Evidence from psychophysics. *Frontiers in Neuroscience* 9:214. doi:10.3389/fnins.2015.00214.

Chee, Michael W. L., Edsel W. L. Tan, and Thorsten Thiel. 1999. Mandarin and English single word processing studied with functional magnetic resonance imaging. *Journal of Neuroscience* 19 (8): 3050–3056.

Cheour, Marie, Kimmo Alho, Kimmo Sainio, Kalevi Reinikainen, Martin Renlund, Olli Aaltonen, Osmo Eerola, and Risto Näätänen. 1997. The mismatch negativity to changes in speech sounds at the age of three months. *Developmental Neuropsychology* 13 (2): 167–174.

Cheour, Marie, Rita Čeponiene, Paavo Leppänen, Kimmo Alho, Teija Kujala, Martin Renlund, Vineta Fellman, and Risto Näätänen. 2002. The auditory sensory memory trace decays rapidly in newborns. *Scandinavian Journal of Psychology* 43 (1): 33–39.

Cheour, Marie, Rita Čeponiene, Anne Lehtokoski, Aavo Luuk, Jüri Allik, Kimmo Alho, and Risto Näätänen. 1998. Development of language-specific phoneme representations in the infant brain. *Nature Neuroscience* 1 (5): 351–353.

Cheour-Luhtanen, Marie, Kimmo Alho, Teija Kujala, Kimmo Sainio, Kalevi Reinikainen, Martin Renlund, Olli Aaltonen, Osmo Eerola, and Risto Näätänen. 1995. Mismatch negativity indicates vowel discrimination in newborns. *Hearing Research* 82 (1): 53–58.

Cheour-Luhtanen, Marie, Kimmo Alho, Kimmo Sainio, Teemu Rinne, Kalevi Reinikainen, M. Pohjavouri, Martin Renlund, Olli Aaltonen, Osmo Eerola, and Risto Näätänen. 1996. The ontogenetically earliest discriminative response of the human brain. *Psychophysiology* 33 (4): 478–481.

Chi, Je Geung, Elizabeth C. Dooling, and Floyd H. Gilles. 1977. Left-right asymmetries of the temporal speech areas of the human fetus. *Archives of Neurology* 34 (6): 346–348.

Chiarello, Christine, Robert Knight, and Mark Mandel. 1982. Aphasia in a prelingually deaf woman. *Brain* 105:29–51.

Chomsky, Noam. 1957. *Syntactic Structures*. The Hague: Mouton & Co.

Chomsky, Noam. 1965. *Aspects of the Theory of Syntax*. Cambridge, MA: MIT Press.

Chomsky, Noam. 1980. On binding. *Linguistic Theory* 11 (1): 1–46.

Chomsky, Noam. 1981. *Lectures on Government and Binding. The Pisa Lectures*. Dordrecht: Foris.

Chomsky, Noam. 1986. *Knowledge of Language: Its Nature, Origin and Use*. New York: Praeger.

Chomsky, Noam. 1995a. Language and nature. *Mind* 104 (413): 1–61.

Chomsky, Noam. 1995b. *The Minimalist Program*. MIT Press.

Chomsky, Noam. 2007. Approaching UG from below. In *Interfaces + Recursion = Language?* ed. Uli Sauerland and Hans-Martin Gärtner, 1–29. Berlin: Mouton de Gruyter.

Chomsky, Noam. 2013. Problems of projection. *Lingua* 130:33–49.

Chwilla, Dorothee J., Colin M. Brown, and Peter Hagoort. 1995. The N400 as a function of the level of processing. *Physiophysiology* 32 (3): 274–285.

Citron, Francesca M. M., Regine Oberecker, Angela D. Friederici, and Jutta L. Mueller. 2011. Mass counts: ERP correlates of non-adjacent dependency learning under different exposure conditions. *Neuroscience Letters* 487 (3): 282–286.

Clark, Herbert H. 1996. *Using Language*. Cambridge: Cambridge University Press.

Clos, Mareike, Katrin Amunts, Angela R. Laird, Peter T. Fox, and Simon B. Eickhoff. 2013. Tackling the multifunctional nature of Broca's region meta-analytically: Co-activation-based parcellation of area 44. *NeuroImage* 83:174–188.

Cohen, Laurent, Antoinette Jobert, Denis Le Bihan, and Stanislas Dehaene. 2004. Distinct unimodal and multimodal regions for word processing in the left temporal cortex. *NeuroImage* 23 (4): 1256–1270.

Collins, Allan M., and Elizabeth F. Loftus. 1975. Spreading activation theory of semantic processing. *Physiological Reviews* 82 (6): 407–428.

Crinion, Jennifer T., Matthew A. Lambon-Ralph, Elizabeth A. Warburton, David Howard, and Richard J. S. Wise. 2003. Temporal lobe regions engaged during normal speech comprehension. *Brain* 126 (5): 1193–1201.

Curtiss, Susan, ed. 1977. *Genie. Psycholinguistic Study of a Modern-day "Wild Child"*. London: Academic Press Inc.

Dapretto, Mirella, and Susan Y. Bookheimer. 1999. Form and content: Dissociating syntax and semantics in sentence comprehension. *Neuron* 24 (2): 427–432.

Darwin, Charles. 1871. *The Descent of Man, and Selection in Relation to Sex*. 1st ed. London: John Murray.

David, Olivier, Burkhard Maess, Korinna Eckstein, and Angela D. Friederici. 2011. Dynamic causal modeling of subcortical connectivity of language. *Journal of Neuroscience* 31 (7): 2712–2717.

Davis, Mark H. 2015. The neurobiology of lexical access. In *Neurobiology of Language Volume*, ed. Steven L. Small and Gregory G. Hickok, 541–557. Elsevier Press.

Davis, Mark H., and Gareth M. Gaskell. 2009. A complementary systems account of word learning: Neural and behavioural evidence. *Philosophical Transactions of the Royal Society of London. Series B, Biological Sciences* 364 (1536): 3773–3800.

Dax, Marc. 1836. Lesions de la moitie gauche de l'encephale soincidant avec l'oubli des signes de la pensee. Congres meridional tenu a Montpellier.

De Diego Balaguer, Ruth, Juan Manuel Toro., Antoni Rodriguez-Fornells, and Anne-Catherine Bachoud-Levi. 2007. Different neurophysiological mechanisms underlying word and rule extraction from speech. *PLoS One* 2 (11): e1175. doi:10.1371/journal.pone.0001175.

de Saussure, Ferdinand. 1916. Cours de linguistique générale, ed. Charles Bally and Albert Sechehaye. Lausanne and Paris: Payot. (Trans. Wade Baskin. 1977. Course in General Linguistics, Glasgow: Fontana/Collins).

Dehaene, Stanislas, and Laurent Cohen. 2007. Cultural recycling of cortical maps. *Neuron* 56 (2): 384–398.

Dehaene, Stanislas, Laurent Cohen, Mariano Sigman, and Fabien Vinckier. 2005. The neural code for written words: A proposal. *Trends in Cognitive Sciences* 9 (7): 335–341.

Dehaene-Lambertz, Ghislaine. 2000. Cerebral specialization for speech and non-speech stimuli in infants. *Journal of Cognitive Neuroscience* 12:449–460.

Dehaene-Lambertz, Ghislaine, and Sylvaine Baillet. 1998. A phonological representation in the infant brain. *Neuroreport* 9 (8): 1885–1888.

Dehaene-Lambertz, Ghislaine, and Stanislas Dehaene. 1994. Speed and cerebral correlates of syllable discrimination in infants. *Nature* 370 (6487): 292–295.

Dehaene-Lambertz, Ghislaine, Stanislas Dehaene, and Lucie Hertz-Pannier. 2002. Functional neuroimaging of speech perception in infants. *Science* 298 (5600): 2013–2015.

Dehaene-Lambertz, Ghislaine, Emmanuel Dupoux, and Ariel Gout. 2000. Electrophysiological correlates of phonological processing: A cross-linguistic study. *Journal of Cognitive Neuroscience* 12 (4): 635–647.

Dehaene-Lambertz, Ghislaine, Lucie Hertz-Pannier, Jessica Dubois, Sebastien Meriaux, Alexis Roche, Mariano Sigman, and Stanislas Dehaene. 2006. Functional organization of perisylvian activation during presentation of sentences in preverbal infants. *Proceedings of the National Academy of Sciences of the United States of America* 103 (38): 14240–14245.

Dehaene-Lambertz, Ghislaine, Christophe Pallier, Willy Serniclaes, Liliane Sprenger-Charolles, Antoinette Jobert, and Stanislas Dehaene. 2005. Neural correlates of switching from auditory to speech perception. *NeuroImage* 24 (1): 21–33.

Dehaene-Lambertz, Ghislaine, and Marcela Peña. 2001. Electrophysiological evidence for automatic phonetic processing in neonates. *Neuroreport* 12 (14): 3155–3158.

Del Prato, Paul, and Liina Pylkkänen. 2014. MEG evidence for conceptual combination but not numeral quantification in the left anterior temporal lobe during language production. *Frontiers in Psychology* 5:524. doi:10.3389/fpsyg.2014.00524.

Dell, Gary S. 1986. A spreading-activation theory of retrieval in sentence production. *Psychological Review* 93 (3): 283–321.

Démonet, Jean-François, François Chollet, Stuart Ramsay, Dominique Cardebat, Jean-Luc Nespoulous, Richard Wise, André Rascol, and Richard Frackowiak. 1992. The anatomy of phonological and semantic processing in normal subjects. *Brain* 115:1753–1768.

Démonet, Jean-François, Guillaume Thierry, and Dominique Cardebat. 2005. Renewal of the neurophysiology of language: Functional neuroimaging. *Physiological Reviews* 85 (1): 49–95.

den Ouden, Dirk-Bart, Dorothee Saur, Wolfgang Mader, Bjoern Schelter, Sladjana Lukic, Eisha Wali, Jens Timmer, and Cynthia K. Thompson. 2012. Network modulation during complex syntactic processing. *NeuroImage* 59 (1): 815–823.

Deutsch, Avital, and Shlomo Bentin. 2001. Syntactic and semantic factors in processing gender agreement in Hebrew: Evidence from ERPs and eye movements. *Journal of Memory and Language* 45 (2): 200–224.

DeWitt, Iain, and Josef P. Rauschecker. 2012. Phoneme and word recognition in the auditory ventral stream. *Proceedings of the National Academy of Sciences of the United States of America* 109 (8): 2709–2710.

Dick, Anthony Steven, and Iris Broce. 2015. The neurobiology of gesture and its development. In *Neurobiology of language*, ed. Gregory Hickok and Steven L. Small, 389–398. San Diego: Elsevier.

Dick, Anthony Steven, Eva H. Mok, Anjali Raja Beharelle, Susan Goldin-Meadow, and Steven L. Small. 2014. Frontal and temporal contributions to understanding the iconic co-speech gestures that accompany speech. *Human Brain Mapping* 35 (3): 900–917.

Diesch, Eugen, Carsten Eulitz, Scott Hampson, and Bernhard Ross. 1996. The neurotopography of vowels as mirrored by evoked magnetic field measurements. *Brain and Language* 53 (2): 143–168.

Dikker, Suzanne, Hugh Rabagliati, and Liina Pylkkänen. 2009. Sensitivity to syntax in visual cortex. *Cognition* 110 (3): 293–321.

Ding, Nai, Lucia Melloni, Hang Zhang, Xing Tian, and David Poeppel. 2015. Cortical tracking of hierarchical linguistic structures in connected speech. *Nature Neuroscience* 19 (1): 158–164.

Dittmar, Miriam, Kirsten Abbot-Smith, Elena Lieven, and Michael Tomasello. 2008. Young German children's early syntactic competence: A preferential looking study. *Developmental Science* 11 (4): 575–582.

Dronkers, Nina F. 1996. A new brain region for coordinating speech articulation. *Nature* 384 (6605): 159–161.

Dronkers, Nina F., David P. Wilkins, Robert D. Van Valin, Jr., Brenda B. Redfern, and Jeri J. Jaeger. 2004. Lesion analysis of the brain areas involved in language comprehension. *Cognition* 92 (1–2): 145–177.

Dubois, Jessica, Lucie Hertz-Pannier, Arnaud Cachia, Jean-Francois Mangin, Denis Le Bihan, and Ghislaine Dehaene-Lambertz. 2009. Structural asymmetries in the infant language and sensorimotor networks. *Cerebral Cortex* 19:414–423.

Dubois, Jessica, Cyril Poupon, Bertrand Thirion, Hina Simonnet, Sofya Kulikova, Francois Leroy, Lucie Hertz-Pannier, and Ghislaine Dehaene-Lambertz. 2016. Exploring the early organization and maturation of linguistic pathways in the human infant brain. *Cerebral Cortex* 26 (5): 2283–2298.

Duffau, Hugues. 2008. The anatomo-functional connectivity of language revisited new insights provided by electrostimulation and tractography. *Neuropsychologia* 46 (4): 927–934.

Duffau, Hugues, Peggy Gatignol, Sylvie Moritz-Gasser, and Emmanuel Mandonnet. 2009. Is the left uncinate fasciculus essential for language? *Journal of Neurology* 256 (3): 382–389.

Dunbar, Robin Ian MacDonald. 1996. *Grooming, Gossip and the Evolution of Language*. London: Faber and Faber.

Eckstein, Korinna, and Angela D. Friederici. 2006. It's early: Event-related potential evidence for initial interaction of syntax and prosody in speech comprehension. *Journal of Cognitive Neuroscience* 18 (10): 1696–1711.

Emmorey, Karen, Stephen McCullough, and Diane Brentari. 2003. Categorical perception in American Sign Language. *Language and Cognitive Processes* 18 (1): 21–45.

Ethofer, Thomas, Johannes Bretscher, Markus Gschwind, Benjamin Kreifelts, Dirk Wildgruber, and Patrik Vuilleumier. 2012. Emotional voice areas: Anatomic location, functional properties, and structural connections revealed by combined fMRI/diffusion tensor imaging. *Cerebral Cortex* 22 (1): 191–200.

Fabbri-Destro, Maddalena, and Giacomo Rizzolatti. 2008. Mirror neurons and mirror systems in monkeys and humans. *Physiology (Bethesda, MD)* 23:171–179.

Falk, Dean. 1980. Hominid brain evolution: The approach from paleoneurology. *American Journal of Physical Anthropology* 23 (S1): 93–107.

Falk, Dean. 2014. Interpreting sulci on hominin endocasts: Old hypotheses and new findings. *Frontiers in Human Neuroscience* 8:1–11. doi:10.3389/fnhum.2014.00134.

Fedorenko, Evelina, Michael K. Behr, and Nancy Kanwisher. 2011. Functional specificity for high-level linguistic processing in the human brain. *Proceedings of the National Academy of Sciences of the United States of America* 108 (39): 16428–16433.

Fedorenko, Evelina, and Sharon L. Thompson-Schill. 2014. Reworking the language network. *Trends in Cognitive Sciences* 18 (3): 120–126.

Fell, Juergen, Eva Ludowig, Bernhard P. Staresina, Tobias Wagner, Thorsten Kranz, Christian E. Elger, and Nikolai Axmacher. 2011. Medial temporal theta/alpha power enhancement precedes successful memory encoding: Evidence based on intracranial EEG. *Journal of Neuroscience* 31 (14): 5392–5397.

Felser, Claudia, Harald Clahsen, and Thomas F. Munte. 2003. Storage and integration in the processing of filler-gap dependencies: An ERP study in of topicalization and WH-movement in German. *Brain and Language* 87 (3): 345–354.

Fengler, Anja, Lars Meyer, and Angela D. Friederici. 2016. How the brain attunes to sentence processing: Relating behavior, structure, and function. *NeuroImage* 129:268–278.

Fernandez-Miranda, Juan C., Yibao Wang, Sudhir Pathak, Lucia Stefaneau, Timothy Verstynen, and Fang-Cheng Yeh. 2015. Asymmetry, connectivity, and segmentation of the arcuate fascicle in the human brain. *Brain Structure & Function* 220 (3): 1665–1680.

Ferstl, Evelyn C. 2015. Inferences during text comprehension: What neuroscience can (or cannot) contribute. In *Inferences During Reading*, ed. Edward J. O'Brien, Anne E. Cook and Robert F. Lorch, Jr., 230–259. Cambridge: Cambridge University Press.

Ferstl, Evelyn C., Thomas Guthke, and D. Yves von Cramon. 2002. Text comprehension after brain injury: Left prefrontal lesions affect inference processes. *Neuropsychology* 16 (3): 292–308.

Fiebach, Christian J., Matthias Schlesewsky, and Angela D. Friederici. 2001. Syntactic working memory and the establishment of filler-gap dependencies: Insights from ERPs and fMRI. *Journal of Psycholinguistic Research* 30 (3): 321–338.

Fiebach, Christian J., Matthias Schlesewsky, and Angela D. Friederici. 2002. Separating syntactic memory costs and syntactic integration costs during parsing: The processing of German WH-questions. *Journal of Memory and Language* 47 (2): 250–272.

Fiez, Julie A. 1997. Phonology, semantics, and the role of the left inferior prefrontal cortex. *Human Brain Mapping* 5 (2): 79–83.

Finkl, Theresa, Alfred Anwander, Angela D. Friederici, Johannes Gerber, Dirk Mürbe, and Anja Hahne. Submitted. Differential modulation of the brain networks for speech and language in deaf signers.

Fisher, Simon E., Cecilia S. L. Lai, and Anthony P. Monaco. 2003. Deciphering the genetic basis of speech and language disorders. *Annual Review of Neuroscience* 26:57–80.

Fisher, Simon E., and Sonja C. Vernes. 2015. Genetics and the Language Sciences. *Annual Review of Linguistics* 1:289–310. doi:10.1146/annurev-linguist-030514-125024.

Fitch, W. Tecumseh. 2010. *The Evolution of Language*. Cambridge: Cambridge University Press.

Fitch, W. Tecumseh, and Marc D. Hauser. 2004. Computational constraints on syntactic processing in a nonhuman primate. *Science* 303 (5656): 377–380.

Flechsig, Paul E. 1896. *Gehirn und Seele*. Leipzig: Veit & comp.

Flinker, Adeen, Anna Korzeniewska, Avgusta Y. Shestyuk, Piotr J. Franaszczuk, Nina F. Dronkers, Robert T. Knight, and Nathan E. Crone. 2015. Redefining the role of Broca's area in speech. *Proceedings of the National Academy of Sciences of the United States of America* 112 (9): 2871–2875.

Fodor, Jerry Alan. 1983. *The Modularity of Mind*. Cambridge, MA: MIT Press.

Fogassi, Leonardo, Pier Francesco Ferrari, Benno Gesierich, Stefano Rozzi, Fabian Chersi, and Giacomo Rizzolatti. 2005. Parietal lobe: From action organization to intention understanding. *Science* 29 (5722): 662–667.

Fonteneau, Elisabeth, and Heather K. J. van der Lely. 2008. Electrical brain responses in language-impaired children reveal grammar-specific deficits. *PLoS One* 3 (3): e1832. doi:10.1371/journal.pone.0001832.

Fox, Michael D., and Marcus E. Raichle. 2007. Spontaneous fluctuations in brain activity observed with functional magnetic resonance imaging. *Nature Reviews. Neuroscience* 8 (9): 700–711.

Frazier, Lyn, and Janet Dean Fodor. 1978. The sausage machine: A new two-stage parsing model. *Cogniton* 6 (4): 291–325.

Frey, Stephen, Jennifer S. W. Campbell, G. Bruce Pike, and Michael Petrides. 2008. Dissociating the human language pathways with high angular resolution diffusion fiber tractography. *Journal of Neuroscience* 28 (45): 11435–11444.

Friederici, Angela D. 1982. Syntactic and semantic processes in aphasic deficits: The availability of prepositions. *Brain and Language* 15 (2): 249–258.

Friederici, Angela D. 1983. Children's sensitivity to function words during sentence comprehension. *Linguistics* 21 (5): 717–739.

Friederici, Angela D. 1985. Levels of processing and vocabulary types: Evidence from online comprehension in normals and agrammatics. *Cognition* 19 (2): 133–166.

Friederici, Angela D. 1990. On the properties of cognitive modules. *Psychological Research* 52 (2–3): 175–180.

Friederici, Angela D. 2002. Towards a neural basis of auditory sentence processing. *Trends in Cognitive Sciences* 6 (2): 78–84.

Friederici, Angela D. 2005. Neurophysiological markers of early language acquisition: From syllables to sentences. *Trends in Cognitive Sciences* 9 (10): 481–488.

Friederici, Angela D. 2009a. Pathways to language: Fiber tracts in the human brain. *Trends in Cognitive Sciences* 13 (4): 175–181.

Friederici, Angela D. 2009b. Allocating function to fiber tracts: Facing its indirectness. *Trends in Cognitive Sciences* 13 (9): 370–371.

Friederici, Angela D. 2011. The brain basis of language processing: From structure to function. *Physiological Reviews* 91 (4): 1357–1392.

Friederici, Angela D. 2012a. Language development and the ontogeny of the dorsal pathway. *Frontiers in Evolutionary Neuroscience* 4:3. doi:10.3389/fnevo.2012.00003.

Friederici, Angela D. 2012b. The cortical language circuit: From auditory perception to sentence comprehension. *Trends in Cognitive Sciences* 16:262–268.

Friederici, Angela. D. 2015. White matter pathways for speech and language processing. In *The Human Auditory System, Volume 129: Fundamental Organization and Clinical Disorders (Handbook of Clinical Neurology)*, ed. Gastone G. Celesia and Gregory Hickok, 177–186. San Diego: Elsevier.

Friederici, Angela D. 2016. The neuroanatomical pathway model of language: Syntactic and semantic networks. In *Neurobiology of Language*, ed. Steven L. Small and Gregory Hickok, 349–356. San Diego: Elsevier.

Friederici, Angela D., and Kai Alter. 2004. Lateralization of auditory language functions: A dynamic dual pathway model. *Brain and Language* 89 (2): 267–276.

Friederici, Angela D., Jörg Bahlmann, Stefan Heim, Riccardo I. Schubotz, and Alfred Anwander. 2006. The brain differentiates human and non-human grammars: Functional localization and structural connectivity. *Proceedings of the National Academy of Sciences of the United States of America* 103 (7): 2458–2463.

Friederici, Angela D., Christian J. Fiebach, Matthias Schlesewsky, Ina Bornkessel, and D. Yves von Cramon. 2006. Processing linguistic complexity and grammaticality in the left frontal cortex. *Cerebral Cortex* 16 (12): 1709–1717.

Friederici, Angela D., Manuela Friedrich, and Anne Christophe. 2007. Brain responses in 4-month-old infants are already language specific. *Current Biology* 17 (14): 1208–1211.

Friederici, Angela D., Manuela Friedrich, and Christiane Weber. 2002. Neural manifestation of cognitive and precognitive mismatch detection in early infancy. *Neuroreport* 13 (10): 1251–1254.

Friederici, Angela D., and Stefan Frisch. 2000. Verb-argument structure processing: The role of verb-specific and argument-specific information. *Journal of Memory and Language* 43 (3): 476–507.

Friederici, Angela D., and Sarah M. E. Gierhan. 2013. The language network. *Current Opinion in Neurobiology* 23 (2): 250–254.

Friederici, Angela D., Thomas C. Gunter, Anja Hahne, and Kerstin Mauth. 2004. The relative timing of syntactic and semantic processes in sentence comprehension. *Neuroreport* 15 (1): 165–169.

Friederici, Angela D., Anja Hahne, and Axel Mecklinger. 1996. The temporal structure of syntactic parsing: Early and late event-related brain potential effects. *Journal of Experimental Psychology: Learning, Memory, and Cognition* 22 (5): 1219–1248.

Friederici, Angela D., Anja Hahne, and Douglas Saddy. 2002. Distinct neurophysiological patterns reflecting aspects of syntactic complexity and syntactic repair. *Journal of Psycholinguistic Research* 31 (1): 45–63.

Friederici, Angela D., and Sonja A. Kotz. 2003. The brain basis of syntactic processes: Functional imaging and lesion studies. *NeuroImage* 20:S8–S17.

Friederici, Angela D., Sonja A. Kotz, Sophie K. Scott, and Jonas Obleser. 2010. Disentangling syntax and intelligibility in auditory language comprehension. *Human Brain Mapping* 31 (3): 448–457.

Friederici, Angela D., and Willem J. M. Levelt. 1988. Sprache. In *Psychobiologie. Grundlagen des Verhaltens*, ed. Klaus Immelmann, Klaus Scherer, Christian Vogel and Peter Schmoock, 648–671. Stuttgart: Gustav Fischer.

Friederici, Angela D., Michiru Makuuchi, and Jörg Bahlmann. 2009. The role of the posterior superior temporal cortex in sentence comprehension. *Neuroreport* 20 (6): 563–568.

Friederici, Angela D., and Claudia Männel. 2014. Neural correlates of the development of speech perception and comprehension. In *The Oxford Handbook of Cognitive Neuroscience*, ed. Kevin N. Ochsner and Stephen Kosslyn, 171–192. New York: Oxford University Press.

Friederici, Angela D., Martin Meyer, and D. Yves von Cramon. 2000. Auditory language comprehension: An event-related fMRI study on the processing of syntactic and lexical information. *Brain and Language* 74 (2): 289–300.

Friederici, Angela D., Jutta L. Mueller, and Regine Oberecker. 2011. Precursors to natural grammar learning: Preliminary evidence from 4-month-old infants. *PLoS One* 6 (3): e17920. doi:10.1371/journal.pone.0017920.

Friederici, Angela D., Jutta L. Mueller, Bernhard Sehm, and Patrick Ragert. 2013. Language learning without control: The role of the PFC. *Journal of Cognitive Neuroscience* 25 (5): 814–821.

Friederici, Angela D., Erdmut Pfeifer, and Anja Hahne. 1993. Event-related brain potentials during natural speech processing: Effects of semantic, morphological and syntactic violations. *Brain Research. Cognitive Brain Research* 1 (3): 183–192.

Friederici, Angela D., Shirley-Ann Rüschemeyer, Anja Hahne, and Christian J. Fiebach. 2003. The role of left inferior frontal and superior temporal cortex in sentence comprehension: Localizing syntactic and semantic processes. *Cerebral Cortex* 13 (2): 170–177.

Friederici, Angela D., and Wolf Singer. 2015. Grounding language processing on basic neurophysiological principles. *Trends in Cognitive Sciences* 19 (6): 329–338.

Friederici, Angela D., Karsten Steinhauer, and Erdmut Pfeifer. 2002. Brain signatures of artificial language processing: Evidence challenging the critical period hypothesis. *Proceedings of the National Academy of Sciences of the United States of America* 99 (1): 529–534.

Friederici, Angela D., D. Yves von Cramon, and Sonja A. Kotz. 1999. Language related brain potentials in patients with cortical and subcortical left hemisphere lesions. *Brain* 122:1033–1047.

Friederici, Angela D., D. Yves von Cramon, and Sonja A. Kotz. 2007. Role of the corpus callosum in speech comprehension: Interfacing syntax and prosody. *Neuron* 53 (1): 135–145.

Friederici, Angela D., Yunhua Wang, Christoph S. Herrmann, Burkhard Maess, and Ulrich Oertel. 2000. Localization of early syntactic processes in frontal and temporal cortical areas: A magnetoencephalographic study. *Human Brain Mapping* 11 (1): 1–11.

Friederici, Angela D., and Isabell Wartenburger. 2010. Language and Brain. (Overview) *Wiley Interdisciplinary Reviews: Cognitive Science*. doi: 10.1002/WCS.9.

Friederici, Angela D., and Jürgen Weissenborn. 2007. Mapping sentence form onto meaning: The syntax-semantic interface. *Brain Research* 1146:50–58.

Friederici, Angela D., and Jeanine M. I. Wessels. 1993. Phonotactic knowledge of word boundaries and its use in infant speech perception. *Perception & Psychophysics* 54 (3): 287–295.

Friedrich, Claudia K., and Sonja A. Kotz. 2007. ERP evidence of form and meaning coding during online speech recognition. *Journal of Cognitive Neuroscience* 19 (4): 594–604.

Friedrich, Claudia K., Ulrike Schild, and Brigitte Röder. 2009. Electrophysiological indices of word fragment priming allow characterizing neuronal stages of speech recognition. *Biological Psychology* 80 (1): 105–113.

Friedrich, Manuela, and Angela D. Friederici. 2005a. Lexical priming and semantic integration reflected in the ERP of 14-month-olds. *Neuroreport* 16 (6): 653–656.

Friedrich, Manuela, and Angela D. Friederici. 2005b. Phonotactic knowledge and lexical-semantic processing in one-year-olds: Brain responses to words and nonsense words in picture contexts. *Journal of Cognitive Neuroscience* 17 (11): 1785–1802.

Friedrich, Manuela, and Angela D. Friederici. 2005c. Semantic sentence processing reflected in the event-related potentials of one- and two-year-old children. *Neuroreport* 16 (6): 1801–1804.

Friedrich, Manuela, and Angela D. Friederici. 2010. Maturing brain mechanisms and developing behavioral language skills. *Brain and Language* 114:66–71.

Friedrich, Manuela, Ines Wilhelm, Jan Born, and Angela D. Friederici. 2015. Generalization of word meanings during infant sleep. *Nature Communications* 6:6004. doi:10.1038/ncomms7004.

Frisch, Stefan, Anja Hahne, and Angela D. Friederici. 2004. Word category and verb-argument structure information in the dynamics of parsing. *Cognition* 91 (3): 191–219.

Frisch, Stefan, Sonja A. Kotz, D. Yves von Cramon, and Angela D. Friederici. 2003. Why the P600 is not just a P300: The role of the basal ganglia. *Clinical Neurophysiology* 114 (2): 336–340.

Frisch, Stefan, and Matthias Schlesewsky. 2001. The N400 reflects problems of thematic hierarchizing. *Neuroreport* 12 (15): 3391–3394.

Friston, Karl J., Christian Buechel, Gereon R. Fink, John S. Morris, Edmund Rolls, and Ray J. Dolan. 1997. Psychophysiological and modulatory interactions in neuroimaging. *NeuroImage* 6 (3): 218–229.

Friston, Karl J., Lee M. Harrison, and Will Penny. 2003. Dynamic causal modelling. *NeuroImage* 19 (4): 1273–1302.

Frith, Uta, and Christopher D. Frith. 2003. Development and neurophysiology of mentalizing. *Philosophical Transactions of the Royal Society of London. Series B, Biological Sciences* 358 (1431): 459–473.

Fromkin, Victoria A., ed. 1973. Speech errors as linguistic evidence. The Hague: Mouton.

Frühholz, Sascha, Markus Gschwind, and Didier Grandjean. 2015. Bilateral dorsal and ventral fiber pathways for the processing of affective prosody identified by probabilistic fiber tracking. *NeuroImage* 109:27–34.

Fuster, Joaquin M. 2002. Frontal lobe and cognitive development. *Journal of Neurocytology* 31 (3): 373–385.

Galaburda, Albert M., Marjorie LeMay, Thomas L. Kemper, and Norman Geschwind. 1978. Right-left asymmetries in the brain. *Science* 199 (4331): 852–856.

Galantucci, Sebastiano, Maria Carmela Tartaglia, Stephen M. Wilson, Maya L. Henry, Massimo Filippi, Federica Agosta, Nina F. Dronkers, et al. 2011. White matter damage in primary progressive aphasias: A diffusion tensor tractography study. *Brain* 134:3011–3029.

Gandour, Jackson, Yunxia Tong, Donald Wong, Thomas Talavage, Mario Dzemidzic, Yisheng Xu, Xiaojian Li, and Mark Lowe. 2004. Hemispheric roles in the perception of speech prosody. *NeuroImage* 23 (1): 344–357.

Gandour, Jack, Donald Wong, Li Hsieh, Bret Weinzapfel, Diana Van Lancker, and Gary D. Hutchins. 2000. A cross-linguistic PET study of tone perception. *Journal of Cognitive Neuroscience* 12 (1): 207–222.

Gandour, Jack, Donald Wong, Mark Lowe, Mario Dzemidzic, Nakarin Satthamnuwong, Yunxia Tong, and Xiaojian Li. 2002. A cross-linguistic fMRI study of spectral and temporal cues underlying phonological processing. *Journal of Cognitive Neuroscience* 14 (7): 1076–1087.

Garfield, Jay L., ed. 1987. *Modularity in Knowledge Representation and Natural-Language Understanding.* Cambridge, MA: MIT Press.

Garrett, Merrill F. 1980. Levels of processing in sentence production. In *Language Production, Volume 1: Speech and Talk*, ed. Brian Butterworth, 177–220. London: Academic Press.

Garrido, Marta I., James M. Kilner, Klaas E. Stephan, and Karl J. Friston. 2009. The mismatch negativity: A review of underlying mechanisms. *Clinical Neurophysiology* 120 (3): 453–463.

Ge, Jianqiao, Gang Peng, Bingjiang Lyu, Yi Wang, Yan Zhuo, Zhendong Niu, Hai Tan Li, Alexander P. Leff, and Jia-Hong Gao. 2015. Cross-language differences in the brain network subserving intelligible speech. *Proceedings of the National Academy of Sciences of the United States of America* 112 (10): 2972–2977.

Gentner, Timothy Q., Kimberly M. Fenn, Daniel Margoliash, and Howard C. Nusbaum. 2006. Recursive syntactic pattern learning by songbirds. *Nature* 440 (7088): 1204–1207.

Geschwind, Norman, and Walter Levitsky. 1968. Human brain: Left-right asymmetries in temporal speech region. *Science* 161 (3837): 186–187.

Giedd, Jay N., Jonathan Blumenthal, Neal O. Jeffries, F. Xavier Castellanos, Hong Liu, Alex Zijdenbos, Tomas Paus, Alan C. Evans, and Judith L. Rapoport. 1999. Brain development during childhood and adolescence: A longitudinal MRI study. *Nature Neuroscience* 2 (10): 861–863.

Gierhan, Sarah M. E. 2013. Connections for auditory language in the human brain. *Brain and Language* 127 (2): 205–221.

Giraud, Anne-Lise, Andreas Kleinschmidt, David Poeppel, Torben E. Lund, Richard S. J. Frackowiak, and Helmut Laufs. 2007. Endogenous cortical rhythms determine cerebral specialization for speech perception and production. *Neuron* 56 (6): 1127–1134.

Giraud, Anne-Lise, and David Poeppel. 2012. Cortical oscillations and speech processing: Emerging computational principles and operations. *Nature Neuroscience* 15 (4): 511–517.

Giraud, Anne-Lise, and Cathy J. Price. 2001. The constraints functional neuroimaging places on classical models of auditory word processing. *Journal of Cognitive Neuroscience* 13 (6): 754–765.

Glasser, Matthew F., and James K. Rilling. 2008. DTI tractography of the human brain's language pathways. *Cerebral Cortex* 18 (11): 2471–2482.

Gleitman, Lila R., and Eric Wanner. 1982. Language acquisition: The state of the state of the art. In *Language Acquisition: The State of the Art*, ed. Eric Wanner and Lila R. Gleitman, 3–48. Cambridge: Cambridge University Press.

Gogtay, Nitin, Jay N. Giedd, Leslie Lusk, Kiralee M. Hayashi, Deanna Greenstein, A. Catherine Vaituzis, Tom F. Nugent, et al. 2004. Dynamic mapping of human cortical development during childhood through early adulthood. *Proceedings of the National Academy of Sciences of the United States of America* 101 (21): 8174–8179.

Gómez, Rebecca L., and LouAnn Gerken. 1999. Artificial grammar learning by 1-year-olds leads to specific and abstract knowledge. *Cognition* 70 (2): 109–135.

Goodglass, Harold, and Edith Kaplan. 1972. *The Assessment of Aphasia and Related Disorders*. Philadelphia: Lea & Febiger.

Goucha, Tomás, Alfred Anwander, and Angela D. Friederici. Submitted. Language differences shape a universal brain network.

Goucha, Tomás, Alfred Anwander, and Angela D. Friederici. 2015. How language shapes the brain: Cross-linguistic differences in structural connectivity. Poster presented at 7th Annual Meeting of the Society for the Neurobiology of Language, Chicago, IL, USA (2015) and 45th Annual Meeting of the Society for Neuroscience (SfN 2015), Chicago, IL, USA (2015).

Goucha, Tomás B., and Angela D. Friederici. 2015. The language skeleton after dissecting meaning: A functional segregation within Broca's area. *NeuroImage* 114 (6): 294–302.

Green, Antonia, Benjamin Straube, Susanne Weis, Andreas Jansen, Klaus Willmes, Kerstin Konrad, and Tilo Kircher. 2009. Neural integration of iconic and unrelated coverbal gestures: A functional MRI study. *Human Brain Mapping* 30 (10): 3309–3324.

Grewe, Tanja, Ina Bornkessel, Stefan Zysset, Richard Wiese, and D. Yves von Cramon. 2005. The emergence of the unmarked: A new perspective on the language-specific function of Broca's area. *Human Brain Mapping* 26 (3): 178–190.

Griffiths, John D., William D. Marslen-Wilson, Emmanuel A. Stamatakis, and Lorraine K. Tyler. 2013. Functional organization of the neural language system: Dorsal and ventral pathways are critical for syntax. *Cerebral Cortex* 23 (1): 139–147.

Griffiths, Timothy D., and Jason D. Warren. 2002. The planum temporale as a computational hub. *Trends in Neurosciences* 25 (7): 348–353.

Grodzinsky, Yosef. 2000. The neurology of syntax: Language use without Broca's area. *Behavioral and Brain Sciences* 23 (1): 1–21.

Grodzinsky, Yosef, and Angela D. Friederici. 2006. Neuroimaging of syntax and syntactic processing. *Current Opinion in Neurobiology* 16 (2): 240–246.

Grodzinsky, Yosef, and Andrea Santi. 2008. The battle for Broca's region. *Trends in Cognitive Sciences* 12:474–480.

Grossman, Murray, Ayanna Cooke, Chris DeVita, David Alsop, John Detre, Willis Chen, and James Gee. 2002. Age-related changes in working memory during sentence comprehension: An fMRI study. *NeuroImage* 15 (2): 302–317.

Grossman, Murray, Corey McMillan, Peachie Moore, Lijun Ding, Guila Glosser, Melissa Work, and James Gee. 2004. What's in a name: Voxel-based morphometric analyses of MRI and naming difficulty in Alzheimer's disease, frontotemporal dementia and corticobasal degeneration. *Brain* 127:628–649.

Gross, Joachim, Andreas A. Ioannides, Jürgen Dammers, Burkhard Maess, Angela D. Friederici, and Hans-Wilhelm Müller-Gärtner. 1998. Magnetic field tomography analysis of continuous speech. *Brain Topography* 10 (4): 273–281.

Gruber, Oliver, and D. Yves von Cramon. 2003. The functional neuroanatomy of human working memory revisited—Evidence from 3-T fMRI studies using classical domain-specific interference tasks. *NeuroImage* 19 (3): 797–809.

Guasti, Maria Teresa. 2002. *Language Acquisition. The Growth of Grammar*. Cambridge, MA: MIT Press.

Guenther, Frank H. 1995. Speech sound acquisition, coarticulation, and rate effects in a neural network model of speech production. *Psychological Review* 102 (3): 594–621.

Guenther, Frank H. 2016. *Neural Control of Speech*. Cambridge, MA: MIT Press.

Guenther, Frank H., and Gregory Hickok. 2015. Role of the auditory system in speech production. In *Handbook of Clinical Neurology*, ed. Michael J. Aminoff, François Boller and Dick F. Swaab, 161–175. Amsterdam: Elsevier.

Gunter, Thomas C., Angela D. Friederici, and Anja Hahne. 1999. Brain responses during sentence reading: Visual input affects central processes. *Neuroreport* 10 (15): 3175–3178.

Gunter, Thomas C., Angela D. Friederici, and Herbert Schriefers. 2000. Syntactic gender and semantic expectancy: ERPs reveal early autonomy and late interaction. *Journal of Cognitive Neuroscience* 12 (4): 556–568.

Gunter, Thomas C., Leon Kroczek, Henning Holle, and Angela D. Friederici. 2013. Neural correlates of gesture-syntax interaction. *Poster presented at the 5th Annual Meeting of the Society for the Neurobiology of Language*. San Diego, CA.

Haarmann, Henk J., and Katherine A. Cameron. 2005. Active maintenance of sentence meaning in working memory: Evidence from EEG coherences. *International Journal of Psychophysiology* 57 (2): 115–128.

Haarmann, Henk J., Katherine A. Cameron, and Daniel S. Ruchkin. 2002. Neural synchronization mediates on-line sentence processing: EEG coherence evidence from filler-gap constructions. *Psychophysiology* 39 (6): 820–825.

Hagoort, Peter. 2003. Interplay between syntax and semantics during sentence comprehension: ERP effects of combining syntactic and semantic violations. *Journal of Cognitive Neuroscience* 15 (6): 883–899.

Hagoort, Peter. 2005. On Broca, brain, and binding: A new framework. *Trends in Cognitive Sciences* 9 (9): 416–423.

Hagoort, Peter. 2008. The fractionation of spoken language understanding by measuring electrical and magnetic brain signals. *Philosophical Transactions of the Royal Society of London. Series B, Biological Sciences* 363 (1493): 1055–1069.

Hagoort, Peter, Giosuè Baggio, and Roel M. Willems. 2009. Semantic unification. In *The Cognitive Neurosciences*. 4th ed., ed. Michael S. Gazzaniga, 819–836. Cambridge, MA: MIT Press.

Hagoort, Peter, Colin Brown, and Jolanda Groothusen. 1993. The syntactic positive shift (SPS) as an ERP measure of syntactic processing. *Language and Cognitive Processes* 8 (4): 439–483.

Hagoort, Peter, Lea A. Hald, Marcel C. M. Bastiaansen, and Karl Magnus Petersson. 2004. Integration of word meaning and world knowledge in language comprehension. *Science* 304 (5669): 438–441.

Hagoort, Peter, and Peter Indefrey. 2014. The neurobiology of language beyond single words. *Annual Review of Neuroscience* 37:347–362.

Hagoort, Peter, and Jos van Berkum. 2007. Beyond the sentence given. *Philosophical Transactions of the Royal Society of London. Series B, Biological Sciences* 362 (1481): 801–811.

Hahne, Anja. 2001. What's different in second-language processing? Evidence from event-related brain potentials. *Journal of Psycholinguistic Research* 30 (3): 251–266.

Hahne, Anja, Korinna Eckstein, and Angela D. Friederici. 2004. Brain signatures of syntactic and semantic processes during children's language development. *Journal of Cognitive Neuroscience* 16 (7): 1302–1318.

Hahne, Anja, and Angela D. Friederici. 2002. Differential task effects on semantic and syntactic processes as revealed by ERPs. *Brain Research. Cognitive Brain Research* 13 (3): 339–356.

Hahne, Anja, and Angela D. Friederici. 1999. Electrophysiological evidence for two steps in syntactic analysis: Early automatic and late controlled processes. *Journal of Cognitive Neuroscience* 11 (2): 194–205.

Hahne, Anja, and Angela D. Friederici. 2001. Processing a second language: Late learners' comprehension strategies as revealed by event-related brain potentials. *Bilingualism: Language and Cognition* 4 (2): 123–141.

Halgren, Eric, Rupali P. Dhond, Natalie Christensen, Cyma Van Petten, Ksenija Marinkovic, Jeffrey D. Lewine, and Anders M. Dale. 2002. N400-like magnetoencephalography responses modulated by semantic context, word frequency, and lexical class in sentences. *NeuroImage* 17 (3): 1101–1116.

Hall, Deborah A., Ingrid S. Johnsrude, Mark P. Haggard, Alan R. Palmer, Michael A. Akeroyd, and A. Quentin Summerfield. 2002. Spectral and temporal processing in human auditory cortex. *Cerebral Cortex* 12 (2): 140–149.

Haller, Sven, Ernst-Wilhelm Radue, Michael Erb, Wolfgang Grodd, and Tilo Kircher. 2005. Overt sentence production in event-related fMRI. *Neuropsychologia* 43 (5): 807–814.Harris, Julia J., and David Attwell. 2012. The energetics of CNS white matter. *Journal of Neuroscience* 32 (1): 356–371.

Hartwigsen, Gesa, Thomas Golombek, and Jonas Obleser. 2015. Repetitive transcranial magnetic stimulation over left angular gyrus modulates the predictability gain in degraded speech comprehension. *Cortex* 68:100–110.

Hauser, Marc D., Noam Chomsky, and W. Tecumseh Fitch. 2002. The faculty of language: What is it, who has it, and how did it evolve? *Science* 298 (5598): 1569–1579.

He, Chao, Lisa Hotson, and Laurel J. Trainor. 2007. Mismatch responses to pitch changes in early infancy. *Journal of Cognitive Neuroscience* 19 (5): 878–892.

Heilman, Kenneth M., Dawn Bowers, Lynn Speedie, and H. Branch Coslett. 1984. Comprehension of affective and nonaffective prosody. *Neurology* 34 (7): 917–921.

Heilman, Kenneth M., Robert Scholes, and Robert T. Watson. 1975. Auditory affective agnosia—Disturbed comprehension of affective speech. *Journal of Neurology, Neurosurgery, and Psychiatry* 38 (1): 69–72.

Heim, Stefan, Angela D. Friederici, Niels O. Schiller, Shirley-Ann Rueschemeyer, and Katrin Amunts. 2009. The determiner congruency effect in language production investigated with functional MRI. *Human Brain Mapping* 30 (3): 928–940.

Heim, Stefan, Bertram Opitz, and Angela D. Friederici. 2002. Broca's area in the human brain is involved in the selection of grammatical gender for language production: Evidence from event-related functional magnetic resonance imaging. *Neuroscience Letters* 328 (2): 101–104.

Hensch, Takao K. 2005. Critical period plasticity in local cortical circuits. *Nature Reviews. Neuroscience* 6 (11): 877–888.

Herder, Johann Gottfried. 1772. Treatise on the Origin of Language. In *Philosophical Writings*, ed. Michael N. Forster, 65–165. Cambridge: Cambridge University Press.

Hernandez, Arturo E., Juliane Hofmann, and Sonja A. Kotz. 2007. Age of acquisition modulates neural activity for both regular and irregular syntactic functions. *NeuroImage* 36 (3): 912–923.

Herrmann, Bjoern, Burkhard Maess, and Angela D. Friederici. 2011. Violation of syntax and prosody—Disentangling their contributions to the early left anterior negativity (ELAN). *Neuroscience Letters* 490 (2): 116–120.

Herrmann, Bjoern, Burkhard Maess, Anja Hahne, Erich Schröger, and Angela D. Friederici. 2011. Syntactic and auditory spatial processing in the human temporal cortex: An MEG study. *NeuroImage* 57 (2): 624–633.

Herrmann, Bjoern, Burkhard Maess, Anna S. Hasting, and Angela D. Friederici. 2009. Localization of the syntactic mismatch negativity in the temporal cortex: An MEG study. *NeuroImage* 48 (3): 590–600.

Hickok, Gregory. 2012. Computational neuroanatomy of speech production. *Nature Reviews. Neuroscience* 13 (2): 135–145.

Hickok, Gregory, Mark Kritchevsky, Ursula Bellugi, and Edward S. Klima. 1996. The role of the left frontal operculum in sign language aphasia. *Neurocase* 2 (5): 373–380.

Hickok, Gregory, and David Poeppel. 2000. Towards a functional neuroanatomy of speech perception. *Trends in Cognitive Sciences* 4 (4): 131–138.

Hickok, Gregory, and David Poeppel. 2004. Dorsal and ventral streams: A framework for understanding aspects of the functional anatomy of language. *Cognition* 92 (1–2): 67–99.

Hickok, Gregory, and David Poeppel. 2007. The cortical organization of speech perception. *Nature Reviews. Neuroscience* 8:393–402.

Hodges, John R., Karalyn Patterson, Susan Oxbury, and Elaine Funnell. 1992. Semantic dementia—Progressive fluent aphasia with temporal lobe atrophy. *Brain* 115:1783–1806.

Hoeks, John C. J., Laurie A. Stowe, and Gina Doedens. 2004. Seeing words in context: The interaction of lexical and sentence level information during reading. *Brain Research. Cognitive Brain Research* 19 (1): 59–73.

Höhle, Barbara, Michaela Schmitz, Lynn M. Santelmann, and Jürgen Weissenborn. 2006. The recognition of discontinuous verbal dependencies by German 19-month-olds: Evidence for lexical and structural influences on children's early processing capacities. *Language Learning and Development* 2 (4): 277–300.

Hofer, Sabine, and Jens Frahm. 2006. Topography of the human corpus callosum revisited: Comprehensive fiber tractography using diffusion tensor magnetic resonance imaging. *NeuroImage* 32 (3): 989–994.

Holcomb, Phillip J., Sharon A. Coffey, and Helen J. Neville. 1992. Visual and auditory sentence processing: A developmental analysis using event-related brain potentials. *Developmental Neuropsychology* 8 (2–3): 203–241.

Holle, Henning, and Thomas C. Gunter. 2007. The role of iconic gestures in speech disambiguation: ERP evidence. *Journal of Cognitive Neuroscience* 19 (7): 1175–1192.

Holle, Henning, Thomas C. Gunter, Shirley-Ann Rüschemeyer, Andreas Hennenlotter, and Marco Iacoboni. 2008. Neural correlates of the processing of co-speech gestures. *NeuroImage* 39 (4): 2010–2024.

Holle, Henning, Christian Obermeier, Maren Schmidt-Kassow, Angela D. Friederici, Jamie Ward, and Thomas C. Gunter. 2012. Gesture facilitates the syntactic analysis of speech. *Frontiers in Psychology* 3:74. doi:10.3389/fpsyg.2012.00074.

Holle, Henning, Jonas Obleser, Shirley-Ann Rueschemeyer, and Thomas C. Gunter. 2010. Integration of iconic gestures and speech in left superior temporal areas boosts speech comprehension under adverse listening conditions. *NeuroImage* 49 (1): 875–884.

Holloway, Ralph L. 1978. The relevance of endocasts for studying primate brain evolution. In *Sensory systems of primates*, ed. Charles R. Noback, 181–200. New York, NY: Plenum Press.

Holloway, Ralph L., Douglas C. Broadfield, and Michael S. Yuan. 2004. *The Human Fossil Record: Brain Endocasts—The Paleoneurological Evidence*. Hoboken, NJ: Wiley-Liss.

Holloway, Ralph L., Ronald J. Clarke, and Phillip V. Tobias. 2004. Posterior lunate sulcus in Australopithecus africanus: Was Dart right? *Comptes Rendus. Palévol* 3 (4): 287–293.

Homae, Fumitaka, Hama Watanabe, Tamami Nakano, Kayo Asakawa, and Gentaro Taga. 2006. The right hemisphere of sleeping infant perceives sentential prosody. *Neuroscience Research* 54 (4): 276–280.

Houston, Derek M., Peter W. Jusczyk, Cecile Kuijpers, Riet Coolen, and Anne Cutler. 2000. Cross-language word segmentation by 9-month-olds. *Psychonomic Bulletin & Review* 7 (3): 504–509.

Hua, Kegang, Kenichi Oishi, Jiangyang Zhang, Setsu Wakana, Takashi Yoshioka, Weihong Zhang, Kazi Dilruba Akhter, et al. 2009. Mapping of functional areas in the human cortex based on connectivity through association fibers. *Cerebral Cortex* 19 (8): 1889–1895.

Huang, Hao, Jiangyang Zhang, Hangyi Jiang, Setsu Wakana, Lidia Poetscher, Michael I. Miller, Peter C. van Zijl, Argye E. Hillis, Robert Wytik, and Susumu Mori. 2005. DTI tractography based parcellation of white matter: Application to the mid-sagittal morphology of corpus callosum. *NeuroImage* 26 (1): 195–205.

Hubel, David H., and Torsten Nils Wiesel. 1962. Receptive fields, binocular interaction and functional architecture in cats visual cortex. *Journal of Physiology-London* 160 (1): 106–154.

Hublin, Jean-Jacques, Simon Neubauer, and Philipp Gunz. 1663. 2015. Brain ontogeny and life history in Pleistocene hominins. *Philosophical Transactions of the Royal Society of London. Series B, Biological Sciences* 370:20140062. doi:10.1098/rstb.2014.0062.

Humphries, Colin, Jeffrey R. Binder, David A. Medler, and Einat Liebenthal. 2006. Syntactic and semantic modulation of neural activity during auditory sentence comprehension. *Journal of Cognitive Neuroscience* 18 (4): 665–679.

Humphries, Colin, Jeffrey R. Binder, David A. Medler, and Einat Liebenthal. 2007. Time course of semantic processes during sentence comprehension. *NeuroImage* 36 (3): 924–932.

Humphries, Colin, Tracy Love, David Swinney, and Gregory Hickok. 2005. Response of anterior temporal cortex to prosodic and syntactic manipulations during sentence processing. *Human Brain Mapping* 26 (2): 128–138.

Hurford, James R. 2004. Language beyond our grasp: What mirror neurons can, and cannot, do for the evolution of language. In *Evolution of Communication Systems: A Comparative Approach*, ed. D. Kimrough Oller and Ulrike Gabriel, 297–313. Cambridge, MA: MIT Press.

Iacoboni, Marco, Istvan Molnar-Szakacs, Vittorio Gallese, Giovanni Buccino, John C. Mazziotta, and Giacomo Rizzolatti. 2005. Grasping the intentions of others with one's own mirror neuron system. *PLoS Biology* 3 (3): 529–535.

Iacoboni, Marco, Roger P. Woods, Marcel Brass, Harold Bekkering, John C. Mazziotta, and Giacomo Rizzolatti. 1999. Cortical mechanisms of human imitation. *Science* 268 (5449): 2526–2528.

Indefrey, Peter. 2007. Brain imaging studies of language production. In *Oxford Handbook of Psycholinguistics*, ed. Gareth Gaskell, 547–564. Oxford: Oxford University Press.

Indefrey, Peter, Colin M. Brown, Frauke Hellwig, Katrin Amunts, Hans Herzog, Rüdiger J. Seitz, and Peter Hagoort. 2001. A neural correlate of syntactic encoding during speech production. *Proceedings of the National Academy of Sciences of the United States of America* 98 (10): 5933–5936.

Indefrey, Peter, and Willem J. M. Levelt. 2004. The spatial and temporal signatures of word production components. *Cognition* 92 (1–2): 101–144.

Indefrey, Peter, Frauke Hellwig, Hans Herzog, Rüdiger J. Seitz, and Peter Hagoort. 2004. Neural responses to the production and comprehension of syntax in identical utterances. *Brain and Language* 89 (2): 312–319.

Isel, Frédéric, Thomas C. Gunter, and Angela D. Friederici. 2003. Prosody-assisted head-driven access to spoken German compounds. *Journal of Experimental Psychology: Learning, Memory, and Cognition* 29 (2): 277–288.

Isel, Frédéric, Anja Hahne, Burkhard Maess, and Angela D. Friederici. 2007. Neurodynamics of sentence interpretation: ERP evidence from French. *Biological Psychology* 74 (3): 337–346.

Itzhak, Inbal, Efrat Pauker, John E. Drury, Shari R. Baum, and Karsten Steinhauer. 2010. Event-related potentials show online influence of lexical biases on prosodic processing. *Neuroreport* 21 (1): 8–13.

Iverson, Jana M., and Susan Goldin-Meadow. 2005. Gesture paves the way for language development. *Psychological Science* 16 (5): 367–371.

Jacobsen, Roman. 1956. Two aspects of language and two types of aphasic disturbances. In *Fundamentals of Language*, ed. Roman Jakobsen and Morris Halle, 69–106. The Hague: Mouton.

Jackendoff, Ray. 1983. *Semantics and Cognition*. Cambridge, MA: MIT Press.

Jackendoff, Ray. 2002. *Foundations of Language (Brain, Meaning, Grammar, Evolution)*. Oxford: Oxford University Press.

Jackendoff, Ray, and Steven Pinker. 2005. The nature of the language faculty and its implications for evolution of language—(Reply to Fitch, Hauser, and Chomsky). *Cognition* 97 (2): 211–225.

Jacquemot, Charlotte, Christophe Pallier, Denis LeBihan, Stanislas Dehaene, and Emmanuel Dupoux. 2003. Phonological grammar shapes the auditory cortex: A functional magnetic resonance imaging study. *Journal of Neuroscience* 23 (29): 9541–9546.

Jasinska, Kaja K., and Laura-Ann Petitto. 2013. How age of bilingual exposure can change the neural systems for language in the developing brain: A functional near infrared spectroscopy investigation of syntactic processing in monolingual and bilingual children. *Developmental Cognitive Neuroscience* 6:87–101.

Jeon, Hyeon-Ae, and Angela D. Friederici. 2013. Two principles of organization in the prefrontal cortex are cognitive hierarchy and degree of automaticity. *Nature Communications* 4:2041. doi:10.1038/ncomms3041.

Jescheniak, Joerg D., and Willem J. M. Levelt. 1994. Word frequency effects in speech production: Retrieval of syntactic information and of phonological form. *Journal of Experimental Psychology: Learning, Memory, and Cognition* 20 (4): 824–843.

Johansen-Berg, Heidi, Tim E. J. Behrens, Matthew D. Robson, Ivana Drobnjak, Matthew F. S. Rushworth, Michael J. Brady, Stephen M. Smith, Desmond J. Higham, and Paul M. Matthews. 2004. Changes in connectivity profiles define functionally distinct regions in human medial frontal cortex. *Proceedings of the National Academy of Sciences of the United States of America* 101 (36): 13335–13340.

Johnsrude, Ingrid S., Anne Lise Giraud, and Richard S. J. Frackowiak. 2002. Functional imaging of the auditory system: The use of positron emission tomography. *Audiology & Neuro-Otology* 7 (5): 251–276.

Jonides, John, Eric H. Schumacher, Edward E. Smith, Robert A. Koeppe, Edward Awh, Patricia A. Reuter-Lorenz, Christy Marshuetz, and Christopher R. Willis. 1998. The role of parietal cortex in verbal working memory. *Journal of Neuroscience* 18 (13): 5026–5034.

Josse, Goulven, Sabine Joseph, Eric Bertasi, and Anne-Lise Giraud. 2012. The brain's dorsal route for speech represents word meaning: Evidence from gesture. *PLoS One* 7 (9): e46108. doi:10.1371/journal.pone.0046108.

Jusczyk, Peter W., Derek M. Houston, and Mary Newsome. 1999. The beginnings of word segmentation in English-learning infants. *Cognitive Psychology* 39 (3–4): 159–207.

Kaan, Edith, Anthony Harris, Edward Gibson, and Phillip Holcomb. 2000. The P600 as an index of syntactic integration difficulty. *Language and Cognitive Processes* 15 (2): 159–201.

Kelly, Spencer D., Corinne Kravitz, and Michael Hopkins. 2004. Neural correlates of bimodal speech and gesture comprehension. *Brain and Language* 89 (1): 253–260.

Kelly, Spencer D., Tara McDevitt, and Megan Esch. 2009. Brief training with co-speech gesture lends a hand to word learning in a foreign language. *Language and Cognitive Processes* 24 (2): 313–334.

Kemeny, Stefan, Frank Q. Ye, Rasmus Birn, and Allan R. Braun. 2005. Comparison of continuous overt speech fMRI using BOLD and arterial spin labelling. *Human Brain Mapping* 24 (3): 173–183.

Kerkhofs, Roel, Wietske Vonk, Herbert Schriefers, and Dorothee J. Chwilla. 2007. Discourse, syntax, and prosody: The brain reveals an immediate interaction. *Journal of Cognitive Neuroscience* 19 (9): 1421–1434.

Kerkhofs, Roel, Wietske Vonk, Herbert Schriefers, and Dorothee J. Chwilla. 2008. Sentence processing in the visual and auditory modality: Do comma and prosodic break have parallel functions? *Brain Research* 1224:102–118.

Kim, Albert, and Lee Osterhout. 2005. The independence of combinatory semantic processing: Evidence from event-related potentials. *Journal of Memory and Language* 52 (2): 205–225.

Kim, Karl H. S., Norman R. Relkin, Kyoung-Min Lee, and Joy Hirsch. 1997. Distinct cortical areas associated with native and second languages. *Nature* 388 (6638): 171–174.

Kinno, Ryuta, Mitsuru Kawamura, Seiji Shioda, and Kuniyoshi L. Sakai. 2008. Neural correlates of noncanonical syntactic processing revealed by a picture-sentence matching task. *Human Brain Mapping* 29 (9): 1015–1027.

Klein, Denise, Robert J. Zatorre, Brenda Milner, and Viviane Zhao. 2001. A cross-linguistic PET study of tone perception in mandarin Chinese and English speakers. *NeuroImage* 13 (4): 646–653.

Kleist, K. 1914. Aphasie und Geisteskrankheit. *Munchener Medizinische Wochenschrift* 61 (1): 8–12.

Klouda, Gayle V., Donald A. Robin, Neill R. Graff-Radford, and William E. Copper. 1988. The role of callosal connections in speech prosody. *Brain and Language* 35 (1): 154–171.

Knickmeyer, Rebecca C., Martin Styner, Sarah J. Short, Gabriele R. Lubach, Chaeryon Kang, Robert Hamer, Christopher L. Coe, and John H. Gilmore. 2010. Maturational trajectories of cortical brain development through the pubertal transition: Unique species and sex differences in the monkey revealed through structural magnetic resonance imaging. *Cerebral Cortex* 20 (5): 1053–1063.

Knösche, Thomas R., Burkhard Maess, and Angela D. Friederici. 1999. Processing of syntactic information monitored by brain surface current density mapping based on MEG. *Brain Topography* 12 (2): 75–87.

Kolk, Herman H. J., Dorothee J. Chwilla, Marieke van Herten, and Patrick J. W. Oor. 2003. Structure and limited capacity in verbal working memory: A study with event-related potentials. *Brain and Language* 85 (1): 1–36.

Kooijman, Valesca, Peter Hagoort, and Anne Cutler. 2005. Electrophysiological evidence for prelinguistic infants' word recognition in continuous speech. *Brain Research. Cognitive Brain Research* 24 (1): 109–116.

Kooijman, Valesca, Peter Hagoort, and Anne Cutler. 2009. Prosodic structure in early word segmentation: ERP evidence from Dutch ten-month-olds. *Infancy* 14 (6): 591–612.

Kotz, Sonja A., Martin Meyer, Kai Alter, Mireille Besson, Yves von Cramon, and Angela D. Friederici. 2003. On the lateralization of emotional prosody: An event-related functional MR investigation. *Brain and Language* 86 (3): 366–376.

Kotz, Sonja A., and Silke Paulmann. 2011. Emotion, language, and the brain. *Language and Linguistics Compass* 5 (3): 108–125.

Kotz, Sonja A., and Michael Schwartze. 2015. Motor-Timing and Sequencing in Speech Production. In *Neurobiology of Language Volume*, ed. Steven L. Small and Gregory Hickok, 717–724. Amsterdam: Elsevier Press.

Kovelman, Ioulia, Stephanie A. Baker, and Laura-Ann Petitto. 2008. Bilingual and monolingual brains compared: A functional magnetic resonance imaging investigation of syntactic processing and a possible "neural signature" of bilingualism. *Journal of Cognitive Neuroscience* 20 (1): 153–169.

Krause, Johannes, Ludovic Orlando, David Serre, Bence Viola, Kay Prüfer, Michael P. Richards, Jean-Jacques Hublin, Catherine Hanni, Anatoly P. Derevianko, and Svante Pääbo. 2007. Neanderthals in central Asia and Siberia. *Nature* 449 (7164): 902–904.

Kreitewolf, Jens, Angela D. Friederici, and Katharina von Kriegstein. 2014. Hemispheric lateralization of linguistic prosody recognition in comparison to speech and speaker recognition. *NeuroImage* 102 (2): 332–344.

Krekelberg, Bart, Geoffrey M. Boynton, and Richard J. A. van Wezel. 2006. Adaptation: From single cells to BOLD signals. *Trends in Neurosciences* 29 (5): 250–256.

Kubanek, Jan, Peter Brunner, Aysegul Gunduz, David Poeppel, and Gerwin Schalk. 2013. The tracking of speech envelope in the human cortex. *PLoS One* 8 (1): e53398. doi:10.1371/journal.pone.0053398.

Kubota, Mikio, Paul Ferrari, and Timothy P. L. Roberts. 2003. Magnetoencephalography detection of early syntactic processing in humans: Comparison between L1 speakers and L2 learners of English. *Neuroscience Letters* 353 (2): 107–110.

Kumar, Sukhbinder, Klaas E. Stephan, Jason D. Warren, Karl J. Friston, and Timothy D. Griffiths. 2007. Hierarchical processing of auditory objects in humans. *PLoS Computational Biology*, 3 (6): e100, 977–985.

Kuperberg, Gina R., David Caplan, Tatiana Sitnikova, Marianna Eddy, and Phillip J. Holcomb. 2006. Neural correlates of processing syntactic, semantic, and thematic relationships in sentences. *Language and Cognitive Processes* 21 (5): 489–530.

Kuperberg, Gina R., Phillip J. Holcomb, Tatiana Sitnikova, Douglas Greve, Anders M. Dale, and David Caplan. 2003. Distinct patterns of neural modulation during the processing of conceptual and syntactic anomalies. *Journal of Cognitive Neuroscience* 15 (2): 272–293.

Kuperberg, Gina R., Donna A. Kreher, Tatiana Sitnikova, David Caplan, and Phillip J. Holcomb. 2007. The role of animacy and thematic relationships in processing active English sentences: Evidence from event-related potentials. *Brain and Language* 100 (3): 223–238.

Kuperberg, Gina R., Philip K. McGuire, Ed T. Bullmore, Michael J. Brammer, Sophia Rabe-Hesketh, I. C. Wright, David J. Lythgoe, Steven C. R. Williams, and Anthony S. David. 2000. Common and distinct neural substrates for pragmatic, semantic, and syntactic processing of spoken sentences: An fMRI study. *Journal of Cognitive Neuroscience* 12 (2): 321–341.

Kurowski, Kathleen, Eric Hazen, and Sheila E. Blumstein. 2003. The nature of speech production impairments in anterior aphasics: An acoustic analysis of voicing in fricative consonants. *Brain and Language* 84 (3): 353–371.

Kutas, Marta, and Kara D. Federmeier. 2000. Electrophysiology reveals semantic memory use in language comprehension. *Trends in Cognitive Sciences* 4 (12): 463–470.

Kutas, Marta, and Steven A. Hillyard. 1980. Reading senseless sentences: Brain potentials reflect semantic incongruity. *Science* 207 (4427): 203–205.

Kutas, Marta, and Steven A. Hillyard. 1983. Event-related brain potentials to grammatical errors and semantic anomalies. *Memory & Cognition* 11 (5): 539–550.

Kutas, Marta, and Cyma Van Petten. 1994. Psycholinguistics electrified: Event-related brain potential investigations. In *Handbook of Psycholinguistics*, ed. Morton A. Gemsbacher, 83–143. San Diego: Academic Press.

Kwon, Hyukchan, Shynia Kuriki, Jin Mok Kim, Jong Ho Lee, Kiwoong Kim, and Kichun Nam. 2005. MEG study on neural activities associated with syntactic and semantic violations in spoken Korean sentences. *Neuroscience Research* 51 (4): 349–357.

Lakoff, George. 1987. *Women, Fire, and Dangerous Things. What Categories Reveal about the Mind*. Chicago: The University of Chicago Press.

Lambon, Ralph Matthew A., and Karalyn Patterson. 2008. Generalization and differentiation in semantic memory: Insights from semantic dementia. *Annals of the New York Academy of Sciences* 1124:61–76.

Larkum, Matthew. 2013. A cellular mechanism for cortical associations: An organizing principle for the cerebral cortex. *Trends in Neurosciences* 36 (3): 141–151.

Lau, Ellen F., Colin Phillips, and David Poeppel. 2008. A cortical network for semantics: (De)constructing the N400. *Nature Reviews. Neuroscience* 9 (12): 920–933.

Lauro, Leonor J. Romero, Janine Reis, Leonardo G. Cohen, Carlo Cecchetto, and Costanza Papagno. 2010. A case for the involvement of phonological loop in sentence comprehension. *Neuropsychologia* 48 (14): 4003–4011.

Leaver, Amber M., and Josef P. Rauschecker. 2010. Cortical representation of natural complex sounds: Effects of acoustic features and auditory object category. *Journal of Neuroscience* 30 (22): 7604–7612.

Lebel, Catherine, Lindsay Walker, Alexander Leemans, Linda Phillips, and Christian Beaulieu. 2008. Microstructural maturation of the human brain from childhood to adulthood. *NeuroImage* 40 (3): 1044–1055.

Leigh, Steven R. 2004. Brain growth, life history, and cognition in primate and human evolution. *American Journal of Primatology* 62 (3): 139–164.

Lenneberg, Eric H. 1967. Biological foundations of language. New York: John Wiley and Sons, Inc.

Leitman, David I., Daniel H. Wolf, J. Daniel Ragland, Petri Laukka, James Loughead, Jeffrey N. Valdez, Daniel C. Javitt, Bruce I. Turetsky, and Ruben C. Gur. 2010. "It's not what you say, but how you say it": A reciprocal temporo-frontal network for affective prosody. *Frontiers in Human Neuroscience* 4:19. doi:10.3389/fnhum.2010.00019.

Leppänen, Paavo H. T., Elina Pihko, Kenneth M. Eklund, and Heikki Lyytinen. 1999. Cortical responses of infants with and without a genetic risk for dyslexia: II. Group effects. *Neuroreport* 10 (5): 969–973.

Leppänen, Paavo H. T., Ulla Richardson, and Heikki Lyytinen. 1997. Brain ERPs to changes of speech segment durations in six-month-olds. *International Journal of Psychophysiology* 25 (1): 17–84.

Leroy, Francois, Herve Glasel, Jessica Dubois, Lucie Hertz-Pannier, Bertrand Thirion, Jean-Francois Mangin, and Ghislaine Dehaene-Lambertz. 2011. Early maturation of the linguistic dorsal pathway in human infants. *Journal of Neuroscience* 31 (4): 1500–1506.

Levelt, Willem J. M. 1989. *Speaking: From Intention to Articulation*. Cambridge, MA: The MIT Press.

Levelt, Willem J. M. 2013. *A history of Psycholinguistics: The Pre-Chomskyan Era*. Oxford: Oxford University Press.

Levelt, Willem J. M., Ardi Roelofs, and Antje S. Meyer. 1999. Multiple perspectives on word production. *Behavioral and Brain Sciences* 22 (1): 61–75.

Levelt, Willem J. M., Herbert Schriefers, Dirk Vorberg, Antje S. Meyer, Thomas Pechmann, and Jaap Havinga. 1991. The time course of lexical access in speech production—a study of picture naming. *Psychological Review* 98 (1): 122–142.

Levinson, Stephen C. 1983. *Pragmatics. Cambridge Textbooks in Linguistics.* Cambridge: Cambridge University Press.

Lewis, Ashley G., and Marcel Bastiaansen. 2015. A predictive coding framework for rapid neural dynamics during sentence-level language comprehension. *Cortex* 68:155–168.

Lewis, Ashley G., Jan-Mathijs Schoffelen, Herbert Schriefers, and Marcel Bastiaansen. 2016. A predictive coding perspective on beta oscillations during sentence-level language comprehension. *Frontiers in Human Neuroscience* 10:85. doi:10.3389/fnhum.2016.00085.

Li, Xuesong, Hua Shu, Youyi Liu, and Ping Li. 2006. Mental representation of verb meaning: Behavioral and electrophysiological evidence. *Journal of Cognitive Neuroscience* 18 (10): 1774–1787.

Li, Weijun, and Yufang Yang. 2009. Perception of prosodic hierarchical boundaries in Mandarin Chinese sentences. *Neuroscience* 158 (4): 1416–1425.

Liberman, Alvin M., Katherine Safford Harris, Howard S. Hoffman, and Belver C. Griffith. 1957. The discrimination of speech sounds within and across phoneme boundaries. *Journal of Experimental Psychology* 54 (5): 358–368.

Liu, Fang, Aniruddh D. Patel, Adrian Fourcin, and Lauren Stewart. 2010. Intonation processing in congenital amusia: Discrimination, identification and imitation. *Brain* 133:1682–1693.

Lohmann, Gabriele, Kerstin Erfurth, Karsten Muller, and Robert Turner. 2012. Critical comments on dynamic causal modeling. *NeuroImage* 59 (3): 2322–2329.

Lohmann, Gabriele, Stefanie Hoehl, Jens Brauer, Claudia Danielmeier, Ina Bornkessel-Schlesewsky, Joerg Bahlmann, Robert Turner, and Angela D. Friederici. 2010. Setting the frame: The human brain activates a basic low-frequency network for language processing. *Cerebral Cortex* 20 (6): 1286–1292.

Lu, Lisa H., Christiana M. Leonard, Paul M. Thompson, E. Kan, J. Jolley, S. E. Welcome, Arthur Toga, and E. R. Sowell. 2007. Normal developmental changes in inferior frontal gray matter are associated with improvement in phonological processing: A longitudinal MRI analysis. *Cerebral Cortex* 17 (5): 1092–1099.

Lyn, Heidi, Peter Pierre, Allyson J. Bennett, Scott Fears, Roger Woods, and William D. Hopkins. 2011. Planum temporale grey matter asymmetries in chimpanzees (Pan troglodytes), vervet (Chlorocebus aethiops sabaeus), rhesus (Macaca mulatta) and bonnet (Macaca radiata) monkeys. *Neuropsychologia* 49 (7): 2004–2012.

Macedonia, Manuela, Karsten Müller, and Angela D. Friederici. 2011. The impact of iconic gestures on foreign language word learning and its neural substrate. *Human Brain Mapping* 32 (6): 982–998.

MacDonald, Maryellen C., Neal J. Pearlmutter, and Mark S. Seidenberg. 1994. Lexical nature of syntactic ambiguity resolution. *Psychological Review* 101 (4): 676–703.

MacGregor, Lucy J., Friedemann Pulvermüller, Marten van Casteren, and Yury Shtyrov. 2012. Ultra-rapid access to words in the brain. *Nature Communications* 3:711. doi:10.1038/ncomms1715.

MacSweeney, Mairead, Cheryl M. Capek, Ruth Campbell, and Bencie Woll. 2008. The signing brain: The neurobiology of sign language. *Trends in Cognitive Sciences* 12 (11): 432–440.

MacWhinney, Brian, Elizabeth Bates, and Reinhold Kliegl. 1984. Cue validity and sentence interpretation in English, German, and Italian. *Journal of Verbal Learning and Verbal Behavior* 23 (2): 127–150.

Maess, Burkhard, Angela D. Friederici, Markus Damian, Antje S. Meyer, and Willem J. M. Levelt. 2002. Semantic category interference in overt picture naming: Sharpening current density localization by PCA. *Journal of Cognitive Neuroscience* 14 (3): 455–462.

Männel, Claudia, and Angela D. Friederici. 2009. Pauses and intonational phrasing: ERP studies in 5-month-old German infants and adults. *Journal of Cognitive Neuroscience* 21 (10): 1988–2006.

Männel, Claudia, and Angela D. Friederici. 2011. Intonational phrase structure processing and syntactic knowledge in childhood: ERP studies in 2-, 3-, and 6-year-old children. *Developmental Science* 14 (4): 786–798.

Männel, Claudia, Christine Schipke, and Angela D. Friederici. 2013. The role of pause as prosodic boundary marker: Language ERP studies in German 3- and 6-year-olds. *Developmental Cognitive Neuroscience* 5:86–94.

Mahmoudzadeh, Mahdi, Ghislaine Dehaene-Lambertz, Marc Fournier, Guy Kongolo, Sabrina Goudjil, Jessica Dubois, Reinhard Grebe, and Fabrice Wallois. 2013. Syllabic discrimination in premature human infants prior to complete formation of cortical layers. *Proceedings of the National Academy of Sciences of the United States of America* 110 (12): 4846–4851.

Makris, Nikos, David N. Kennedy, Sean McInerney, A. Gregory Sorensen, Ruopeng Wang, Verne S. Caviness, and Deepak N. Pandya. 2005. Segmentation of subcomponents within the superior longitudinal fascicle in humans: A quantitative, in vivo, DT-MRI study. *Cerebral Cortex* 15 (6): 854–869.

Makris, Nikos, and Deepak N. Pandya. 2009. The extreme capsule in humans and rethinking of the language circuitry. *Brain Structure & Function* 213 (3): 343–358.

Makuuchi, Michiru, Joerg Bahlmann, Alfred Anwander, and Angela D. Friederici. 2009. Segregating the core computational faculty of human language from working memory. *Proceedings of the National Academy of Sciences of the United States of America* 106 (20): 8362–8367.

Makuuchi, Michiru, Joerg Bahlmann, and Angela D. Friederici. 2012. An approach to separating the levels of hierarchical structure building in language and mathematics. *Philosophical Transactions of the Royal Society of London. Series B, Biological Sciences* 367 (1598): 2033–2045.

Makuuchi, Michiru, and Angela D. Friederici. 2013. Hierarchical functional connectivity between the core language system and the working memory system. *Cortex* 49 (9): 2416–2423.

Mampe, Birgit, Angela D. Friederici, Anne Christophe, and Kristiane Wermke. 2009. Newborns' cry melody is shaped by their native language. *Current Biology* 19 (23): 1994–1997.

Marcus, Gary F., S. Vijayan, Shoba Bandi Rao, and Peter M. Vishton. 1999. Rule learning by seven-month-old infants. *Science* 283 (5398): 77–80.

Marler, Peter. 1970. Birdsong and speech development—could there be parallels. *American Scientist* 58 (6): 669–673.

Marslen-Wilson, William. 1987. Functional parallelism in spoken word-recognition. *Cognition* 25 (1–2): 71–102.

Marslen-Wilson, William, and Lorraine K. Tyler. 1980. The temporal structure of spoken language understanding. *Cognition* 8 (1): 1–71.

Marslen-Wilson, William, and Lorraine K. Tyler. 1987. Against modularity. In *Modularitiy in Knowledge Representation and Natural-Language Understanding*, ed. Jay L. Garfield, 37–62. Cambridge, MA: MIT Press.

Marslen-Wilson, William D., Lorraine K. Tyler, Paul Warren, P. Grenier, and C. S. Lee. 1992. Prosodic effects in minimal attachment. *Quarterly Journal of Experimental Psychology* 45 (1): 73–87.

Martynova, Olga, Jarkko Kirjavainen, and Marie Cheour. 2003. Mismatch negativity and late discriminative negativity in sleeping human newborns. *Neuroscience Letters* 340 (2): 75–78.

Matsumoto, Riki, Dileep R. Nair, Eric LaPresto, Imad Najm, William Bingaman, Hiroshi Shibasaki, and Hans O. Lüders. 2004. Functional connectivity in the human language system: a cortico-cortical evoked potential study. *Brain*, 127: 2316–2330.

Matsuo, Koushun, Toshiki Mizuno, Kei Yamada, Kentaro Akazawa, Takashi Kasai, Masaki Kondo, Satoru Mori, Tsunehiko Nishimura, and Masanori Nakagawa. 2008. Cerebral white matter damage in frontotemporal dementia assessed by diffusion tensor tractography. *Neuroradiology* 50 (7): 605–611.

Mazoyer, Bernard M., Nathalie Tzourio, V. Frak, André Syrota, N. Murayama, Olivier Levrier, George Salamon, Stanislas Dehaene, Laurent Cohen, and Jacques Mehler. 1993. The cortical representation of speech. *Journal of Cognitive Neuroscience* 5 (4): 467–479.

McGee, Aaron W., Yupeng Yang, Quentin S. Fischer, Nigel W. Daw, and Stephen M. Strittmatter. 2005. Experience-driven plasticity of visual cortex limited by myelin and Nogo receptor. *Science* 305 (5744): 2222–2226.

McLaughlin, Judith, Lee Osterhout, and Albert Kim. 2004. Neural correlates of second-language word learning: Minimal instruction produces rapid change. *Nature Neuroscience* 7 (7): 703–704.

Meisel, Juergen M. 2011. *First and Second Language Acquisition: Parallels and Differences*. Cambridge: Cambridge University Press.

Mellem, Monika S., Rhonda B. Friedman, and Andrei V. Medvedev. 2013. Gamma- and theta-band synchronization during semantic priming reflect local and long-range lexical-semantic networks. *Brain and Language* 127 (3): 440–451.

Menenti, Laura, Sarah M. E. Gierhan, Katrien Segaert, and Peter Hagoort. 2011. Shared language: Overlap and segregation of the neuronal infrastructure for speaking and listening revealed by functional MRI. *Psychological Science* 22 (9): 1173–1182.

Menenti, Laura, Katrien Segaert, and Peter Hagoort. 2012. The neuronal infrastructure of speaking. *Brain and Language* 122 (2): 71–80.

Meringer, R., and C. Mayer. 1895. *Versprechen und Verlesen: Eine psychologisch-linguistische Studie*. Stuttgart: G. J. Göschensche Verlagshandlung.

Mestres-Misse, Anna, Antoni Rodriguez-Fornells, and Tomas F. Münte. 2010. Neural differences in the mapping of verb and noun concepts onto novel words. *NeuroImage* 49 (3): 2826–2835.

Mesulam, M. Marsel, Cynthia K. Thompson, Sandra Weintraub, and Emily J. Rogalski. 2015. The Wernicke conundrum and the anatomy of language comprehension in primary progressive aphasia. *Brain* 138:2423–2437.

Meyer, Lars, Maren Grigutsch, Nora Schmuck, Phoebe Gaston, and Angela D. Friederici. 2015. Frontal-posterior theta oscillations reflect memory retrieval during sentence comprehension. *Cortex* 71:205–218.

Meyer, Lars, Molly J. Henry, Phoebe Gaston, Noura Schmuck, and Angela D. Friederici. 2016. Linguistic bias modulates interpretation of speech via neural delta-band oscillations. *Cerebral Cortex*; Advance online publication. doi:10.1093/cercor/bhw228.

Meyer, Lars, Jonas Obleser, Alfred Anwander, and Angela D. Friederici. 2012. Linking ordering in Broca's area to storage in left temporo-parietal regions: The case of sentence processing. *NeuroImage* 62 (3): 1987–1998.

Meyer, Lars, Jonas Obleser, and Angela D. Friederici. 2013. Left parietal alpha enhancement during working memory-intensive sentence processing. *Cortex* 49 (3): 711–721.

Meyer, Martin, Kai Alter, Angela D. Friederici, Gabriele Lohmann, and D. Yves von Cramon. 2002. FMRI reveals brain regions mediating slow prosodic modulations in spoken sentences. *Human Brain Mapping* 17 (2): 73–88.

Meyer, Martin, Karsten Steinhauer, Kai Alter, Angela D. Friederici, and D. Yves von Cramon. 2004. Brain activity varies with modulation of dynamic pitch variance in sentence melody. *Brain and Language* 89 (2): 277–289.

Meyer, Martin, Ulrike Toepel, Joerg Keller, Daniela Nussbaumer, Stefan Zysset, and Angela D. Friederici. 2007. Neuroplasticity of sign language: Implications from structural and functional brain imaging. *Restorative Neurology and Neuroscience* 25 (3–4): 335–351.

Meyer, Patric, Axel Mecklinger, Thomas Grunwald, Juergen Fell, Christian E. Elger, and Angela D. Friederici. 2005. Language processing within the human medial temporal lobe. *Hippocampus* 15 (4): 451–459.

Milberg, William P., Sheila E. Blumstein, and Barbara Dworetzky. 1987. Processing of lexical ambiguities in aphasia. *Brain and Language* 31 (1): 138–150.

Miller, George A. 1978. Semantic relations among words. In *Linguistic Theory and Psychological Reality*, ed. Morris Halle, Joan Bresnan and George A. Miller, 60–118. Cambridge, MA: MIT Press.

Miller, Daniel J., Tetyana Duka, Cheryl D. Stimpson, Steven J. Schapiro, Wallace B. Baze, Marc J. McArthur, Archibald J. Fobbs, et al. 2012. Prolonged myelination in human neocortical evolution. *Proceeding of the National Academy of Sciences of the United States of America* 109 (41): 16480–16485.

Mills, Debra L., Sharon Coffey-Corina, and Helen J. Neville. 1997. Language comprehension and cerebral specialization from 13 to 20 months. *Developmental Neuropsychology* 13 (3): 397–446.

Milne, Alice E., Jutta L. Mueller, Claudia Männel, Adam Attaheri, Angela D. Friederici, and Chris I. Petkov. 2016. Evolutionary origins of non-adjacent sequence processing in primate brain potentials. *Scientific Reports* 6:36259. doi:10.1038/srep36259.

Milner, Brenda, Larry R. Squire, and Eric R. Kandel. 1998. Cognitive neuroscience and the study of memory. *Neuron* 20 (3): 445–468.

Minagawa-Kawai, Yasuyo, Alejandrina Cristia, and Emmanuel Dupoux. 2011. Cerebral lateralization and early speech acquisition: A developmental scenario. *Developmental Cognitive Neuroscience* 1 (3): 217–232.

Mitchell, Don C. 1994. Sentence parsing. In *Handbook of Psycholinguistics*, ed. Morton Ann Gernsbacher, 375–409. San Diego: Academic Press.

Mitchell, Rachel L. C., Rebecca Elliott, Martin Barry, Alan Cruttenden, and Peter W. R. Woodruff. 2003. The neural response to emotional prosody, as revealed by functional magnetic resonance imaging. *Neuropsychologia* 41 (10): 1410–1421.

Molinaro, Nicola, Pedro M. Paz-Alonso, Jon Andoni Duñabeitia, and Manuel Carreiras. 2015. Combinatorial semantics strengthens angular-anterior temporal coupling. *Cortex* 65:113–127.

Montgomery, James W., Beula M. Magimairaj, and Michelle H. O'Malley. 2008. Role of working memory in typically developing children's complex sentence comprehension. *Journal of Psycholinguistic Research* 37 (5): 331–354.

Moore-Parks, Erin Nicole, Erin L. Burns, Rebecca Bazzill, Sarah Levy, Valerie Posada, and Ralph-Axel Muller. 2010. An fMRI study of sentence-embedded lexical-semantic decision in children and adults. *Brain and Language* 114 (2): 90–100.

Morford, Marolyn, and Susan Goldin-Meadow. 1992. Comprehension and production of gesture in combination with speech in one-word speakers. *Journal of Child Language* 19 (3): 559–580.

Mori, Susumu, and Peter C. M. Zijl. 2002. Fiber tracking: Principles and strategies—a technical review. *NMR in Biomedicine* 15 (7–8): 468–480.

Morillon, Benjamin, Katia Lehongre, Richard S. J. Frackowiak, Antoine Ducorps, Andreas Kleinschmidt, David Poeppel, and Anne-Lise Giraud. 2010. Neurophysiological origin of human brain asymmetry for speech and language. *Proceeding of the National Academy of Sciences of the United States of America* 107 (43): 18688–18693.

Moro, Andrea. 2014. On the similarity between syntax and actions. *Trends in Cognitive Sciences* 18 (3): 109–110.

Morosan, Patricia, Jörg Rademacher, Axel Schleicher, Katrin Amunts, Thorsten Schormann, and Karl Zilles. 2001. Human primary auditory cortex: Cytoarchitectonic subdivisions and mapping into a spatial reference system. *NeuroImage* 13 (4): 684–701.

Morosan, Patricia, Jörg Rademacher, Nicola Palomero-Gallagher, and Karl Zilles. 2005. Anatomical organization of the human auditory cortex: Cytoarchitecture and transmitter receptors. In *Auditory Cortex: A Synthesis of Human and Animal Research*, ed. Peter Heil, Henning Scheich, Eike Budinger and Reinhard König, 27–50. Mahwah, NJ: Lawrence Erlbaum.

Morr, Mara L., Valerie L. Shafer, Judith A. Kreuzer, and Diane Kurtzberg. 2002. Maturation of mismatch negativity in typically developing infants and preschool children. *Ear and Hearing* 23 (2): 118–136.

Mueller, Jutta L., Angela D. Friederici, and Claudia Männel. 2012. Auditory perception at the root of language learning. *Proceedings of the National Academy of Sciences of the United States of America* 109 (39): 15953–15958.

Mueller, Jutta L., Regine Oberecker, and Angela D. Friederici. 2009. Syntactic learning by mere exposure—An ERP study in adult learners. *BMC Neuroscience* 10:89. doi:10.1186/1471-2202-10-89.

Mummery, Catherine J., John Ashburner, Sophie K. Scott, and Richard J. S. Wise. 1999. Functional neuroimaging of speech perception in six normal and two aphasic subjects. *Journal of the Acoustical Society of America* 106 (1): 449–457.

Muralikrishnan, Ramasamy, Matthias Schlesewsky, and Ina Bornkessel-Schlesewsky. 2015. Animacy-based predictions in language comprehension are robust: Contextual cues modulate but do not nullify them. *Brain Research* 1608:108–137.

Musso, Mariacristina, Andrea Moro, Volkmar Glauche, Michel Rijntjes, Jürgen Reichenbach, Christian Büchel, and Cornelius Weiller. 2003. Broca's area and the language instinct. *Nature Neuroscience* 6 (7): 774–781.

Nakamura, Akinori, Burkhard Maess, Thomas R. Knösche, Thomas C. Gunter, Patric Bach, and Angela D. Friederici. 2004. Cooperation of different neuronal systems during hand sign recognition. *NeuroImage* 23 (1): 25–34.

Nan, Yun, and Angela D. Friederici. 2013. Differential roles of right temporal cortex and Broca's area in pitch processing: Evidence from music and Mandarin. *Human Brain Mapping* 34 (9): 2045–2054.

Nan, Yun, Yanan Sun, and Isabelle Peretz. 2010. Congenital amusia in speakers of a tone language: Association with lexical tone agnosia. *Brain* 133:2635–2642.

Narain, Charvy, Sophie K. Scott, Richard J. S. Wise, Stuart Rosen, Alexander Leff, S. D. Iversen, and Paul M. Matthews. 2003. Defining a left-lateralized response specific to intelligible speech using fMRI. *Cerebral Cortex* 13 (12): 1362–1368.

Näätänen, R., and Kimmo Alho. 1997. Higher-order processes in auditory change detection. *Trends in Cognitive Sciences* 1 (2): 44–45.

Näätänen, Risto, Anne Lehtokoski, Mietta Lennes, Marie Cheour, Minna Huotilainen, Antti Iivonen, Martti Vainio, et al. 1997. Language-specific phoneme representations revealed by electric and magnetic brain responses. *Nature* 385 (6615): 432–434.

Näätänen, Risto, Petri Paavilainen, Teemu Rinne, and Kimmo Alho. 2007. The mismatch negativity (MMN) in basic research of central auditory processing: A review. *Clinical Neurophysiology* 118 (12): 2544–2590.

Näätänen, Risto, Mari Tervaniemi, Elyse Sussman, Petri Paavilainen, and István Winkler. 2001. 'Primitive intelligence' in the auditory cortex. *Trends in Neurosciences* 24 (5): 283–288.

Neubert, Franz-Xaver, Rogier B. Mars, Adam G. Thomas, Jerome Sallet, and Matthew F. S. Rushworth. 2014. Comparison of human ventral frontal cortex areas for cognitive control and language with areas in monkey frontal cortex. *Neuron* 81 (3): 700–713.

Neville, Helen J., and Daphne Bavelier. 1998. Neural organization and plasticity of language. *Current Opinion in Neurobiology* 8 (2): 254–258.

Neville, Helen J., Daphne Bavelier, David Corina, Josef Rauschecker, Avi Karni, Anil Lalwani, Allen Braun, Vince Clark, Peter Jezzard, and Robert Turner. 1998. Cerebral organization for language in deaf and hearing subjects: Biological constraints and effects of experience. *Proceedings of the National Academy of Sciences of the United States of America* 95 (3): 922–929.

Neville, Helen J., Sharon A. Coffey, Donald S. Lawson, Andrew Fischer, Karen Emmorey, and Ursula Bellugi. 1997. Neural systems mediating American sign language: Effects of sensory experience and age of acquisition. *Brain and Language* 57 (3): 285–308.

Neville, Helen J., Janet L. Nicol, Andrew Barss, Kenneth I. Forster, and Merrill F. Garrett. 1991. Syntactically based sentence processing classes: Evidence from event-related brain potentials. *Journal of Cognitive Neuroscience* 3 (2): 151–165.

Newman, Aaron J., Daphne Bavelier, David Corina, Peter Jezzard, and Helen J. Neville. 2002. A critical period for right hemisphere recruitment in American Sign Language processing. *Nature Neuroscience* 5 (1): 76–80.

Newman, Sharlene D., Toshikazu Ikuta, and Thomas Burns, Jr. 2010. The effect of semantic relatedness on syntactic analysis: An fMRI study. *Brain and Language* 113 (2): 51–58.

Ng, Wai Pui, Nicholas Cartel, John Roder, Arthur Roach, and Andres Lozano. 1996. Human central nervous system myelin inhibits neurite outgrowth. *Brain Research* 720 (1–2): 17–24.

Nosarti, Chiara, Teresa M. Rushe, Peter W. R. Woodruff, Ann L. Stewart, Larry Rifkin, and Robin M. Murray. 2004. Corpus callosum size and very preterm birth: Relationship to neuropsychological outcome. *Brain* 127:2080–2089.

Nunez, S. Christopher, Mirella Dapretto, Tami Katzir, Ariel Starr, Jennifer Bramen, Eric Kan, Susan Bookheimer, and Elizabeth R. Sowell. 2011. FMRI of syntactic processing in typically developing children: Structural correlates in the inferior frontal gyrus. *Developmental Cognitive Neuroscience* 1 (3): 313–323.

Oberecker, Regine, and Angela D. Friederici. 2006. Syntactic event-related potential components in 24-month-olds' sentence comprehension. *Neuroreport* 17 (10): 1017–1021.

Oberecker, Regine, Manuela Friedrich, and Angela D. Friederici. 2005. Neural correlates of syntactic processing in two-year-olds. *Journal of Cognitive Neuroscience* 17 (10): 1667–1678.

Obermeier, Christian, Thomas Dolk, and Thomas C. Gunter. 2012. The benefit of gestures during communication: Evidence from hearing and hearing impaired persons. *Cortex* 48 (7): 857–870.

Obleser, Jonas, and Sonja A. Kotz. 2010. Expectancy constraints in degraded speech modulate the language comprehension network. *Cerebral Cortex* 20 (3): 633–640.

Obleser, Jonas, Aditi Lahiri, and Carsten Eulitz. 2003. Auditory-evoked magnetic field codes place of articulation in timing and topography around 100 milliseconds post syllable onset. *NeuroImage* 20 (3): 1839–1847.

Obleser, Jonas, Lars Meyer, and Angela D. Friederici. 2011. Dynamic assignment of neural resources in auditory comprehension of complex sentences. *NeuroImage* 56 (4): 2310–2320.

Obleser, Jonas, Sophie K. Scott, and Carsten Eulitz. 2006. Now you hear it, now you don't: Transient traces of consonants and their nonspeech analogues in the human brain. *Cerebral Cortex* 16 (8): 1069–1076.

Obleser, Jonas, Richard J. S. Wise, M. Alex Dresner, and Sophie K. Scott. 2007. Functional integration across brain regions improves speech perception under adverse listening conditions. *Journal of Neuroscience* 27 (9): 2283–2289.

Obleser, Jonas, Jonas Zimmermann, John Van Meter, and Josef P. Rauschecker. 2007. Multiple stages of auditory speech perception reflected in event-related FMRI. *Cerebral Cortex* 17 (10): 2251–2257.

Özyürek, Asli, Roel M. Willems, Sotaro Kita, and Peter Hagoort. 2007. On-line integration of semantic information from speech and gesture: Insights from event-related brain potentials. *Journal of Cognitive Neuroscience* 19 (4): 605–616.

Ohta, Shinri, Naoki Fukui, and Kuniyoshi L. Sakai. 2013. Syntactic computation in the human brain: The degree of merger as a key factor. *PLoS One* 8 (2): e56230. doi:10.1371/journal.pone.0056230.

Okanoya, Kazuo. 2004. The Bengalese finch: A window on the behavioral neurobiology of birdsong syntax. *Annals of the New York Academy of Sciences* 1016:724–735.

Olshausen, Bruna A., and David J. Field. 2004. Sparse coding of sensory inputs. *Current Opinion in Neurobiology* 14 (4): 481–487.

Oostenveld, Robert, Pascal Fries, Eric Maris, and Jan-Mathijs Schoffelen. 2011. FieldTrip: Open source software for advanced analysis of MEG, EEG, and Invasive Electrophysiological Data. *Computational Intelligence and Neuroscience*, vol. 2011, Article ID 156869, 9 pages. doi:10.1155/2011/156869.

Opitz, Bertram, and Angela D. Friederici. 2003. Interactions of the hippocampal system and the prefrontal cortex in learning language-like rules. *NeuroImage* 19 (4): 1730–1737.

Opitz, Bertram, and Angela D. Friederici. 2004. Brain correlates of language learning: The neuronal dissociation of rule-based versus similarity-based learning. *Journal of Neuroscience* 24 (39): 8436–8440.

Opitz, Bertram, and Angela D. Friederici. 2007. Neural basis of processing sequential and hierarchical syntactic structures. *Human Brain Mapping* 28 (7): 585–592.

Osterhout, Lee, and Phillip J. Holcomb. 1992. Event-related brain potentials elicited by syntactic anomaly. *Journal of Memory and Language* 31 (6): 785–806.

Osterhout, Lee, Phillip J. Holcomb, and David A. Swinney. 1994. Brain potentials elicited by garden-path sentences: Evidence of the application of verb information during parsing. *Journal of Experimental Psychology: Learning, Memory, and Cognition* 20 (4): 786–803.

Osterhout, Lee, and Linda A. Mobley. 1995. Event-related brain potentials elicited by failure to agree. *Journal of Memory and Language* 34 (6): 739–773.

Osterhout, Lee, Judith McLaughlin, Ilona Pitkanen, Cheryl Frenck-Mestre, and Nicola Molinaro. 2006. Novice learners, longitudinal designs, and event-related potentials: A means for exploring the neurocognition of second language processing. *Language Learning* 56 (Suppl 1): 199–230.

Osterhout, Lee, and Janet Nicol. 1999. On the distinctiveness, independence, and time course of the brain responses to syntactic and semantic anomalies. *Language and Cognitive Processes* 14 (3): 283–317.

Pakulak, Eric, and Helen J. Neville. 2011. Maturational constraints on the recruitment of early processes for syntactic processing. *Journal of Cognitive Neuroscience* 23 (10): 2752–2765.

Pallier, Christophe, Anne-Dominique Devauchelle, and Stanislas Dehaene. 2011. Cortical representation of the constituent structure of sentences. *Proceedings of the National Academy of Sciences of the United States of America* 108 (6): 2522–2527.

Pannekamp, Ann, Ulrike Toepel, Kai Alter, Anja Hahne, and Angela D. Friederici. 2005. Prosody-driven sentence processing: An event-related brain potential study. *Journal of Cognitive Neuroscience* 17 (3): 407–421.

Pannekamp, Ann, Christiane Weber, and Angela D. Friederici. 2006. Prosodic processing at the sentence level in infants. *Neuroreport* 17 (6): 675–678.

Papoutsi, Marina, Emmanuel A. Stamatakis, John Griffiths, William D. Marslen-Wilson, and Lorraine K. Tyler. 2011. Is left fronto-temporal connectivity essential for syntax? Effective connectivity, tractography and performance in left-hemisphere damaged patients. *NeuroImage* 58 (2): 656–664.

Park, Hae-Jeong, Jae Jin Kim, Seung-Koo Lee, Jeong Ho Seok, Jiwon Chun, Dong Ik Kim, and Jong Doo Lee. 2008. Corpus callosal connection mapping using cortical gray matter parcellation and DT-MRI. *Human Brain Mapping* 29 (5): 503–516.

Parker, Geoffrey J. M., Simona Luzzi, Daniel C. Alexander, Claudia A. M. Wheeler-Kingshott, Olga Ciccarelli, and Matthew A. Lambon Ralph. 2005. Lateralization of ventral and dorsal auditory language pathways in the human brain. *NeuroImage* 24 (3): 656–666.

Patel, Aniruddh D. 2008. *Music, Language, and the Brain*. New York: Oxford University Press.

Patterson, Karalyn, and Matthew A. Lambon Ralph. 2016. The hub-and-spoke hypothesis of semantic memory. In *Neurobiology of Language*, ed. Steven L. Small and Gregory Hickok, 765–775. Elsevier Press.

Patterson, Karalyn, Peter J. Nestor, and Timothy T. Rogers. 2007. Where do you know what you know? The representation of semantic knowledge in the human brain. *Nature Reviews. Neuroscience* 8 (12): 976–987.

Paulmann, Silke, and Marc D. Pell. 2010. Contextual influences of emotional speech prosody on face processing: How much is enough? *Cognitive, Affective & Behavioral Neuroscience* 10 (2): 230–242.

Paulmann, Silke, Debra Titone, and Marc D. Pell. 2012. How emotional prosody guides your way: Evidence from eye movements. *Speech Communication* 54 (1): 92–107.

Pearce, Joshua M. 1987. Communication and language. In *An Introduction to Animal Cognition*, 251–283. Hillsdale, NJ: Lawrence Erlbaum Associates.

Pell, Marc D., and Shari R. Baum. 1997. Unilateral brain damage, prosodic comprehension deficits, and the acoustic cues to prosody. *Brain and Language* 57 (2): 195–214.

Peña, Marcela, Atsushi Maki, Damir Kovacic, Ghislaine Dehaene-Lambertz, Hideaki Koizumi, Furio Bouquet, and Jacques Mehler. 2003. Sounds and silence: An optical topography study of language recognition at birth. *Proceedings of the National Academy of Sciences of the United States of America* 100 (20): 11702–11705.

Peña, Marcela, Janet F. Werker, and Ghislaine Dehaene-Lambertz. 2012. Earlier speech exposure does not accelerate speech acquisition. *Journal of Neuroscience* 32 (33): 11159–11163.

Penke, Martina, Helga Weyerts, Matthias Gross, Elke Zander, Thomas F. Münte, and Harald Clahsen. 1997. How the brain processes complex words: An event-related potential study of German verb inflections. *Brain Research. Cognitive Brain Research* 6 (1): 37–52.

Perani, Daniela, Eraldo Paulesu, Nuria Sebastian Galles, Emmanuel Dupoux, Stanislas Dehaene, Valentino Bettinardi, Stefano F. Cappa, Ferruccio Fazio, and Jacques Mehler. 1998. The bilingual brain. Proficiency and age of acquisition of the second language. *Brain* 121 (10): 1841–1852.

Perani, Daniela, Maria C. Saccuman, Paola Scifo, Alfred Anwander, Danilo Spada, Cristina Baldoli, Antonella Poloniato, Gabriele Lohmann, and Angela D. Friederici. 2011. The neural language networks at birth. *Proceedings of the National Academy of Sciences of the United States of America* 108 (38): 16056–16061.

Peretz, Isabelle. 1993. Auditory atonalia for melodies. *Cognitive Neuropsychology* 10 (1): 21–56.

Perkins, Judy M., Jane A. Baran, and Jack Gandour. 1996. Hemispheric specialization in processing intonation contours. *Aphasiology* 10 (4): 343–362.

Petersson, Karl-Magnus, Vasiliki Folia, and Peter Hagoort. 2012. What artificial grammar learning reveals about the neurobiology of syntax. *Brain and Language* 120 (2): 83–95.

Petersson, Karl-Magnus, Christian Forkstam, and Martin Ingvar. 2004. Artificial syntactic violations activate Broca's region. *Cognitive Science* 28 (3): 383–407.

Petitto, Laura Ann, and Paula F. Marentette. 1991. Babbling in the manual mode—evidence for the ontogeny of language. *Science* 251 (5000): 1493–1496.

Petitto, Laura Ann, Siobhan Holowka, Lauren E. Sergio, and David Ostry. 2013. Language rhythms in baby hand movements. *Nature* 413 (6851): 35–36.

Petitto, Laura Ann, Robert J. Zatorre, Kristine Gauna, E. J. Nikelski, Deanna Dostie, and Alan C. Evans. 2000. Speech-like cerebral activity in profoundly deaf people processing signed languages: Implications for the neural basis of human language. *Proceedings of the National Academy of Sciences of the United States of America* 97 (25): 13961–13966.

Petkov, Christopher I., and Erich D. Jarvis. 2012. Birds, primates, and spoken language origins: Behavioral phenotypes and neurobiological substrates. *Frontiers in Evolutionary Neuroscience* 4:12. doi:10.3389/fnevo.2012.00012.

Petrides, Michael, and Deepak N. Pandya. 1984. Projections to the frontal cortex from the posterior parietal region in the rhesus monkey. *Journal of Comparative Neurology* 228 (1): 105–116.

Petrides, Michael, and Depaak N. Pandya. 2009. Distinct parietal and temporal pathways to the homologues of Broca's area in the monkey. *PLoS Biology* 7 (8): e1000170. doi:10.1371/journal.pbio.1000170.

Phillips, Colin. 2001. Levels of representation in the electrophysiology of speech perception. *Cognitive Science* 25 (5): 711–731.

Phillips, Colin, Thomas Pellathy, Alec Marantz, Elron Yellin, Kenneth Wexler, David Poeppel, Martha McGinnis, and Timothy Roberts. 2000. Auditory cortex accesses phonological categories: An MEG mismatch study. *Journal of Cognitive Neuroscience* 12 (6): 1038–1055.

Pick, Arnold. 1913. *Die agrammatischen Sprachstörungen; Studien zur psychologischen Grundlegung der Aphasielehre*. Berlin: Springer.

Picton, Terry W., Shlomo Bentin, Patrick Berg, Emanuel Donchin, Steven A. Hillyard, Ray Johnson, Jr., Gregory A. Miller, et al. 2000. Guidelines for using human event-related potentials to study cognition: Recording standards and publication criteria. *Psychophysiology* 37 (2): 127–152.

Pierce, Lara J., Denise Klein, Jen-Kai Chen, Audrey Delcenserie, and Fred Genesee. 2014. Mapping the unconscious maintenance of a lost first language. *Proceedings of the National Academy of Sciences of the United States of America* 111 (48): 17314–17319.

Pihko, Elina, Paavo H. T. Leppänen, Kenneth M. Eklund, Marie Cheour, Tomi K. Guttorm, and Heikki Lyytinen. 1999. Cortical responses of infants with and without a genetic risk for dyslexia: I. Age effects. *Neuroreport* 10 (5): 901–905.

Pinard, Minola, Howard Chertkow, Sandra Black, and Isabelle Peretz. 2002. A case study of pure word deafness: Modularity in auditory processing? *Neurocase* 8 (1–2): 40–55.

Pinker, Steven. 1994. *The Language Instinct*. New York: Harper Perennial Modern Classics.

Plante, Elena, Marlena Creusere, and Cynthia Sabin. 2002. Dissociating sentential prosody from sentence processing: Activation interacts with task demands. *NeuroImage* 17 (1): 401–410.

Poeppel, David, Colin Phillips, Elron Yellin, Howard A. Rowley, Timothy P. L. Roberts, and Alec Marantz. 1997. Processing of vowels in supratemporal auditory cortex. *Neuroscience Letters* 221 (2–3): 145–148.

Poizner, Howard, Edward S. Klima, and Ursula Bellugi. 1987. *What the Hands Reveal about the Brain*. Cambridge, MA: Bradford Books/MIT Press. (Reprinted in paperback, 1990.)

Pollard, Carl, and Ivan A. Sag. 1994. *Head-Driven Phrase Structure Grammar*. Chicago; Stanford, CA: University of Chicago Press and CSLI Publications.

Pollmann, Stefan, Marianne Maertens, D. Yves von Cramon, Joeran Lepsien, and Kenneth Hughdahl. 2002. Dichotic listening in patients with posterior CC and nonsplenial callosal lesions. *Neuropsychologia* 16 (1): 56–64.

Polka, Linda, and Megha Sundara. 2012. Word segmentation in monolingual infants acquiring Canadian-English and Canadian-French: Native language, cross-language and cross-dialect comparisons. *Infancy* 17 (2): 198–232.

Price, Cathy J. 2010. The anatomy of language: A review of 100 fMRI studies published in 2009. *Annals of the New York Academy of Sciences* 1191:62–88.

Price, Amy R., Michael F. Bonner, Jonathan E. Peelle, and Murray Grossman. 2015. Converging evidence for the neuroanatomic basis of combinatorial semantics in the angular gyrus. *Journal of Neuroscience* 35 (7): 3276–3284.

Prüfer, Kay, Fernando Racimo, Nick Patterson, Flora Jay, Sriram Sankararaman, Susanna Sawyer, Anja Heinze, et al. 2014. The complete genome sequence of a Neanderthal from the Altai Mountains. *Nature* 505 (7481): 43–49.

Pulvermüller, Friedemann, and Luciano Fadiga. 2010. Active perception: Sensorimotor circuits as a cortical basis for language. *Nature Reviews. Neuroscience* 11:351–360.

Pulvermüller, Friedemann, Werner Lutzenberger, and Hubert Preissl. 1999. Nouns and verbs in the intact brain: Evidence from event-related potentials and high-frequency cortical responses. *Cerebral Cortex* 9 (5): 497–506.

Pustejovsky, James. 1995. *The Generative Lexicon*. Cambridge, MA: MIT Press.

Quian Quiroga, Rodrigo, Leila Reddy, Gabriel Kreiman, Christof Koch, and Itzhak Fried. 2005. Invariant visual representation by single neurons in the human brain. *Nature* 435:1102–1107.

Raettig, Tim, Sonja A. Kotz, Alfred Anwander, D. Yves von Cramon, and Angela D. Friederici. Submitted. [The language connection: a structural connectivity-based parcellation of the left superior temporal cortex.]

Rauschecker, Josef P., and Sophie K. Scott. 2009. Maps and streams in the auditory cortex: Nonhuman primates illuminate human speech processing. *Nature Neuroscience* 12 (6): 718–724.

Rauschecker, Josef P., and Biao Tian. 2000. Mechanisms and streams for processing of "what" and "where" in auditory cortex. *Proceedings of the National Academy of Sciences of the United States of America* 97 (22): 11800–11806.

Richardson, Fiona M., Michael S. C. Thomas, Roberto Filippi, Helen Harth, and Cathy J. Price. 2010. Contrasting effects of vocabulary knowledge on temporal and parietal brain structure across lifespan. *Journal of Cognitive Neuroscience* 22 (5): 943–954.

Rilling, James K., Sarah K. Barks, Lisa A. Parr, Todd M. Preuss, Tracy L. Faber, Giuseppe Pagnoni, J. Douglas Bremner, and John R. Votaw. 2007. A comparison of resting-state brain activity in humans and chimpanzees. *Proceedings of the National Academy of Sciences of the United States of America* 104 (43): 17146–17151.

Rilling, James K., Matthew F. Glasser, Saad Jbabdi, Jesper Andersson, and Todd M. Preuss. 2012. Continuity, divergence, and the evolution of brain language pathways. *Frontiers in Evolutionary Neuroscience* 3:11. doi:10.3389/fnevo.2011.00011.

Rilling, James K., Matthew F. Glasser, Todd M. Preuss, Xiangyang Ma, Tiejun Zhao, Xiaoping Hu, and Timothy E. J. Behrens. 2008. The evolution of the arcuate fasciculus revealed with comparative DTI. *Nature Neuroscience* 11 (4): 426–428.

Rilling, James K., and Rebecca A. Seligman. 2002. A quantitative morphometric comparative analysis of the primate temporal lobe. *Journal of Human Evolution* 42 (5): 505–533.

Rivera-Gaxiola, Maritza, Juan Silva-Pereyra, and Patricia K. Kuhl. 2005. Brain potentials to native and nonnative speech contrasts in 7-and 11-month-old American infants. *Developmental Science* 8 (2): 162–172.

Rizzolatti, Giacomo, and Michael A. Arbib. 1998. Language within our grasp. *Trends in Neurosciences* 21 (5): 188–194.

Rizzolatti, Giacomo, Luciano Fadiga, Massimo Matelli, Valentino Bettinardi, Eraldo Paulesu, Daniela Perani, and Ferruccio Fazio. 1996. Localization of grasp representations in humans by PET. 1. Observation versus execution. *Experimental Brain Research* 111 (2): 246–252.

Rodd, Jennifer M., Matthew H. Davis, and Ingrid S. Johnsrude. 2005. The neural mechanisms of speech comprehension: fMRI studies of semantic ambiguity. *Cerebral Cortex* 15 (8): 1261–1269.

Röder, Brigitte, Oliver Stock, Helen Neville, Siegfried Bien, and Frank Rösler. 2002. Brain activation modulated by the comprehension of normal and pseudo-word sentences of different processing demands: A functional magnetic resonance imaging study. *NeuroImage* 15 (4): 1003–1014.

Rösler, Frank. 2011. *Psychophysiologie der Kognition: Eine Einführung in die Kognitive Neurowissenschaft*. Heidelberg: Spektrum Akademischer Verlag.

Rogalsky, Corianne. 2015. The role of the anterior temporal lobe in sentence processing. In *Neurobiology of Language Volume*, ed. Steven L. Small and Gregory Hickok, 587–596. Elsevier Press.

Rogalsky, Corianne, and Gregory Hickok. 2011. The role of Broca's area in sentence comprehension. *Journal of Cognitive Neuroscience* 23 (7): 1664–1680.

Rogalsky, Corianne, William Matchin, and Gregory Hickok. 2008. Broca's area, sentence comprehension, and working memory: An fMRI Study. *Frontiers in Human Neuroscience* 2:14. doi:10.3389/neuro.09.014.2008.

Roland, Per E., and Karl Zilles. 1998. Structural divisions and functional fields in the human cerebral cortex. *Brain Research. Brain Research Reviews* 26 (2–3): 87–105.

Rolheiser, Tyler, Emmanuel A. Stamatakis, and Lorraine K. Tyler. 2011. Dynamic processing in the human language system: Synergy between the arcuate fascicle and extreme capsule. *Journal of Neuroscience* 31 (47): 16949–16957.

Román, Patricia E., Julio González, Noelia Ventura-Campos, Aina Rodriguez-Pujadas, Ana Sanjuán, and Cesar Avila. 2015. Neural differences between monolinguals and early bilinguals in their native language during comprehension. *Brain and Language* 150:80–89.

Rossi, Sonja, Manfred F. Gugler, Angela D. Friederici, and Anja Hahne. 2006. The impact of proficiency on syntactic second language processing of German and Italian: Evidence from event-related potentials. *Journal of Cognitive Neuroscience* 18 (12): 2030–2048.

Rumsey, Judith M., Manuel Casanova, Glenn B. Mannheim, Nicholas Patronas, Nathan DeVaughn, Susan D. Hamburger, and Tracy Aquino. 1996. Corpus callosum morphology, as measured with MRI, in dyslexic men. *Biological Psychiatry* 39 (9): 769–775.

Rüschemeyer, Shirley-Ann, Christian J. Fiebach, Vera Kempe, and Angela D. Friederici. 2005. Processing lexical semantic and syntactic information in first and second language: fMRI evidence from German and Russian. *Human Brain Mapping* 25 (2): 266–286.

Rymarczyk, Krystyna, and Anna Grabowska. 2007. Sex differences in brain control of prosody. *Neuropsychologia* 45 (5): 921–930.

Saffran, Jenny R., Richard N. Aslin, and Elissa L. Newport. 1996. Statistical learning by 8-month-old infants. *Science* 274 (5294): 1926–1928.

Sahin, Ned T., Steven Pinker, Sydney S. Cash, Donald Schomer, and Eric Halgren. 2009. Sequential processing of lexical, grammatical, and phonological information within Broca's area. *Science* 326 (5951): 445–449.

Sajin, Stanislav M., and Cynthia M. Connine. 2014. Semantic richness: The role of semantic features in processing spoken words. *Journal of Memory and Language* 70:13–35.

Sakai, Tomoko, Mie Matsui, Akichika Mikami, Ludise Malkova, Yuzuru Hamada, Masaki Tomonaga, Juri Suzuki, et al. 2013. Developmental patterns of chimpanzee cerebral tissues provide important clues for understanding the remarkable enlargement of the human brain. *Proceedings of the National Academy of Sciences of the United States of America* 280 (1753): 20122398.

Sakai, Tomoko, Akichika Mikami, Masaki Tomonaga, Mie Matsui, Juri Suzuki, Yuzuru Hamada, Masayuki Tanaka, et al. 2011. Differential prefrontal white matter development in chimpanzees and humans. *Current Biology* 21 (16): 1397–1402.

Sammler, Daniela, Sonja A. Kotz, Korinna Eckstein, Derek V. M. Ott, and Angela D. Friederici. 2010. Prosody meets syntax: The role of the corpus callosum. *Brain* 133:2643–2655.

Sammler, Daniela, Marie-Helene Grosbras, Alfred Anwander, Patricia E. G. Bestelmeyer, and Pascal Belin. 2015. Dorsal and ventral pathways for prosody. *Current Biology* 25 (23): 3079–3085.

Sanides, Friedrich. 1962. The architecture of the human frontal lobe and the relation to its functional differentiation. *International Journal of Neurology* 5:247–261.

Santelmann, Lynn M., and Peter W. Jusczyk. 1998. Sensitivity to discontinuous dependencies in language learners: Evidence for limitations in processing space. *Cognition* 69 (2): 105–134.

Santi, Andrea, and Yosef Grodzinsky. 2010. fMRI adaptation dissociates syntactic complexity dimensions. *NeuroImage* 51 (4): 1285–1293.

Sarnthein, Johannes, Hellmuth Petsche, Peter Rappelsberger, Gordon L. Shaw, and Astrid von Stein. 1998. Synchronization between prefrontal and posterior association cortex during human working memory. *Proceedings of the National Academy of Sciences of the United States of America* 95 (12): 7092–7096.

Sarubbo, Silvio, Alessandro De Benedictis, Igor L. Maldonado, Gianpaolo Basso, and Hugues Duffau. 2013. Frontal terminations for the inferior fronto-occipital fascicle: Anatomical dissection, DTI study and functional considerations on a multi-component bundle. *Brain Structure & Function* 218 (1): 21–37.

Saur, Dorothee, Annette Baumgaertner, Anja Moehring, Christian Büchel, Matthias Bonnesen, Michael Rose, Mariachristina Musso, and Jürgen M. Meisel. 2009. Word order processing in the bilingual brain. *Neuropsychologia* 47 (1): 158–168.

Saur, Dorothee, Björn W. Kreher, Susanne Schnell, Dorothee Kümmerer, Philipp Kellmeyer, Magnus-Sebastian Vry, Roza Umarova, et al. 2008. Ventral and dorsal pathways for language. *Proceedings of the National Academy of Sciences of the United States of America* 105 (46): 18035–18040.

Saur, Dorothee, Björn Schelter, Susanne Schnell, David Kratochvil, Hanna Küpper, Philipp Kellmeyer, Dorothee Kümmerer, et al. 2010. Combining functional and anatomical connectivity reveals brain networks for auditor language comprehension. *NeuroImage* 49 (4): 3187–3197.

Sauseng, Paul, Wolfgang Klimesch, Manuel Schabus, and Michael Doppelmayr. 2005. Fronto-parietal EEG coherence in theta and upper alpha reflect central executive functions of working memory. *International Journal of Psychophysiology* 57 (2): 97–103.

Saxe, Rebecca, and Nancy Kanwisher. 2003. People thinking about thinking people—The role of the temporo-parietal junction in "theory of mind". *NeuroImage* 19 (4): 1835–1842.

Scharff, Constance, and Jana Petri. 2011. Evo-devo, deep homology and FoxP2: Implications for the evolution of speech and language. *Philosophical Transactions of the Royal Society of London. Series B, Biological Sciences* 366 (1574): 2124–2140.

Scheibel, Arnold B. 1984. A dendritic correlate of human speech. In *Cerebral dominance: The biological foundations*, ed. Norman Geschwind and Albert M. Galaburda, 43–52. Cambridge, MA: Harvard University Press.

Schenker, Natalie M., William D. Hopkins, Muhammad A. Spocter, Amy R. Garrison, Cheryl D. Stimpson, Joseph M. Erwin, Patrick R. Hof, and Cher C. Sherwood. 2010. Broca's area homologue in chimpanzees (pan troglodytes): Probabilistic mapping, asymmetry, and comparison to humans. *Cerebral Cortex* 20 (3): 730–742.

Schipke, Christine S., Angela D. Friederici, and Regine Oberecker. 2011. Brain responses to case-marking violations in German preschool children. *Neuroreport* 22 (16): 850–854.

Schipke, Christine S., Lisa J. Knoll, Angela D. Friederici, and Regine Oberecker. 2012. Preschool children's interpretation of object-initial sentences: Neural correlates of their behavioral performance. *Developmental Science* 15 (6): 762–774.

Schirmer, Annett, Nicolas Escoffier, Stefan Zysset, Dirk Koester, Tricia Striano., and Angela D. Friederici. 2008. When vocal processing gets emotional. On the role of social orientation in relevance detection by the human amygdala. *NeuroImage* 40 (3): 1402–1410.

Schirmer, Annett, and Sonja A. Kotz. 2006. Beyond the right hemisphere: Brain mechanisms mediating vocal emotional processing. *Trends in Cognitive Sciences* 10 (1): 24–30.

Schirmer, Annett, Sonja A. Kotz, and Angela D. Friederici. 2002. Sex differentiates the role of emotional prosody during word processing. *Brain Research. Cognitive Brain Research* 14 (2): 228–233.

Schirmer, Annett, Sonja A. Kotz, and Angela D. Friederici. 2005. On the role of attention for the processing of emotions in speech: Sex differences revisited. *Brain Research. Cognitive Brain Research* 24 (3): 442–452.

Schmahmann, Jeremy D., Depaak N. Pandya, Ruopeng Wang, Guangping Dai, Helen E. D'Arceuil, Alex J. de Crespigny, and Van J. Wedeen. 2007. Association fibre pathways of the brain: Parallel observations from diffusion spectrum imaging and autoradiography. *Brain* 130:630–653.

Scott, Sophie K., Catrin Blank, Stuart Rosen, and Richard J. S. Wise. 2000. Identification of a pathway for intelligible speech in the left temporal lobe. *Brain* 123:2400–2406.

Scott, Sophie K., and Ingrid S. Johnsrude. 2003. The neuroanatomical and functional organization of speech perception. *Trends in Neurosciences* 26 (2): 100–107.

Seidl, Amanda. 2007. Infants' use and weighting of prosodic cues in clause segmentation. *Journal of Memory and Language* 57 (1): 24–48.

Service, Elisabet, Paeivi Helenius, Sini Maury, and Riitta Salmelin. 2007. Localization of syntactic and semantic brain responses using magnetoencephalography. *Journal of Cognitive Neuroscience* 19 (7): 1193–1205.

Shannon, Robert V., Fang-Gang Zeng, Vivek Kamath, John Wygonski, and Michael Ekelid. 1995. Speech recognition with primarily temporal cues. *Science* 270 (5234): 303–304.

Shestakova, Anna, Elvira Brattico, Alexei Soloviev, Vasily Klucharev, and Minna Huotilainen. 2004. Orderly cortical representation of vowel categories presented by multiple exemplars. *Brain Research. Cognitive Brain Research* 21 (3): 342–350.

Shtyrov, Yury, Friedemann Pulvermuller, Risto Näätänen, and Risto J. Ilmoniemi. 2003. Grammar processing outside the focus of attention: An MEG study. *Journal of Cognitive Neuroscience* 15 (8): 1195–1206.

Shukla, Mohinish, Katherine S. White, and Richard N. Aslin. 2011. Prosody guides the rapid mapping of auditory word forms onto visual objects in 6-mo-old infants. *Proceedings of the National Academy of Sciences of the United States of America* 108 (15): 6038–6043.

Siapas, Anthanassios G., Evgueniy V. Lubenov, and Matthew A. Wilson. 2005. Prefrontal phase locking to hippocampal theta oscillations. *Neuron* 46 (1): 141–151.

Siebörger, Florian Th., Evelyn C. Ferstl, and D. Yves von Cramon. 2007. Making sense of nonsense: An fMRI study of task induced coherence processes. *Brain Research* 1166:77–91.

Sigurdsson, Torfi, Kimberly L. Stark, Maria Karayiorgou, Joseph A. Gogos, and Joshua A. Gordon. 2010. Impaired hippocampal-prefrontal synchrony in a genetic mouse model of schizophrenia. *Nature* 464 (7289): 763–767.

Silva-Pereyra, Juan F., and Manuel Carreiras. 2007. An ERP study of agreement features in Spanish. *Brain Research* 1185:201–211.

Silva-Pereyra, Juan F., Lindsay Klarman, Lotus Jo-Fu Lin, and Patricia K. Kuhl. 2005. Sentence processing in 30-month-old children: An event-related potential study. *Neuroreport* 16 (6): 645–648.

Silva-Pereyra, Juan F., Maritza Rivera-Gaxiola, and Patricia K. Kuhl. 2005. An event-related brain potential study of sentence comprehension in preschoolers: Semantic and morphosyntactic processing. *Brain Research. Cognitive Brain Research* 23 (2–3): 247–258.

Simonds, Roderick J., and Arnold B. Scheibel. 1989. The postnatal development of the motor speech area: A preliminary study. *Brain and Language* 37 (1): 42–58.

Singer, Tania, and Claus Lamm. 2009. The social neuroscience of empathy. *Annals of the New York Academy of Sciences* 1156:81–96.

Skeide, Michael A., Jens Brauer, and Angela D. Friederici. 2014. Syntax gradually segregates from semantics in the developing brain. *NeuroImage* 100:206–211.

Skeide, Michael A., Jens Brauer, and Angela D. Friederici. 2016. Brain functional and structural predictors of language performance. *Cerebral Cortex* 26 (5): 2127–2139.

Skeide, Michael A., and Angela D. Friederici. 2016. The ontogeny of the cortical language network. *Nature Reviews. Neuroscience* 17:323–332.

Smith, G. Elliot. 1904. The morphology of the occipital region of the cerebral hemisphere in man and the apes. [Annals of Anatomy] *Anatomischer Anzeiger* 24:436–451.

Smith, Edward E., and John Jonides. 1999. Storage and executive processes in the frontal lobes. *Science* 283 (5408): 1657–1661.

Snijders, Tineke M., Theo Vosse, Gerard Kempen, Jos J. A. Van Berkum, Karl Magnus Petersson, and Peter Hagoort. 2009. Retrieval and unification of syntactic structure in sentence comprehension: An fMRI study using word-category ambiguity. *Cerebral Cortex* 19 (7): 1493–1503.

Song, Xindong, Michael S. Osmanski, Yueqi Guo, and Xiaoqin Wang. 2016. Complex pitch perception mechanisms are shared by humans and a New World monkey. *Proceedings of the National Academy of Sciences of the United States of America* 113 (3): 781–786.

Sowell, Elizabeth R., Bradley S. Peterson, Paul M. Thompson, Suzanne E. Welcome, Amy L. Henkenius, and Arthur W. Toga. 2003. Mapping cortical change across the human life span. *Nature Neuroscience* 6 (3): 309–315.

Starkstein, Sergio E., J. Paul Federoff, Thomas R. Price, Ramón C. Leiguarda, and Robert G. Robinson. 1994. Neuropsychological and neuroradiologic correlates of emotional prosody comprehension. *Neurology* 44 (3): 515–522.

Steinhauer, Karsten. 2003. Electrophysiological correlates of prosody and punctuation. *Brain and Language* 86 (1): 142–164.

Steinhauer, Karsten, Kai Alter, and Angela D. Friederici. 1999. Brain potentials indicate immediate use of prosodic cues in natural speech processing. *Nature Neuroscience* 2 (2): 191–196.

Steinhauer, Karsten, and John E. Drury. 2012. On the early left-anterior negativity (ELAN) in syntax studies. *Brain and Language* 120 (2): 135–162.

Steinhauer, Karsten, and Angela D. Friederici. 2001. Prosodic boundaries, comma rules, and brain responses: The Closure Positive Shift in the ERPs as a universal marker for prosodic phrasing in listeners and readers. *Journal of Psycholinguistic Research* 30 (3): 267–295.

Steinmann, Saskia, Gregor Leicht, Matthias Ertl, Christina Andreou, Nenad Polomac, Rene Westerhausen, Angela D. Friederici, and Christoph Mulert. 2014. Conscious auditory perception related to long-range synchrony of gamma oscillations. *NeuroImage* 100:435–443.

Steinmetz, Helmuth, Jörg Rademacher, Yanxiong Huang, Harald Hefter, Karl Zilles, Armin Thron, and Hans-Joachim Freund. 1989. Cerebral asymmetry—MR planimetry of the human planum temporale. *Journal of Computer Assisted Tomography* 13 (6): 996–1005.

Strauss, Antje, Sonja A. Kotz, Matthias Scharinger, and Jonas Obleser. 2014. Alpha and theta brain oscillations index dissociable processes in spoken word recognition. *NeuroImage* 97:387–395.

Strelnikov, Kuzma N., V. A. Vorobyev, Tatiana V. Chernigovskaya, and Svyatoslav V. Medvedev. 2006. Prosodic clues to syntactic processing: A PET and ERP study. *NeuroImage* 29 (4): 1127–1134.

Strotseva-Feinschmidt, Anna, Christine S. Schipke, Thomas C. Gunter, Angela D. Friederici, and Jens Brauer. Submitted. The use of syntactic and semantic cues during sentence comprehension in early childhood. *Developmental Cognitive Neuroscience*.

Stroud, Claire, and Colin Phillips. 2012. Examining the evidence for an independent semantic analyzer: An ERP study in Spanish. *Brain and Language* 120 (2): 108–126.

Stowe, Laurie A., Cees A. J. Broere, Anne M. J. Paans, Albertus A. Wijers, Gijsbertus Mulder, Wim Vaalburg, and Frans Zwarts. 1998. Localizing components of a complex task: Sentence processing and working memory. *Neuroreport* 9 (13): 2995–2999.

Summerfield, Christopher, and Jennifer A. Mangels. 2005. Coherent theta-band EEG activity predicts item-context binding during encoding. *NeuroImage* 24 (3): 692–703.

Suzuki, Kei, and Kuniyoshi L. Sakai. 2003. An event-related fMRI study of explicit syntactic processing of normal/anomalous sentences in contrast to implicit syntactic processing. *Cerebral Cortex* 13 (5): 517–526.

Tadayonnejad, Reza, Shaolin Yang, Anand Kumar, and Olusola Ajilore. 2015. Clinical, cognitive, and functional connectivity correlations of resting-state intrinsic brain activity alterations in unmedicated depression. *Journal of Affective Disorders* 172:241–250.

Takashima, Atsuko, Iske Bakker, Janet G. van Hell, Gabriele Janzen, and James M. McQueen. 2014. Richness of information about novel words influences how episodic and semantic memory networks interact during lexicalization. *NeuroImage* 84:265–278.

Tan, Arlene A., and Dennis L. Molfese. 2009. ERP Correlates of noun and verb processing in preschool-age children. *Biological Psychology* 8 (1): 46–51.

Teffer, Kate, and Katerina Semendeferi. 2012. Human prefrontal cortex: Evolution, development, and pathology. *Progress in Brain Research* 195:191–218.

ten Cate, Carel, and Kazuo Okanoya. 2012. Revisiting the syntactic abilities of non-human animals: Natural vocalizations and artificial grammar learning. *Philosophical Transactions of the Royal Society of London. Series B, Biological Sciences* 367 (1598): 1984–1994.

Tettamanti, Marco, Hatem Alkhadi, Andrea Moro, Dorothea Weniger, Daniela Perani, Spyros Kollias, and Ferruccio Fazio. 2001. Neural correlates of learning new grammatical rules: A fMRI study. *NeuroImage* 13 (6): S616–S616.

Tettamanti, Marco, Hatem Alkadhi, Andrea Moro, Daniela Perani, Spyros Kollias, and Dorothea Weniger. 2002. Neural correlates for the acquisition of natural language syntax. *NeuroImage* 17 (2): 700–709.

Thiebaut de Schotten, Michel, Flavio Dell'Acqua, Romain Valabregue, and Marco Catani. 2012. Monkey to human comparative anatomy of the frontal lobe association tracts. *Cortex* 48 (1): 82–96.

Thiebaut de Schotten, Michel, Dominic H. Ffytche, Alberto Bizzi, Flavio Dell'Acqua, Matthew Allin, Muriel Walshe, Robin Murray, Steven C. Williams, Declan G. M. Murphy, and Marco Catani. 2011. Atlasing location, asymmetry and inter-subject variability of white matter tracts in the human brain with MR diffusion tractography. *NeuroImage* 54 (1): 49–59.

Thierry, Guillaume, Marilyn Vihman, and Marc Roberts. 2003. Familiar words capture the attention of 11-month-olds in less than 250 ms. *Neuroreport* 14 (18): 2307–2310.

Thompson, Sandra A. 1978. Modern English from a typological point of view: Some implications of the function of word order. *Linguistische Berichte Braunschweig* 54:19–35.

Thompson-Schill, Sharon L., Mark D'Esposito, Geoffrey K. Aguirre, and Martha J. Farah. 1997. Role of left inferior prefrontal cortex in retrieval of semantic knowledge: A reevaluation. *Proceedings of the National Academy of Sciences of the United States of America* 94 (26): 14792–14797.

Tillmann, Barbara, Denis Burnham, Sebastien Nguyen, Nicolas Grimault, Nathalie Gosselin, and Isabelle Peretz. 2011. Congenital amusia (or tone-deafness) interferes with pitch processing in tone languages. *Frontiers in Psychology* 2:120. doi:10.3389/fpsyg.2011.00120.

Tincoff, Ruth, Lynn M. Santelmann, and Peter W. Jusczyk. 2000. Auxiliary verb learning and 18-month-olds' acquisition of morphological relationships. In *Proceedings of the 24th Annual Boston University Conference on Language Development*, Vol. 2, ed. S. Catherine Howell, Sarah A. Fish, and Thea Keith-Lucas, 726–737. Somerville, MA: Cascadilla Press.

Tomasello, Michael. 1999. *The Cultural Origins of Human Cognition*. Cambridge, MA: Harvard University Press.

Tomasello, Michael. 2008. *Origins of Human Communication*. Cambridge, MA: MIT Press.

Tomasello, Michael, Malinda Carpenter, Josep Call, Tanya Behne, and Henrike Moll. 2005. Understanding and sharing intentions: The origins of cultural cognition. *Behavioral and Brain Sciences* 28 (5): 675–691.

Tomasi, Dardo, and Nora D. Volkow. 2012. Resting functional connectivity of language networks: Characterization and reproducibility. *Molecular Psychiatry* 17 (8): 841–854.

Trainor, Laurel, Melissa McFadden, Lisa Hodgson, Lisa Darragh, Jennifer Barlow, Laura Matsos, and Ranil Sonnadara. 2003. Changes in auditory cortex and the development of mismatch negativity between 2 and 6 months of age. *International Journal of Psychophysiology* 51 (1): 5–15.

Travis, Katherine E., Matthew K. Leonard, Timothy T. Brown, Donald J. Hagler, Megan Curran, Anders M. Dale, Jeffrey L. Elman, and Eric Halgren. 2011. Spatiotemporal neural dynamics of word understanding in 12- to 18-month-old-infants. *Cerebral Cortex* 21 (8): 1832–1839.

Tune, Sarah, Matthias Schlesewsky, Steven L. Small, Anthony J. Sanford, Jason Bohan, Jona Sassenhagen, and Ina Bornkessel-Schlesewsky. 2014. Cross-linguistic variation in the neurophysiological response to semantic processing: Evidence from anomalies at the borderline of awareness. *Neuropsychologia* 56:147–166.

Turner, Robert. 2015. Myelin Imaging. *Brain Mapping* 1:137–142.

Turkeltaub, Peter E., Guinevere F. Eden, Karen M. Jones, and Thomas A. Zeffiro. 2002. Meta-analysis of the functional neuroanatomy of single-word reading: Method and validation. *NeuroImage* 16 (3): 765–780.

Turken, And U., and Nina F. Dronkers. 2011. The neural architecture of the language comprehension network: Converging evidence from lesion and connectivity analyses. *Frontiers in Systems Neuroscience* 5:1. doi:10.3389/fnsys.2011.00001.

Tyler, Lorraine K., and William D. Marslen-Wilson. 2008. Fronto-temporal brain systems supporting spoken language comprehension. *Philosophical Transactions of the Royal Society of London. Series B, Biological Sciences* 363 (1493): 1037–1054.

Tyler, Lorraine K., Meredith A. Shafto, Billi Randall, Paul Wright, William D. Marslen-Wilson, and Emmanuel A. Stamatakis. 2010. Preserving syntactic processing across the adult life span: The modulation of the fronto-temporal language system in the context of age-related atrophy. *Cerebral Cortex* 20 (2): 352–364.

Tyler, Lorraine K., Emmanuel A. Stamatakis, Peter Bright, Kadia Acres, Sherief Abdallah, Jennifer M. Rodd, and Helen E. Moss. 2004. Processing objects at different levels of specificity. *Journal of Cognitive Neuroscience* 16 (3): 351–362.

Tzourio-Mazoyer, Nathalie, and Bernard Mazoyer. 2017. Variations of planum temporale asymmetries with Heschl's Gyri duplications and association with cognitive abilities: MRI investigation of 428 healthy volunteers. *Brain Structure and Function*. doi:10.1007/s00429-017-1367-5.

Uddén, Julia, Vasiliki Folia, Christian Forkstam, Martin Ingvar, Guillen Fernandez, Sebastiaan Overeem, Gijs van Elswijk, Peter Hagoort, and Karl Magnus Petersson. 2008. The inferior frontal cortex in artificial syntax processing: An rTMS study. *Brain Research* 1224:69–78.

Uddin, Lucina Q., Kaustubh Supekar, Hitha Amin, Elena Rykhlevskaia, Daniel A. Nguyen, Michael D. Greicius, and Vinod Menon. 2010. Dissociable connectivity within human angular gyrus and intraparietal sulcus: Evidence from functional and structural connectivity. *Cerebral Cortex* 20 (11): 2636–2646.

Ullman, Michael T. 2001. A neurocognitive perspective on language: The declarative/procedural model. *Nature Reviews. Neuroscience* 2 (10): 717–726.

Ullman, Michael T., Suzanne Corkin, Marie Coppola, Gregory Hickok, John H. Growdon, Walter J. Koroshetz, and Steven Pinker. 1997. A neural dissociation within language: Evidence that the mental dictionary is part of declarative memory, and that grammatical rules are processed by the procedural system. *Journal of Cognitive Neuroscience* 9 (2): 266–276.

Upadhyay, Jaymin, Andrew Silver, Tracey A. Knaus, Kristin A. Lindgren, Mathieu Ducros, Dae-Shik Kim, and Helen Tager-Flusberg. 2008. Effective and structural connectivity in the human auditory cortex. *Journal of Neuroscience* 28 (13): 3341–3349.

van der Lely, Heather K. J., and Steven Pinker. 2014. The biological basis of language: Insight from developmental grammatical impairments. *Trends in Cognitive Sciences* 18 (11): 586–595.

van der Lely, Heather K. J., Stuart Rosen, and Alastair McClelland. 1998. Evidence for a grammar-specific deficit in children. *Current Biology* 8 (23): 1253–1258.

Van Lancker, Diana. 1980. Cerebral lateralization of pitch cues in the linguistic signal. *Papers in Linguistics: International Journal of Human Communication* 13 (2): 227–277.

Van Lancker, Diana, and John J. Sidtis. 1992. The identification of affective-prosodic stimuli by left-hemisphere-damaged and right-hemisphere-damaged subjects—all errors are not created equal. *Journal of Speech and Hearing Research* 35 (5): 963–970.

van Turennout, Miranda, Peter Hagoort, and Colin M. Brown. 1998. Brain activity during speaking: From syntax to phonology in 40 milliseconds. *Science* 280 (5363): 572–574.

Vandenberghe, Rik, Anna C. Nobre, and Cathy J. Price. 2002. The response of left temporal cortex to sentences. *Journal of Cognitive Neuroscience* 14 (4): 550–560.

Van den Brink, Danielle, and Peter Hagoort. 2004. The influence of semantic and syntactic context constraints on lexical selection and integration in spoken-word comprehension as revealed by ERPs. *Journal of Cognitive Neuroscience* 16 (6): 1068–1084.

Vernooij, Meike W., Marion Smits, Piotr A. Wielopolski, Gavin C. Houston, Gabriel P. Krestin, and Aad van der Lugt. 2007. Fiber density asymmetry of the arcuate fasciculus in relation to functional hemispheric language lateralization in both right- and left-handed healthy subjects: A combined fMRI and DTI study. *NeuroImage* 35 (3): 1064–1076.

Vigneau, Mathieu, Virginie Beaucousin, Pierre-Yves Herve, Hugues Duffau, Fabrice Crivello, Olivier Houde, Bernard Mazoyer, and Nathalie Tzourio-Mazoyer. 2006. Meta-analyzing left hemisphere language areas: Phonology, semantics, and sentence processing. *NeuroImage* 30 (4): 1414–1432.

Vissiennon, Kodjo, Angela D. Friederici, Jens Brauer, and Chiao-Yi Wu. 2017. Functional organization of the language network in three- and six-year-old children. *Neuropsychologia* 98 (4): 24–33.

Wang, Liping, Lynn Uhrig, Bechir Jarraya, and Stanislas Dehaene. 2015. Representation of numerical and sequential patterns in macaque and human brains. *Current Biology* 25 (15): 1966–1974.

Wake, Hiroaki, Philip R. Lee, and Douglas Fields. 2011. Control of local protein synthesis and initial events in myelination by action potentials. *Science* 333 (6049): 1647–1651.

Warren, Paul, Esther Grabe, and Francis Nolan. 1995. Prosody, phonology and parsing closure ambiguities. *Language and Cognitive Processes* 10 (5): 457–486.

Wartenburger, Isabell, Hauke R. Heekeren, Jubin Abutalebi, Stefano F. Cappa, Arnold Villringer, and Daniela Perani. 2003. Early setting of grammatical processing in the bilingual brain. *Neuron* 37 (1): 159–170.

Watkins, Kate E., Tomas Paus, Jason P. Lerch, Alex Zijdenbos, D. Louis Collins, Peter Neelin, John C. Taylor, Keith J. Worsley, and Alan C. Evans. 2001. Structural asymmetries in the human brain: A voxel-based statistical analysis of 142 MRI scans. *Cerebral Cortex* 11 (9): 868–877.

Waxman, Stephen G. 1980. Determinants of conduction velocity in myelinated nerve fibers. *Muscle & Nerve* 3 (2): 141–150.

Weber, Christiane, Anja Hahne, Manuela Friedrich, and Angela D. Friederici. 2004. Discrimination of word stress in early infant perception: Electrophysiological evidence. *Brain Research. Cognitive Brain Research* 18 (2): 149–161.

Weber-Fox, Christine M., and Helen J. Neville. 1996. Maturational constraints on functional specializations for language processing: ERP and behavioral evidence in bilingual speakers. *Journal of Cognitive Neuroscience* 8 (3): 231–256.

Weighall, Anna R., and Gerry T. M. Altmann. 2011. The role of working memory and contextual constraints in children's processing of relative clauses. *Journal of Child Language* 38 (3): 579–605.

Weiller, Cornelius, Mariachristina Musso, Michel Rijntjes, and Dorothee Saur. 2009. Please don't underestimate the ventral pathway in language. *Trends in Cognitive Sciences* 13 (9): 369–370.

Weiller, Cornelius, Tobias Bormann, Dorothee Saur, Mariachristina Musso, and Michel Rijntjes. 2011. How the ventral pathway got lost: And what its recovery might mean. *Brain and Language* 118 (1–2): 29–39.

Weintraub, Sandra, M. Marsel Mesulam, and Lena Kramer. 1981. Disturbances in prosody: A right-hemisphere contribution to language. *Archives of Neurology* 38 (12): 742–744.

Weiss, Sabine, and Peter Rappelsberger. 2000. Long-range EEG synchronization during word encoding correlates with successful memory performance. *Brain Research. Cognitive Brain Research* 9 (3): 299–312.

Wernicke, Carl. 1874. *Der aphasische Symptomencomplex*. Berlin: Springer-Verlag.

Werker, Janet, and Takao K. Hensch. 2015. Critical periods in speech perception: New directions. *Annual Review of Psychology* 66:173–196.

Werker, Janet F., and Richard C. Tees. 1984. Cross language speech perception: Evidence for perceptual reorganization during the first year of life. *Infant Behavior and Development* 25 (1): 121–133.

Westerlund, Masha, and Liina Pylkkänen. 2014. The role of the left anterior temporal lobe in semantic composition vs. semantic memory. *Neuropsychologia* 57:59–70.

Westerlund, Masha, Itamar Kastner, Meera Al Kaabi, and Liina Pylkkänen. 2015. The LATL as locus of composition: MEG evidence from English and Arabic. *Brain and Language* 141:124–134.

Wiesel, Torsten N. 1982. Postnatal-development of the visual-cortex and the influence of environment. *Nature* 299 (5884): 583–591.

Wildgruber, Dirk, Hermann Ackermann, Benjamin Kreifelts, and Thomas Ethofer. 2006. Cerebral processing of linguistic and emotional prosody: fMRI studies. *Progress in Brain Research* 156:249–268.

Wildgruber, Dirk, Axel Riecker, Ingo Hertrich, Michael Erb, Wolfgang Grodd, Thomas Ethofer, and Hermann Ackermann. 2005. Identification of emotional intonation evaluated by fMRI. *NeuroImage* 24 (4): 1233–1241.

Willems, Roel M., Ash Özyürek, and Peter Hagoort. 2007. When language meets action: The neural integration of gesture and speech. *Cerebral Cortex* 17 (10): 2322–2333.

Willems, Roel M., Ash Özyürek, and Peter Hagoort. 2009. Differential roles for left inferior frontal and superior temporal cortex in multimodal integration of action and language. *NeuroImage* 47 (4): 1992–2004.

Wilson, Benjamin, Heather Slater, Yukiko Kikuchi, Alice E. Milne, William D. Marslen-Wilson, Kenny Smith, and Christopher I. Petkov. 2013. Auditory artificial grammar learning in macaque and marmoset monkeys. *Journal of Neuroscience* 33 (48): 18825–18835.

Wilson, Benjamin, Yukiko Kikuchi, Li Sun, David Hunter, Frederic Dick, Kenny Smith, Alexander Thiele, Timothy D. Griffiths, William D. Marslen-Wilson, and Christopher I. Petkov. 2015. Auditory sequence processing reveals evolutionarily conserved regions of frontal cortex in macaques and humans. *Nature Communications* 6:8901. doi:10.1038/ncomms9901.

Wilson, Stephen M., Nina F. Dronkers, Jennifer M. Ogar, Jung Jang, Matthew E. Growdon, Federica Agosta, Maya L. Henry, Bruce L. Miller, and Maria Luisa Gorno-Tempini. 2010. Neural correlates of syntactic processing in the nonfluent variant of primary progressive aphasia. *Journal of Neuroscience* 30 (50): 16845–16854.

Wilson, Stephen M., Sebastiano Galantucci, Maria Carmela Tartaglia, Kindle Rising, Dianne K. Patterson, Maya L. Henry, Jennifer M. Ogar, Jessica DeLeon, Bruce L. Miller, and Maria Luisa Gorno-Tempini. 2011. Syntactic processing depends on dorsal language tracts. *Neuron* 72 (2): 397–403.

Winkler, István, János Horvath, Júlia Weisz, and Leonard J. Trejo. 2009. Deviance detection in congruent audiovisual speech: Evidence for implicit integrated audiovisual memory representations. *Biological Psychology* 82 (3): 281–292.

Wise, Richard J. S. 2003. Language systems in normal and aphasic subjects: Functional imaging studies and inferences from animal studies. *British Medical Bulletin* 65 (1): 95–119.

Wise, Richard J. S., Sophie K. Scott, S. Catrin Blank, Cath J. Mummery, Kevin Murphy, and Elizabeth A. Warburton. 2001. Separate neural subsystems within "Wernicke's area". *Brain* 124:83–95.

Witelson, Sandra F. 1982. Hemisphere specialization from birth. *International Journal of Neuroscience* 17 (1): 54–55.

Wolff, Susann, Matthias Schlesewsky, Masako Hirotani, and Ina Bornkessel-Schlesewsky. 2008. The neural mechanisms of word order processing revisited: Electrophysiological evidence from Japanese. *Brain and Language* 107 (2): 133–157.

Wong, Patrick C. M., Lawrence M. Parsons, Michael Martinez, and Randy L. Diehl. 2004. The role of the insular cortex in pitch pattern perception: The effect of linguistic contexts. *Journal of Neuroscience* 24 (41): 9153–9160.

Wu, Ying Choon, and Seana Coulson. 2007. How iconic gestures enhance communication: An ERP study. *Brain and Language* 101 (3): 234–245.

Xiang, Hua-Dong, Hubert M. Fonteijn, David G. Norris, and Peter Hagoort. 2010. Topographical functional connectivity pattern in the perisylvian language networks. *Cerebral Cortex* 20 (3): 549–560.

Xiao, Yaqiong, Angela D. Friederici, Daniel Margulies, and Jens Brauer. 2016a. Development of a selective left-hemispheric fronto-temporal network for processing syntactic complexity in language. *Neuropsychologia* 83:274–282.

Xiao, Yaqiong, Angela D. Friederici, Daniel Margulies, and Jens Brauer. 2016b. Longitudinal changes in resting-state fMRI from age 5 to age 6 years covary with language development. *NeuroImage* 128:116–124.

Xu, Wei, and Thomas C. Sudhof. 2013. A neural circuit for memory specificity and generalization. *Science* 339 (6125): 1290–1295.

Yeatman, Jason D., Michal Ben-Shachar, Gary H. Glover, and Heidi M. Feldman. 2010. Individual differences in auditory sentence comprehension in children: An exploratory event-related functional magnetic resonance imaging investigation. *Brain and Language* 114 (2): 72–79.

Zaccarella, Emiliano, and Angela D. Friederici. 2015a. Merge in the human brain: A sub-region based functional investigation in the left pars opercularis. *Frontiers in Psychology* 6:1818. doi:10.3389/fpsyg.2015.01818.

Zaccarella, Emiliano, and Angela D. Friederici. 2015b. Syntax in the Brain. In *Brain Mapping: An Encyclopedic Reference*, ed. Arthur W. Toga, 461–468. Academic Press: Elsevier.

Zaccarella, Emiliano, and Angela D. Friederici. 2016. The neuroanatomical network of syntactic merge for language. MPI CBS, Research Report 2014–2016 (p. 68). http://www.cbs.mpg.de/institute/research-reports.

Zaccarella, Emiliano, Lars Meyer, Michiru Makuuchi, and Angela D. Friederici. 2015. Binding by syntax: The neural basis of minimal linguistic structures. *Cerebral Cortex*. doi:10.1093/cercor/bhv234.

Zaehle, Tino, Torsten Wüstenberg, Martin Meyer, and Lutz Jäncke. 2004. Evidence for rapid auditory perception as the foundation of speech processing: A sparse temporal sampling fMRI study. *European Journal of Neuroscience* 20 (9): 2447–2456.

Zatorre, Robert J., Pascal Belin, and Virginia B. Penhune. 2002. Structure and function of auditory cortex: Music and speech. *Trends in Cognitive Sciences* 6 (1): 37–46.

Zatorre, Robert J., Alan C. Evans, Ernst Meyer, and Albert Gjedde. 1992. Lateralization of phonetic and pitch discrimination in speech processing. *Science* 256 (5058): 846–849.

Zilles, Karl, and Katrin Amunts. 2009. Receptor mapping: Architecture of the human cerebral cortex. *Current Opinion in Neurology* 22 (4): 331–339.

Zilles, Karl, Maraike Bacha-Trams, Nicola Palomero-Gallagher, Katrin Amunts, and Angela D. Friederici. 2014. Common molecular basis of the sentence comprehension network revealed by neurotransmitter receptor fingerprints. *Cortex* 63:79–89.

Zilles, Karl, Nicola Palomero-Gallagher, and Axel Schleicher. 2004. Transmitter receptors and functional anatomy of the cerebral cortex. *Journal of Anatomy* 205 (6): 417–432.

Zurif, Edgar B., Alfonso Caramazza, and Roger Myerson. 1972. Grammatical judgments of agrammatic aphasics. *Neuropsychologia* 10 (4): 405–417.

Zou, Qihong, Thomas J. Ross, Hong Gu, Xiujuan Geng, Xi-Nian Zuo, L. Elliot Hong, Jia-Hong Gao, Elliot A. Stein, Yu-Feng Zang, and Yihong Yang. 2013. Intrinsic resting-state activity predicts working memory brain activation and behavioral performance. *Human Brain Mapping* 34 (12): 3204–3215.

Zwitserlood, Pienie. 1989. The locus of the effects of sentential-semantic context in spoken-word processing. *Cognition* 32 (1): 25–64.

Index

AG. *See* Angular gyrus (AG)
Age of acquisition, 143, 146–152, 155
Agrammatism, 7
Angular gyrus, 56, 57, 59–62, 111, 119, 137, 139–141, 160, 166. *See also* AG
Animacy, 63
Anterior insula (aINS), 91, 92, 96
Anterior superior temporal gyrus (aSTG), 6, 25, 38, 40, 43, 82, 110, 111, 113, 116, 136–138, 140, 189, 190, 226
Anterior temporal lobe, 29, 30, 37, 50, 57, 60–62, 95, 109, 111, 118, 119, 137, 140, 162
Arcuate fasciculus (AF), 105–107, 109–111, 113, 114, 120, 129, 139–141, 164, 165, 189, 192–195, 198, 214, 218, 224–227, 229–231
Argument structure, 28, 63, 65, 71
Articulation, 8, 86–92, 94, 157, 192, 203
Artificial grammar learning (AGL), 40, 154, 178, 209
Artificial grammars, 40, 42, 207–209
Associative learning, 2, 183
Attention, 2, 45, 96, 171, 205
Axons, 5, 103, 122, 156

BA 44, 6, 8–10, 24, 40, 42–55, 82, 92–94, 98, 99, 107–109, 112, 114–118, 120, 122–124, 126, 128, 132, 135, 137–141, 153, 156, 161, 165, 189–191, 193–196, 199, 201, 204, 206–208, 211–215, 217, 218, 224–231. *See also* Pars opercularis
BA 45, 6, 8, 9, 10, 24, 46, 48–57, 60, 62, 81, 92–94, 107, 110, 112, 114, 115, 119, 120, 122, 123, 126, 128, 135, 137–141, 165, 189, 190, 194, 195, 199, 211, 212, 217, 218, 225–228, 230. *See also* Pars triangularis
BA 47, 6, 24, 50, 52–54, 56, 57, 60, 62, 82, 107, 110, 112, 115, 119, 120, 122, 123, 126, 135, 137–141, 165, 225–227. *See also* Pars orbitalis
Basal ganglia, 37, 70, 78, 79, 90, 102, 126, 127, 164
Bilingualism, 146, 147, 150
Blood-oxygen-level dependent (BOLD), 18, 19, 126, 190.
BOLD. *See* Blood-oxygen-level dependent (BOLD)

Brain lesion, 6–8, 17, 37, 87, 94, 95
BROCANTO, 148, 153, 154
Broca, Paul, 5, 87, 94
Broca's aphasia, 7, 87, 88
Broca's area, 6–11, 41–47, 51, 53, 54, 81, 91–94, 103–109, 111, 112, 115, 117, 118, 120, 126, 127, 129, 131, 138, 149, 150, 153, 154, 158, 160, 164, 186, 187, 192, 193, 201, 204, 207, 211–214, 217, 218, 222, 224, 225, 227–229, 230, 231
Brodmann area (BA), 6, 8, 55, 109, 123, 214, 226

Canonical subject-first word order, 44
Carl Wernicke, 6
Case marking, 62, 63, 65, 66, 185, 194
Center-embedded sentences, 123, 189
Central sulcus, 9, 214, 218
Cerebellum, 90, 91, 152, 189
Cloze probability, 69, 137
Closure Positive Shift (CPS), 72, 73, 82, 176–178
Combinatorial processes, 40, 61, 110–112, 115, 137
Communication, 1–5, 11, 13, 70, 85, 95–97, 99, 101, 111, 124, 134, 156, 209, 210
Complexity, 43–47, 53, 54, 98, 116, 117, 127, 128, 138, 140, 150, 157, 188, 209
Conceptual-semantic, 4, 29, 30
Conceptual-semantic representations, 4
Conduction aphasia, 88
Connectivity-based parcellation, 9, 10, 24, 26
Content words, 7, 38, 50, 51, 54, 55, 132
Coordinate system, 8
Corpus callosum, 76–78, 81, 83, 114, 115, 161, 225
Co-speech gestures, 96–98
CPS. *See* Closure Positive Shift (CPS)
Critical period, 2, 143, 145–148, 155, 213
Cytoarchitectonic, 6, 8, 48, 51, 131, 207, 211, 212, 217, 227, 230

Dendrite, 5, 122, 212, 213, 227
Derivational morphology, 50, 51
Diffusion tensor imaging, 19, 104, 107, 192, 218
Dorsal fiber tracts, 108, 109, 113, 115, 135

Dorsal pathway, 6, 105, 107–115, 120, 124, 129, 135, 140, 141, 161, 192, 195, 206, 207, 213, 214, 217, 218, 226, 229, 230
Dynamic causal modeling, 127–129, 140, 153
Dysarthria, 87

Early left anterior negativity (ELAN), 18, 34–38, 67–68, 82, 83, 147, 148, 164, 180, 181, 184, 197
EEG. *See* Electroencephalography (EEG)
Effective connectivity, 125, 127
ELAN. *See* Early left anterior negativity (ELAN)
Electroencephalography (EEG), 17, 19, 20, 33, 37, 56, 57, 72, 125, 129, 130, 165, 215
Embedding, 44, 46–48, 116
Emotional prosody, 71, 78–81, 95, 112, 113
ERP. *See* Event-related brain potentials (ERP)
Event-related brain potentials (ERP), 17, 18, 20, 21, 27, 28, 34–37, 49, 57, 58, 59, 63–65, 67, 69, 70, 72, 73, 76, 78–80, 82, 83, 97, 98, 115, 131, 147–149, 153, 159, 162, 164, 165, 167–186, 197, 199, 215, 216

Feature-based semantic theories, 31
Fiber bundles, 3, 5, 11, 19, 24, 103–106, 108, 113, 115, 116, 121, 155, 156, 188, 213, 225
FMRI. *See* Functional magnetic resonance imaging (FMRI)
FOP, 6, 10, 75, 109, 120, 135, 226
Fractional anisotropy, 103, 156, 189, 194, 195
Frequency band, 17, 23, 130, 131, 132
Frontal lobe, 5, 6, 8, 77, 83, 165
Frontal operculum, 6, 9, 10, 38, 40–43, 82, 107–110, 112, 113, 115–117, 120, 126, 135, 137, 138, 165, 214, 217, 224–226, 228. *See also* FOP
Functional connectivity, 60, 78, 81, 102, 121, 125–129, 133, 134, 152, 153, 191–194, 196, 217, 225, 229
Functional magnetic resonance imaging (FMRI), 17–19, 22, 29, 38, 40, 44, 47, 49, 50, 51, 56, 57, 59, 60, 61, 69, 70, 74, 79, 80, 91–98, 104, 108, 117–119, 122, 125–127, 129, 132, 136–138, 147, 149, 150–154, 159, 165, 166, 176, 182, 183, 186, 188, 189, 194, 196, 211, 216
Function words, 7, 38, 40, 50, 54, 55, 87–89, 131, 204
Fusiform gyrus, 60, 61, 91, 92, 128

Gap, 44–47, 49, 184
Gestures, 1, 3, 95–99, 158, 204, 208
Grammar, 3, 7, 8, 40–42, 53, 85, 86, 88, 94, 108, 109, 124, 148, 149, 153, 154, 163, 164, 178, 179, 203, 206–209, 214
Grammatical encoding, 86, 88, 92–94
Gray matter, 5, 121, 155–157, 188, 189, 191, 210, 211, 219, 224

Head-driven phrase structure grammar, 3
Heschl's gyrus, 6, 21, 22, 24, 136, 161, 187, 210. *See also* HG
HG, 6, 161
Hierarchical embedding, 44, 47, 48
Hierarchically embedded, 47
Hierarchical structure, 4, 53, 117, 164, 206, 207, 209
Hippocampus, 30–32, 149, 153, 154

Iconic gestures, 1, 96, 97, 158
IFG. *See* Inferior frontal gyrus (IFG)
IFOF, 107, 110, 118, 120, 138, 189, 192, 194, 195, 198, 225, 226. *See also* Inferior fronto-occipital fasciculus
IFS, 123, 128, 129, 214, 218. *See also* Inferior frontal sulcus
Inferior frontal gyrus (IFG), 6, 8–10, 43, 44, 46, 47, 50, 53, 54, 56, 57, 60–62, 71, 87, 91, 96–98, 107, 110, 113–115, 117, 119, 128, 129, 132, 134, 135, 138–141, 150, 152, 156, 159, 160, 165, 186, 188, 189, 192, 194–199, 210, 216–218, 224, 226–229
Inferior frontal sulcus (IFS), 47, 48, 123, 127–129, 140, 196, 214, 218
Inferior fronto-occipital fasciculus, 107, 111, 113, 120, 165, 192, 195, 198, 225, 226
Inflectional morphology, 33, 49, 50
Intonation, 15, 16, 72, 75, 76, 79, 115
Intonational phrase boundary, 72, 73, 76, 176, 178
Intracranial recordings, 31, 224

Korbian Brodmann, 8

LAN, 63, 65–67, 71, 82. *See also* Left anterior negativity
Language faculty, 1, 2, 206, 219, 221, 222, 227, 229, 231
Language production, 5–7, 13, 85–95, 152
Layers, 8, 227
Left anterior negativity, 18, 34, 63, 180, 181. *See also* LAN
Left hemisphere, 5, 6, 8–11, 16, 23, 24, 26, 48, 50, 56, 71, 74–76, 78, 79, 81–83, 87, 95, 101, 105, 112–115, 117, 120, 123, 124, 135, 152, 158, 159, 166, 167, 172, 182, 183, 193, 196–198, 210, 212, 213, 225, 227, 228. *See also* LH
Lexical functional grammar, 3
Lexical selection, 86, 92
Lexical-semantic, 7, 18, 27–33, 35, 36, 57, 58, 119, 136, 137, 141, 151, 172–175, 186, 197, 212, 225
Lexicon, 8, 20, 27, 28, 30, 32, 85, 86, 88, 89, 93, 94, 131, 132, 158
LH, 6, 9, 10, 16, 25, 26, 39, 41, 42, 45, 46, 75, 83, 107, 120, 123, 135, 154, 167, 187, 192, 193, 198, 212, 218, 226
Linguistic prosody, 71, 74, 75, 81, 112–115
Linguistic-semantic representations, 4

Index

Magnetoencephalography, 17–19, 31, 37, 38, 53, 56, 57, 59, 60, 61, 78, 92, 125, 129, 130, 136, 165. *See also* MEG
Marc Dax, 5
MEG, 20, 33, 69
Merge, 3, 4, 40, 42, 52, 55, 157, 209, 221, 224
Middle temporal gyrus, 6, 8, 9, 22, 54, 57, 60–62, 70, 82, 91–93, 97, 98, 108, 110, 112, 115, 119, 120, 126, 128, 129, 135, 137–139, 141, 162, 165, 198, 225. *See also* MTG
Milliseconds (ms), 13, 15, 17, 18, 20, 34, 38, 73, 83
Mirror neuron, 133, 134, 204
Mismatch negativity, 20, 21, 167, 168, 215
Mismatch negativity paradigm, 167
Morphemes, 87, 88, 89
Morphology, 33, 49, 50, 51, 77, 151
Motor cortex, 8, 90, 94, 123, 133, 212, 227, 228
Movement, 45, 46, 96, 98, 158, 169, 204
MTG, 6, 57, 61, 120, 128, 135, 162, 198, 214, 218
Myelin, 5, 103, 156, 157, 162, 189, 191, 230
Myelination, 103, 105, 118, 156, 191, 229

N100, 18, 20–22, 82, 83
N400, 18, 28, 36, 57–59, 63–65, 71, 77, 78, 80, 82, 83, 97, 133, 137, 147–149, 159, 164, 171–175, 185, 197
Near-infrared spectroscopy, 17, 19, 165, 166, 167, 176, 182. *See also* NIRS
Neuron, 5, 9, 30–32, 91, 101, 121, 122, 124, 125, 133, 134, 156, 196, 204, 211, 212, 223, 224, 227
Neurotransmitters, 5, 9, 43, 101, 121, 122, 124, 228
NIRS, 177
Non-canonical, 44, 47, 116–118, 137, 151, 184–186, 189, 199
Number marking, 65, 66

Occipital lobe, 6, 83, 219
Ontogeny, 12, 143, 163, 196, 218, 219, 229, 230
Ordering, 44, 47, 139, 199
Oscillations, 17, 23, 125, 129–134
Oscillatory frequency, 53

P600, 18, 63, 65, 67–70, 78, 82, 83, 98, 147–149, 164, 180, 181, 184, 185, 197
Paragrammatism, 7, 158
Parietal lobe, 6, 77, 83, 106, 107, 126, 152, 176, 189
Pars opercularis, 6, 9, 10, 82, 98, 120, 126, 128, 135, 137, 138, 189, 190, 194, 195, 212, 213, 226. *See also* BA 44
Pars orbitalis, 6, 120, 126, 135, 137. *See also* BA 47
Pars triangularis, 6, 9, 10, 97, 120, 126, 135, 137, 189, 190, 195, 226. *See also* BA 45
Phonemes, 20–23, 27, 71, 90, 91, 136, 159, 168, 172, 183, 225
Phonological encoding, 86, 92, 94
Phonology, 15, 152, 161, 164, 165, 223

Phrase structure grammar, 3, 41, 109, 148, 208, 209
Phylogeny, 2, 116, 217–219, 224, 229, 230
Pitch, 23, 71–75, 79–81, 113, 176, 178, 182, 209, 215, 216
Planum polare, 21, 22, 25
Planum temporale, 21, 22, 25, 82, 159, 160, 164, 210, 211
Pragmatics, 95, 99
Precentral gyrus, 91, 92, 94, 98, 107, 192, 212, 218
Precentral sulcus, 9, 159, 160, 214, 218
Premotor cortex (PMC), 6, 92, 94, 97, 98, 104, 106–108, 110, 112–115, 120, 128, 135, 153, 165, 192, 206, 226, 227
Primary auditory cortex (PAC), 6, 21, 23, 25, 38, 82, 123, 131, 135, 136, 140, 161, 164, 225
Processing pitch, 80, 81, 182, 209
Proficiency, 146–151, 155
Progressive non-fluent aphasia, 87
Prosodic boundary, 72, 176
Prosody, 16, 23, 71–81, 83, 95, 112–115, 169, 175–177, 183, 204

Readiness potential, 90, 91
Receptorarchitecture, 122
Receptorarchitectonic parcellation, 9
Receptors, 122
Resting-state functional connectivity, 81, 126, 127, 196
Resting-state functional magnetic resonance (rfMRI), 125, 152, 194, 196
Right hemisphere, 8, 11, 16, 23, 25, 71, 72, 74–79, 81–83, 95, 101, 105, 112–115, 124, 137, 159, 161, 164–167, 170, 176, 177, 182, 183, 186, 210, 212, 213, 217, 225, 227, 228, 230

Scrambling, 44, 45
Second language learning, 11, 96, 145, 146, 148–150, 153, 155, 158
Segmental, 16, 71, 72, 74, 75, 112, 166, 167, 176, 182
Selectional restrictions, 63, 65, 221
Semantic combinatorics, 29, 60, 61
Semantic network, 29, 56, 57, 61, 115, 118, 119, 140
Semantics, 11, 15, 29, 30, 37, 49, 52, 56, 60–62, 83, 119, 120, 146, 149, 164, 172, 173, 175, 186, 188, 221, 223, 230, 231
Sensitive period, 143, 145, 147, 148, 155, 156, 205
Sign language, 2, 8, 143, 146, 157, 158, 159, 160, 162
Single cell recordings, 30
Specific language impairment (SLI), 164
Speech errors, 87, 89, 90, 94
Structural magnetic resonance imaging (sMRI), 19, 104, 108, 165
Subject-verb agreement, 65, 66, 71, 178, 180
Superior longitudinal fasciculus (SLF), 106, 107, 109, 111, 113, 114, 120, 139, 140, 141, 160, 165, 192, 193, 226

Superior temporal gyrus (STG), 6, 8, 10, 21, 22, 24–27, 38, 40, 43, 44, 50, 53–55, 57, 59, 69, 70, 81, 82, 91, 92, 94, 106, 108–113, 115–117, 119, 120, 123, 124, 126, 127, 129, 131, 132, 135, 136–140, 152, 156, 159, 160, 162, 165, 188–190, 192, 193, 194, 196–199, 204, 206, 210, 213, 217, 225–227, 229, 230
Superior temporal sulcus (STS), 6, 10, 21, 22, 24, 26, 44, 46, 50, 52–55, 57, 59, 69, 70, 97, 110, 113, 114, 117, 120, 123, 124, 128, 135, 136, 138–140, 156, 159, 165, 193, 196, 198, 199, 204, 213, 214, 217, 218, 226
Supramarginal gyrus, 119, 129, 186, 188
Suprasegmental, 15, 16, 23, 71, 74, 75, 114, 166, 167, 176, 182, 186
Syllable, 5, 15, 20, 23, 25, 27, 28, 40, 41, 50, 53, 72, 73, 75, 90, 108, 131–133, 169, 170, 176, 177, 178, 215
Synapses, 5
Syntactic complexity, 44, 45, 47, 54, 98, 117, 127, 128, 138, 140, 150, 188
Syntactic movement, 45
Syntactic networks, 116, 165
Syntactic phrases, 33, 71, 132, 136, 197
Syntax, 2–4, 15, 32–34, 36, 38, 39, 43, 44, 48–52, 54, 55, 60, 70, 72, 75, 76, 78, 81, 83, 88, 91, 98, 99, 109, 110, 115, 117–119, 122, 124, 137, 139, 140, 145, 146, 149, 150, 153, 155, 156, 158, 161, 164, 175–178, 183, 184, 186, 188, 191–193, 199, 203–206, 211, 215, 217, 219, 221, 223, 226–231

Temporal lobe, 6, 8, 21, 29, 30, 32, 37, 50, 56, 57, 60–62, 95, 101, 111, 113, 115, 118, 119, 133, 136–138, 140, 158, 165, 197, 210
Temporal pole, 29, 54, 61, 138
Temporo-parietal junction (TPJ), 95, 96, 99
Thalamus, 78, 90, 91, 102, 126, 161
Thematic relations, 19, 28, 32, 43, 51, 62, 65, 86, 224
Thematic role assignment, 59, 62, 63, 65, 67, 69, 71, 82, 185
Thematic roles, 51, 62, 63, 65, 66, 67, 71, 185
Tractography-based parcellation, 24, 77

Uncinate fasciculus (UF), 110, 111, 116, 118, 120, 138, 160, 165, 198, 225, 226

Ventral fiber tracts, 107, 110, 138
Verb-argument, 28, 36, 44, 53, 54, 58, 63, 65, 69, 70, 71, 117, 140, 141

Wernicke's aphasia, 7, 88, 89
Wernicke's area, 6–9, 11, 103, 105, 111, 120, 126, 127, 158, 210
White matter, 5, 6, 9, 10, 12, 19, 24, 55, 77, 81, 101, 103–105, 111, 113, 115, 116, 118, 121, 124, 134, 155–157, 160, 162, 188, 189, 191, 205, 210, 213, 219, 223–225, 229

White matter fiber tracts, 6, 10, 12, 101, 105, 111, 115, 124, 132, 156, 162, 191, 205, 214, 218, 224
Word categories, 33
Word category, 27, 28, 32–38, 68, 86, 89, 115, 136, 147, 180, 181, 183, 184, 197, 225
Word order, 44, 45, 49, 50, 62, 63, 65, 66, 76, 116, 185, 194
Word substitutions, 88
Working memory (WM), 47–49, 55, 63, 117, 128–130, 132, 139, 188, 189
World knowledge, 15, 16, 58, 59, 86